MEMOIRS OF A MAVERICK MATHEMATICIAN

Zoltan Paul Dienes

MINERVA PRESS
ATLANTA LONDON SYDNEY

MEMOIRS OF A MAVERICK MATHEMATICIAN
Copyright © Zoltan Paul Dienes 1999

All Rights Reserved

No part of this book may be reproduced in any form
by photocopying or by any electronic or mechanical means,
including information storage or retrieval systems,
without permission in writing from both the copyright
owner and the publisher of this book.

ISBN 0 75410 350 1

First Published 1999 by
MINERVA PRESS
315–317 Regent Street
London W1R 7YB

Printed in Great Britain for Minerva Press

MEMOIRS OF A MAVERICK MATHEMATICIAN

Acknowledgements

Zoltan Paul Dienes and Paul Ernest would like to thank Maurice Ash and the Trustees of the Elmgrant Foundation for supporting this publication. The University of Exeter has also provided resources and been supportive and Paul Ernest would also like to thank Mrs Julie Williams for redrawing the figures in the text.

Foreword

I first met Zoltan Dienes in the summer of 1985. My colleague David Burghes invited me to lunch with Dr and Mrs Dienes at the University of Exeter's Crossmead Centre. I had long known of his work, especially through his 1960 masterpiece *Building Up Mathematics*, so I was delighted to meet the great man at last.

We ate and chatted and then Zoltan Dienes produced a pack of cards which he had had specially printed and asked, 'I wonder if you can solve this problem?' The pack consisted of forty-eight cards with pictures of various numbers of flowers and leaves on the front and a brief text on the back such as 'two houses, a tree and a mountain'. Zoltan Dienes gave me forty-seven cards and said, 'Tell me which one is missing.' It is hard to describe my emotions. I was puzzled and intrigued, but I also felt that, as a lecturer in mathematics education, my credibility was at stake.

I looked through the cards. There was no obvious rhyme or reason to them. The time was about one thirty. I started to sort out the cards. By trial and error, I began to realise that I had to lay them out on the table. The plates and accoutrements were pushed back. Eventually, I had laid out a six by eight pattern of cards (with one missing) on the table. The rest of the company kept chatting, most diners left and Tessa trotted around, collecting bread rolls, saying, 'You never know when you might need them!' I ignored all these distractions and kept working at the puzzle.

The waitresses removed the remaining lunch leftovers from the table. I kept turning the cards, trying to complete a pattern. I looked up. By now the waitresses had cleared all the tables, removed the lunch buffet and we were the only party left. At last I worked out the complete pattern and was able to identify the missing card correctly. It was now two thirty and the whole restaurant was empty except for us and the waitresses, waiting for us to leave. I had solved the problem, but it was the hardest working lunch I have ever had! As they say, there is no such thing as a free lunch.

At the time Zoltan Dienes was still commuting between Italy and England, as well as spending summers and winters in Canada and making periodic world trips. He would often stay a month or two at Golden Oktober, his house in Totnes, in the late spring or early summer and then return in the autumn. During these visits he would regularly come to the University of Exeter to work with us, inspiring staff and students alike.

He has always been an intrepid world traveller, as the story in these memoirs reveals. He has mastered an intimidating number of languages and remains fluent in seven. I remember his giving me samples of the primary school curriculum he developed in Italy in the 1980s (in all school subjects) and saying to me, 'Look it's easy – Italian is such an easy language – surely you can see what it says?' I had to make it clear that a mere mortal like myself could only cope with one or two languages!

Dr Zoltan Paul Dienes has led a fascinating and fruitful life, which, as I write these words, is far from over! As an introduction, I shall sketch some of the salient features as a thumbnail 'map' to guide the reader. Zoltan Dienes was born in Budapest in 1916, but moved to England in 1932 and attended Dartington School shortly after it was opened. He went on to study mathematics at University College, London and graduated with a BA in 1937 and Ph.D. in

1939, the latter in record time! Following this, he became a mathematics teacher, working for a while at Highgate School and Dartington. From 1944 onwards, he lectured in mathematics at the universities of Southampton, Sheffield and Manchester before taking a permanent post at Leicester in 1950. There he undertook further studies in psychology and then laid the foundations for his major work, including the development of his theory of learning mathematics and his famous teaching apparatus, and publishing books, such as *Building Up Mathematics*, with a preface by Herbert Read. In 1961 he took up a readership at Adelaide University, Australia and later a personal chair. He also worked with Jerome Bruner at Harvard and briefly with Jean Piaget in Geneva and founded the International Study Group for Mathematics Learning and the international *Journal of Structural Learning*.

In 1966 Zoltan was appointed director of a new federal research centre in mathematics education at the University of Sherbrooke in Canada. He worked in Canada till 1978, researching and travelling world-wide to run projects in countries on virtually every continent. From 1978 until 1986 he worked extensively in Italy, developing a primary school curriculum which he wrote in Italian. The University of Siena honoured him with a doctorate in the 1980s, as did the universities of Caen, France and Mount Alison, in New Brunswick.

In 1980 he bought a house at Totnes in Devon and lived there with his wife Tessa for part of the year, working in Devon schools. From the mid-1980s, he contributed to work in education at Exeter University as an honorary research fellow, lecturing and giving demonstrations to student teachers and staff. When the microcomputer appeared in the 1980s he took to it like a duck to water, writing many imaginative teaching programmes. Exeter University awarded him an honorary doctorate in 1995, in recognition of his lifetime's work. I had the honour of giving the public oration at the degree ceremony and I was spoilt for choice in selecting his most important contributions (not to mention colourful episodes from his life) for my eulogy. Dr Dienes now spends most of his time at Wolfville, in Canada and his flow of creativity in writing and designing teaching games continues unabated.

Dr Dienes's reputation rests on two or perhaps three important areas. First of all, there is his very extensive set of publications on the psychology of learning mathematics. His theories and the way he linked these with practice, will remain an enduring contribution for as long as there is interest in the problems of the teaching and learning of mathematics. Every book on the psychology of learning mathematics includes a section on his ideas. Second, he is no mere theorist, for he developed an immense collection of learning games and apparatus to put his theories into practice. In researching the problems of mathematics, he invented the teaching materials called the Multi-base Arithmetic Blocks, known universally as the Dienes's blocks (universally mispronounced as *Deans* blocks, instead of *Dee-enn-ess* blocks); Algebraic Experience Materials, which put the abstract ideas of algebra into a manageable concrete form and the Logiblocs, for teaching classification and sets to learners of all ages. These are just his best-known pieces of apparatus and he has invented literally hundreds more, many of which will be found in well-equipped primary classrooms around the world. Third, he has worked face-to-face with many hundreds of teachers and many thousands of children, on every continent of the globe, communicating understanding and enthusiasm and changing people's perceptions of mathematics permanently. This alone is more than most people achieve, although it is the first two contributions which will guarantee him immortality.

What is marvellous about these memoirs is that they are not the dry-as-dust account of the life of some ordinary academic. There is a rich human thread running through them, describing how Dienes and his family were buffeted hither and thither by some of the major events of twentieth-century world history. The tale is told with frankness, a vivid eye for detail and an understated, dry sense of humour. I have known Zoltan Dienes for over twelve years but I learnt so much from these memoirs, which never came out in our many scores of conversations. I suppose this is because Zoltan Dienes lives in the present, not the past, continually working on new projects, not to mention skiing and swimming in Wolfville whatever the weather. It is my privilege to be able to introduce this fascinating autobiography to you, the reader, the truly memorable memoirs of a maverick mathematician.

Paul Ernest
Exeter
November 1997

Contents

One	Papua New Guinea	13
Two	The Beginnings	20
Three	Getting Re-established in Hungary	32
Four	Some Holiday Experiences During Pre-Teen Years	41
Five	Pre-Teen Life in and Around Budapest	57
Six	The Last Two Years in Hungary	70
Seven	England	95
Eight	The Dartington Years	112
Nine	The Undergraduate Years	129
Ten	Two Decisive Years	159
Eleven	Getting Used to Wartime	180
Twelve	The Peripatetic University Years	198
Thirteen	Leicester	220
Fourteen	Other Happenings During the Leicester Years	251

Fifteen	The Interregnum	289
Sixteen	Getting Used to Australia	308
Seventeen	The World Trips	340
Eighteen	From Australia to Canada	374
Nineteen	Getting Established at Sherbrooke	388
Twenty	Working Internationally from Sherbrooke	410
Twenty-One	About Our Personal Life During the Sherbrooke Years	454
Twenty-Two	The Squeeze	467
Twenty-Three	Winnipeg	477
Twenty-Four	To Europe Again	494
Twenty-Five	The Italy–Devon Period	516
Twenty-Six	The Devon–Wolfville Period	536
Postscript	November 1997	566

Chapter One

Papua New Guinea

'How would you like to have a look at one of our schools?' said the director of education of the then Territory of Papua New Guinea, while we were sitting comfortably in his air-conditioned office.

'Sure thing,' I replied. 'I suppose this is what the minister for territories sent me to do here.'

Port Moresby was then a straggling town, or rather a large village, built along the south coast of Papua. There were not many roads from Port Moresby and those that existed came to abrupt ends in the most unlikely places. If you wanted to go anywhere, you had to go by plane, walk or get a ride in a native canoe!

The director and I drove along the dusty roads for a short distance and pulled up outside a very modern building, which the director assured me was one of their schools. I was duly introduced to the headmaster and some of the teaching staff and taken round the school. Everything seemed to work like clockwork: the children were marching in and out of classrooms to the sound of stirring martial music, all wearing smart but identical uniforms, all looking very serious, not one of them giving us a smile. We looked in on one or two of the classes, in which the teaching was being done in well established, traditional ways, with the teacher up at the front and most of the time the children mechanically repeating everything the teacher said to them without thinking.

As we were being ushered from one classroom to the next, I just managed to whisper to the director, 'I am sure not all your schools are like this! Couldn't we go and see one of your more usual schools?'

'I am sure we can show you some,' said the director. 'We shall excuse ourselves from here as soon as it is politely possible.'

I met the director the next morning and he introduced me to a number of inspectors, who were to accompany us. I had brought some mathematics materials with me, mainly attribute blocks, so that, if the occasion allowed, I could work with the children to test out the terrain a little.

We made our way to a small school, on the periphery of Port Moresby. The building was just a shack. There were not even walls all round, but the roof managed to keep out the rain. I asked the director if he would like me to do something with the children. He told me to help myself and explained to the teacher in charge that I would take over the class for a little while. I put several sets of attribute blocks on the floor and arranged the children in groups around the blocks and told them to do what they liked with the blocks. Not one child stirred. They all sat still, waiting for commands.

Well, I thought, Rome was not built in a day. I got them to form a semicircle and placed one lot of blocks centrally on the floor. I began by asking one child to pick up any two blocks and asking him how they were alike and how they were different. This led to a pairing game in which blocks had to be paired, so that in each pair the two blocks were different from each other in exactly the same way and they were like each other in exactly the same way. This began to motivate the children and they began to argue with each other over the rights and wrongs of individual pairs of blocks. Before long, I was able to put them back into groups and get them to play this game, followed by a number of other games. I began to get smiles, for the children were beginning to enjoy the freedom of thinking out problems for themselves. When the bell went, the children did not want to stop playing. But of course the bell meant that something else was about to happen according to the routine, so they had to stop.

'What do you think all that was about?' said one of the inspectors, who had been watching all the aforementioned developments. They had a brief discussion amongst themselves, to the effect that it was not really desirable to allow children to play, when more serious things ought to be taught to them. They were totally unaware of the incipient logical thinking involved in these games and they thought the whole thing would be extremely bad for discipline. The director did not join in this discussion, but gave me an occasional meaningful look, as much to say that we would talk later.

When we found ourselves alone in the director's office, the inspectors having retired to the local hotel, no doubt with a view to consuming some appropriate beverages, I said to the director, 'Do you not have any schools where inspectors do not inspect?'

'Do you mean you *really* want to do some work? The lesson you showed us was most interesting and if you like, I can show you some schools where inspectors are never likely to go, if you really are anxious to do something here. You see, most people the minister for territories sends us up here enjoy the first-class flight on the plane, have a few days of looking around, then with perhaps a trip to Manus, Goroka or Mount Hagen. Then they utter some platitudes, pocket a good consulting fee and go home and tell hair-raising stories about our Stone Age natives! I assumed you were one of the usual run of official visitors!'

'Mr Barnes, the minister,' I replied, 'told me on the phone before I came that the mathematics education was so bad here that, if I threw it up in the air and let it fall down, it would be probably be an improvement! I told him I had never done anything like this before, but he still encouraged me to come, after he had sent some observers to our experimental schools in Adelaide, who gave him a glowing report. So I am here and I should like to see what can be done and then do it, if I can!'

'That sounds most interesting,' replied the director. 'I'll tell you what we'll do! Be ready about five tomorrow morning, have some of your material to take with you and a jeep will come to the hotel and take you to a school.'

On this cordial note we parted. I was run back to the only decent (and air-conditioned!) hotel in Port Moresby, where I was wined and dined and then went to bed early, asking for a wake-up call at twenty to five in the morning.

At five the next morning I loaded the jeep with a large box of mathematics materials and sat by the native driver, who was detailed to take me to the school. We soon left Port Moresby behind. The road rapidly deteriorated into a dusty track and soon we found ourselves driving along the beach. It was a beautiful morning, just getting light, the sky getting redder and redder by the minute. The sea was as calm as a duck pond. All of a sudden the driver brought the jeep to a stop. On our right was the ocean and in front of us flowed a sizeable river, somewhat muddy-looking, but still quite majestic, as its waters continued to empty into the ocean.

I said to the driver, 'Where is the school?'
'Over river, two mile,' replied the driver.
'How do we cross the river?' I asked
'Walk,' came the laconic reply.
'All right then,' I said. 'Let's go.'
'I no go,' said the driver. 'You go.'

I must say that I was a little taken aback. I had thought the jeep was going to take me to a school. There was not a soul in sight and no sign of any habitation; only a mangrove swamp on our left, the ocean on our right and the river in front of us.

'Why don't you cross the river? This material is very heavy for me to carry for two miles,' I said to him.

'Crocodiles,' said the driver.

Suddenly everything became clear to me. I had asked the director of education to let me work in a school where inspectors never went and now I realised just why they would not do so!

The driver helped me carry the material down to the bank of the river, said goodbye, went back to his jeep, turned round on the sand and drove away. Of course there was nothing else to do but wade across the river. It did not seem very deep, it would probably come up to my waist – but of course I could not be sure. There seemed no suspicious-looking objects floating down the river, so I heaved the box of materials up on to my head and proceeded to wade. It took only a minute to cross the river, but it was one of the longest minutes I had lived up to that point!

I rested a bit on the other side. The sun had risen and it was getting warmer, so it did not seem to matter that all my clothes were wet. So I put my burden on my head again and started off on my two mile hike, following the coast. The driver had been about right, because, after about three-quarters of an hour, I came across some houses; in fact, I soon found myself in the middle of a native village. Some of the houses were built on stilts in the lagoon and some were a little way from the seashore.

It was not hard to find the school because it was the only building that was obviously not built by the natives and it had a corrugated iron roof. I could hear the noise of the children, from which I assumed that school had not yet started, but was about to start. I put down my box and heaved a sigh of relief, then entered the school building. What I saw surprised me a little: a small boy was lying on the floor and a teacher, a European, was administering something to him.

'Good morning,' I said. 'The director of education has sent me.'

'You'd better just stay where you are for the moment and watch me. This child has malaria and if I don't give him a large dose of nivequin he won't be here much longer,' said the teacher, trying to get the child to swallow the dose of nivequin.

That reminded me that I had not taken mine, so I promptly took a pill out of my pocket and swallowed it. All visitors to New Guinea take such quinine-based medication as a prophylactic against malaria.

The boy on the floor was taken back to his parents by one of the native teachers. As it happens, he did survive.

Having done his morning life-saving chores, the teacher said to me, 'How good to see another European face! I am the only one in the village – all my teachers are natives. So what can I do for you?'

This was the start of the New Guinea Mathematics Project, on which I continued to work for more than two years, trying to devise meaningful ways of teaching native children mathematics, using as much local material as possible and trying to make it relevant to their environment.

How the project developed I will relate in another chapter. Now I should like to tell the reader how it happened that I came to be given the job of bringing mathematics to the Stone Age society of Papua New Guinea.

I shall have to go back to the years of the First World War in the Hungarian city of Budapest, where as a child, I had looked at a bird flying across the sky and exclaimed, 'Look! It's just like an aeroplane!'

Chapter Two

The Beginnings

The year 1916 found the world, particularly Europe, in turmoil. The Central Powers and the Allied Powers were locked in mortal combat on the eastern and western fronts, neither side able to achieve a breakthrough, in spite of enormous losses. Although aeroplanes were taking part in the battles, they clearly had only peripheral use, so that, behind the lines, the populations at large were only aware of the carnage through the news published daily in the newspapers and of course, through the ever-heavier losses accruing to families in the belligerent countries. There were some scarcities, but the Allied blockade never really succeeded in starving the populations of the hinterland of the Central Powers. So life went on in a pseudo-normal way, everybody hoping that the killing would soon end and that some sort of magic post-war peace would descend on Europe. Such was the general situation in Budapest, capital of Hungary, which was then part of the Austro-Hungarian Empire.

As fate would have it, I was born into that tumultuous scene. My father was a secondary school teacher of mathematics, with a right to give courses in mathematics at the university. My mother had been through university studying mathematics, philosophy and music, with combined experiences in Budapest and Paris. My parents were friends with prominent members of the Hungarian intelligentsia and being born into such a milieu was bound to have a profound effect on the way my life evolved. Apart from hazy memories of the layout of the apartment in which we lived, together with memories of somewhat personal scenes, such as feeding from my mother's breast and being placed on a pot for lengthy periods for the purposes of toilet-training, my first definite recollections were to do with going to see my father on the top of a mountain. This was because he had had to go into hiding in 1919, as he had taken part in the abortive communist regime of Béla Kun, having been given the job of reorganising the university to enable working-class people to receive tuition there. When the Red Terror was replaced with the White Terror, people who had taken part in the former, however innocently, were hunted down and executed.

I remember telling my friends, 'My father lives on top of the mountain!' and I recall that they were duly impressed! My father had shaved off his beard and was hidden in a friend's apartment, where he was confined to a cupboard, his sole consolation being that a beautiful young girl brought his food to him at regular intervals! He was later to marry this beautiful young girl, which circumstance again had a great deal of influence on what happened to me in the years of my growing up. Eventually, a way was found of getting him out of the country. My mother was introduced to the captain of a riverboat on the Danube which was to make the trip to Vienna. The captain was asking an enormous sum of money in return for smuggling my father out of the country. The only way the money could be raised was by selling everything, including a unique and valuable library. But the money was raised, the captain was paid and my father was smuggled on to the boat at night and hidden for a time in a wine barrel, to avoid detection by customs. As a matter of fact, he stayed in the barrel until the ship sailed away upstream, from the last Hungarian port of call and made for Vienna. He was put ashore in Vienna with nothing but the clothes he was wearing.

It was not long after, that my mother, my brother, who was but two years older than I, left Hungary for Vienna. This was in the autumn of 1920, when the peace treaties had been dictated and signed and the Austro-Hungarian Empire was carved up, Austria being reduced to its present condition of a little country and Hungary too being reduced to one-third of its pre-war size, with the so-called successor states taking over the confiscated territories. Of course, my parents were more concerned with problems of survival than with the eclipse of empires.

My father found employment in crowd work in a film studio, while my mother managed to put us children into a Montessori children's home, where we could stay without fees, in return for which she had promised to give the children dancing lessons, following a system she had elaborated, on the lines of musical scales.

I distinctly remember my first night in the *Kinderheim*. I lay on a small bed and sobbed and sobbed all night, not knowing what was happening. One of the members of staff stayed with me all night, holding my hand and tried to console me, but of course I understood not a word, as it was all in German. But I did realise that she was trying to be kind and I am eternally grateful for her patience with me on that first terrifying night.

Of course children are very resilient and my brother and I soon got used to the place, soon learning to understand and speak German. We saw our mother every day, as she had to come to give dancing lessons. My brother was with the big ones who had lessons and they were learning to read and write. The big ones were pretty contemptuous of the little ones and used to come round teasing us, shouting at us, '*Die Kleinen, die Kleinen!*' which was not helpful in getting me to settle, but children seem to get over most things so we little ones learned to ignore the big ones. We concentrated on learning our nursery rhymes, which we all enjoyed singing. I remember one was about a hunter and a hare, who had a long conversation about whether the hunter should shoot the hare or not and we often seriously discussed the problem faced by the hunter!

Our life in the *Kinderheim* lasted about a year. Then my mother heard through the grapevine that Raymond Duncan, Isadora's brother, had set up some sort of a commune in Nice and he was asking people to join in a 'New Age'-type of social experiment in living, to use modern terminology. It was soon decided that we should go. In the meantime my parents got divorced, since my father wanted to marry the beautiful girl who had looked after him in the cupboard. Soon after their marriage in Austria, he received an invitation to join the University of Wales at its Aberystwyth College, which was a godsend for him and his new wife, so, while they started their new life in Wales, we started ours in Nice.

I have some fairly precise memories about how the commune functioned. The first thing that seemed rather strange to a child was that you were not allowed to call your mother, 'Mother', because everyone had to be addressed by his or her first name. So Mother became Valerie. All the children were 'owned' in common. Whenever we went out, all the children had to go out together, for we were considered to be one unit, not to be separated. Since children are social animals, I did not find this at all difficult and soon it never occurred to me even to want to go out without the other children. We were taken down to the Promenade des Anglais every day and put out to play on the beach. I did not always want to play all the games, nor was I made to. Sometimes I would sit or stand, gazing out to sea and watching intently the waves coming towards the shore, sometimes splashing up into little fountains. I remember wondering why the little fountains simply never happened to form where I was standing! I kept standing in different places on the edge of the water, but every time the little splashing fountains sprang up somewhere else!

Another odd aspect of the commune was the dress. We wore Greek chitons and Greek sandals, adults and children alike. The local inhabitants were quite used to us and after a while took us as much for granted as the postman or the milkman and did not stare. The climate in Nice is mild and the scanty clothing did not seem to result in any physical inconvenience; in fact, for the adults it must have made life a lot simpler. Looking back, another unusual feature was the furniture. The only kind of furniture in the place that I can remember was the bench or form, but without a back. We slept on these forms, we sat on them and we also ate on them, which we did by kneeling up to two forms placed together to make a table. I assume that this was also done in order to simplify the details in life which tend to get too complicated, with no readily observable advantages.

This went on for several months, by which time my brother and I forgot to speak German and had learned to speak French. Then suddenly Raymond decided that we should all move to Paris. He had been on several lecture tours to the United States, where he was often hailed as a messiah and come back with a lot of funds, with which he purchased a property in the rue de Colisée, just off the Champs-Elysées. Whether the communards liked it or not, we all had to pack our bags and move to Paris. All the children were taught to memorise the address, in case they managed to get lost, although the rule of going out together continued to be strictly observed. I still remember the address clearly; it was rue de Colisée trente-quatre.

The building in which the commune was housed was much more extensive than the one where we had lived in Nice. It seemed that there was quite a labyrinth of rooms and passages and there was one enormous big hall with a stage. On this stage Raymond Duncan held forth every Sunday, preaching his version of salvation for the world and the auditorium was always packed. Sometimes these speeches were interspersed with performances, which were invariably taken from the classical Greek period, everyone wearing classical Greek attire. I remember one such performance when my brother Gedeon was given the part of Ganymede, who gave the gods of Olympus their daily nectar ration. The other children just sat in the front row, gazing up at the stage in envy and admiration. Some of the performances involved dance. Raymond Duncan, possibly so as not to be outdone by his much better known sister Isadora, produced what he considered to be the genuine ancient Greek type of dancing, having accumulated evidence on many previous visits to Greece, as well as to museums all over the world, in which he studied ancient vases on which dancing figures had been depicted.

My mother became extremely interested in this and began to incorporate some of these Greek ideas into her evolving system, through which she was attempting to describe the whole gamut of possible human movements. The development of this system of movement, called 'orchestics', was to occupy the major part of the rest of my mother's life, after our eventual return to Hungary.

Life in the commune went on in a very regular way, following a daily and weekly rhythm. I remember being taught to read and write and gaining much pleasure in being able to read stories to myself. I remember once getting a present of a beautiful, big, illustrated edition of the Arabian Nights, while my brother Gedeon received Hans Andersen's tales, in much smaller print and so with many more stories! I thought this was totally unfair, thinking that you could not read the pictures, that they were really a waste of space and that my brother had the best of the bargain. So I used to secretly read the Andersen stories while no one was looking, quickly going back to the Arabian Nights if I felt there were any danger of being detected. The story that impressed me most was called *La vierge des glaciers* and I would read that time and time again whenever I had the chance! I had carefully kept my original reading book, from which I had learned to read. My favourite little story in that book was called *Les premiers pas*, which was about a baby trying haltingly to take his first steps. Gedeon was always jealous of my reading, as he did not read that much and was more of a sociable creature than I ever was. One day he came over while I was looking at the page on which *Les premiers pas* was printed, tore out the page and crumpled it up. I burst into tears, trying to unruffle the page and I found it quite impossible to forgive this most heinous crime.

I found writing more difficult than reading. I remember that I had to keep asking, '*Est-ce qu'il y a une e à la fin?*' We also had to write a weekly letter to our father and I remember writing *Mon cher père* at the top of the page and sitting there staring at the empty page for hours, it seemed to me, not knowing what and how to write! But eventually I did learn to write and so lost my fear of it, soon beginning to enjoy the possibilities it opened up to me. One part of the daily routine involved the children singing Greek songs. These were, I believe, in modern Greek and we all learned to sing the catchy tunes, vocalising the words as best we could, but Duncan forgot to tell us what the words meant and to this day, I remember some of the words but have absolutely no idea what they mean!

One day Duncan's mother died. Duncan wanted the children to learn about death, so we were all ushered into the room where she was laid out and we were asked to sing some of the Greek songs, which we did. I remember hearing some of the adults whispering to each other that it was not right to bring children to a dead body like that, but we just sang on merrily, without of course realising what we were singing! As far as I know, nothing harmful ever resulted from our contact with the dead Mrs Duncan!

One of the weekly routine events was a trip to the Bois de Boulogne. I am not sure what day of the week this happened, but it was probably on Saturdays. On a fine warm day we wore our usual chitons and sandals, but if it was cold we were provided with fur coats to wear over the chitons. A couple of taxis would draw up and the adults whose duty for that day was to mind the children would pile in with us and we all drove up the Champs-Elysées, past the Étoile, to the Bois de Boulogne, where we played on the grass. We created a small sensation in the bois, especially on snowy days, when we still wore our sandals and rolled in the snow and played snowballs, the onlookers shocked to the core at such cruelty to children. In fact, I do not recollect any of us getting even a mild cold as we were fed on very healthy food, on a diet far ahead of the time and we had plenty to do in helping to run the commune.

I do not remember the discipline being very strict, but of course we had to do what we were told. I do not remember ever being punished for anything and it is unlikely that I could have been such a model child that I did not need the occasional correction! The only time I remember witnessing punishment was when Ligois, one of the girls, was caught masturbating and she was beaten. I remember, she kept trying to run away, but Aia, her mother, who was beating her, kept catching her, beating her some more and screaming at her, '*Cela ne se fait pas! Cela ne se fait pas!*' The memory of this somewhat severe chastisement has stayed with me in vivid imagery to this day and I remember feeling so sorry for poor little Ligois, with whom I had played happily on many occasions.

Our life at the commune went on following a regular pattern. The adults were employed in making enormous pictures on canvas, which Duncan himself had designed. They also engaged in spinning and weaving and in making sculptures. These were sold to the public; no doubt, the commune lived on the proceeds. There was some kind of unwritten contract between Duncan and the members of the commune which was to last a lifetime: Duncan would undertake to provide and the members of the commune had to undertake to do the necessary work, so that Duncan could in fact provide. How legal such a contract was nobody quite knew and I do not suppose it would have stood up in a modern court of law.

To cut a long story short, one day our mother called us, namely Gedeon and myself and carrying just a small bag, asked us to come outside with her. This was of course unheard of and we both exclaimed, '*Mais où sont les autres?*'

Our mother replied simply, '*On va faire une petite promenade.*'

This little statement was to become a family password for decades to come and I used to ask my mother, '*C'est toujours la petite promenade?*' and she would answer that indeed it was and that the whole of life was *a 'petite promenade'*, leading us somewhere we cannot yet know but will one day!

We eventually reached a railway station. I remember that there were a number of trains standing at the platforms which were double-deckers and there was just one with ordinary carriages. I fervently hoped that, if we were to travel on a train, we should travel on a double-decker. Of course these were the local commuter trains and the only long-distance train standing at the station was the one which had normal carriages! This is where we went. So much for my dream of a double-decker trip! The train soon moved off but our mother kept strangely silent. No doubt she was worried that she was breaking the contract and that we could be legally made to go back and rejoin the commune! Whenever we asked if this was still the *'petite promenade'*, she just nodded. So we concentrated on looking out of the window and seeing the trees, towns, villages and everything else flash by.

Soon we reached the border between France and Germany. It did not seem to take long to get through the formalities, but as soon as we were in Germany our mother said to us, almost in a whisper, *'Dès maintenant il ne faut pas parler en français!'*

Gedeon and I looked at each other in amazement. Naturally, by this time we could only *'parler en français'*, so, if we could not speak in French, we both logically concluded that we had to keep quiet and wondered why our mother had chosen such a complicated way of telling us!

It must be remembered that this was happening only a few years after the most bloody war in history. Germany had been defeated on the battlefields, humiliated by the Versailles treaty and made to pay 'reparations' out of what little was left after everything had been spent on the war effort. It would have been useless to tell Germans that we were really Hungarian children, Hungary having been a loyal ally of the Germans during the war. How come we were chatting away in French? Clearly, our mother could foresee what might happen and wished to avoid it. But she could not expect her two small boys to understand such twists and turns of world history and how it tied people in knots who had been involved, if only on the sidelines! So we had to be kept quiet!

The end of this particular journey was the Bavarian town of Oberammergau, famed for its passion play of the Crucifixion of Jesus Christ. Here we met our father and his new wife Sari. With the favourable exchange rate between the English pound and the sliding German mark, we were able to rent a large house and live very well while we stayed in that picturesque part of the world. As children, of course we were not aware that the *'menage à trois'* must have been quite difficult for the adults in charge of us, but, as far as I can remember, we were not disturbed at all by any difficulties they may have had. Life in Oberammergau, however, was life in limbo! We did not know whether we would be going back to France or whether it would be possible for us to go back to Hungary. Everything hung in the balance, but it was summer, the country was beautiful, the people were friendly and we children had begun to learn to speak German again!

We were taken to climb the surrounding peaks, which we all thoroughly enjoyed, except for our mother, who preferred to stay at home. There is just one occasion I remember when I was really frightened climbing up a mountain. We were on the last part of the ascent of the Ettaler Mandel, which involved a certain amount of discreet rock climbing. At one point I remember that I could not find another foothold or another handhold with which to pull myself up and I sadly informed my father, who was climbing up behind me, *'Papa, j'y resterai toute ma vie!'*

This was another logical conclusion, drawn from the premise that I could go neither up nor down and that therefore the only possibilities were up or down or staying where I was. I concluded logically that I would stay where I was for the rest of my life! In spite of the correctness of the reasoning, the premise turned out to be false, as, in fact, I did manage to climb up in the end. We all reached the top, which consisted of a few square metres of level rock, surrounded on every side by almost vertical drops. Gedeon and I dangled our feet down, looking down at least one thousand metres on to a beautiful lake below.

This was the time when the German mark went mad. I remember being given 100,000 mark notes to play with, as they were no longer of any use for buying anything. This did not affect us, as we were living on English sterling, earned in Aberystwyth by our father and we never changed more than we would need that day, as the next day the same money would buy less than one-half of what it could buy today!

Time flew, Gedeon and I learned German again, but we tended to speak a kind of mixture of French and German, more German when we were playing with local children, more French when we were alone or with our parents.

It was finally decided that, when our father and Sari had to go back to Wales, we would go back to Hungary and stay with our grandmother in the country town of Pápa. I remember buying my last sugar bun for seven million marks, which would have been only one million marks the day before. Our train steamed off in an easterly direction and Gedeon and I enjoyed looking out of the window, surveying the world as it passed by. We particularly enjoyed some of the little trains on narrow gauge tracks and as soon as one of us spied such a track, we would exclaim, *'Regarde, le ptiten train ses rails'* the made-up word *'ptiten'* was a Germanised version of *'petit'*. Probably the word *'ses'* was instead of the 's' in the German genitive, but I am not sure. But this sentence, in bastardised French, still keeps ringing in my ear, every time I look back on that journey.

We finally arrived in the town of Pápa, where we were taken in a carriage, drawn by two horses, to the apartment occupied by my mother's mother, our grandmother, who welcomed us like lost souls. Fortunately, she could speak French fluently, so we felt welcome and at last relatively safe. So far so good with the *'petite promenade'*!

Chapter Three

Getting Re-established in Hungary

When we returned to Hungary, I was about six years old. I had already spent various periods of this short span of life communicating at first in Hungarian, then in German, then in French, then I had added another sprinkling of German and now here I was having to learn Hungarian again from scratch, as that was now the language of my new environment. The fact that my grandmother spoke French made things less traumatic, but, of course, Hungarian was the only way to communicate with anyone outside the house.

After a decent interval of initial adjustment, my mother decided that I must go to school. She took me along one morning to the local elementary school and introduced me to the grade one teacher, who appeared to me to be a kind man and what is more – miracle of miracles – he could speak French! For a teacher in a small country town in Eastern Europe this was indeed close to a miracle and to this day I do not know how my first teacher in Hungary had learned to speak French!

The teacher sat me down next to another small boy, who was detailed to look after me. The teacher realised that I could not speak Hungarian, so he asked me to come to the blackboard and wrote out some addition problems for me to do, which I promptly did without any errors. The children seemed impressed, as the numbers in question had several digits! Then he asked me if I knew the Lord's Prayer, which our mother had taught us a few days previously, so I said, yes, I knew the Lord's Prayer and recited it with as much meaning as a six year old can give to such a profound prayer! Then he explained to me, in French, of course, that he would now ask the class to recite the Lord's Prayer in Hungarian. I listened intently to this recitation, trying to make out which words might correspond to the words in the prayer I knew in French. At the end of school, the teacher asked my small protector to take me home, having explained to him where to go. So we walked along the street hand in hand, at times looking at each other but not knowing what to say, as we both realised that the other would not understand! Then, almost at the same time, we had an idea: I started reciting the Lord's Prayer in French and when I had finished, he recited it in Hungarian, then I started off again in French and then he continued in Hungarian, until he delivered me to the door of the apartment house where we now lived. We said goodbye, each in his own language.

This was my first day at school upon our return to Hungary. It seemed that the Lord had chosen to sanctify the day by using the prayer that Jesus had taught us so many years ago! It was also the beginning of a somewhat tortuous spiritual journey, which would take me from a simple childish Catholicism to a somewhat more mature one as I grew older, to be rejected as nonsense during my teens. After a long period of atheism and Marxism, I joined the Religious Society of Friends, to which I still belong, although leaning towards an interpretation more inspired by the teachings of Jesus Christ.

We soon settled down to the routine of our new life in the small country town of Pápa. Gedeon and I shared a small room in the apartment, whose window looked out on to a large farmyard, where every week the farmer would kill a pig. I remember being somehow glued, spellbound, to this cruel spectacle of a screaming pig being bled to death in order to satisfy our own need for whitish meat. I remember feeling a terrible sinking feeling in the gut and trying to tear myself away from the spectacle yet not being able to because of morbid fascination with the impending death of the wretched animal. I am not sure if it was on account of these killings, but I became a strict vegetarian at that point. I remember giving the excuse that I could not chew the stuff!

I remember that I was convinced that one of the walls of the room had wolves in it. I used to have nightmares about wolves and I remember that they always howled three times before they came to attack and devour me; of course, just before they did so, I would wake up in a cold sweat and yell for my mother, who would come and comfort me. So I would explain that the wolves were in the wall and that they had nearly got me just then. My mother tried to explain logically that no wall could have wolves in it, that, anyhow, wolves lived in the forest and would not ever come anywhere near houses, let alone come to live inside a wall. Unfortunately, nightmares are not logical so I was not convinced and for years afterwards I regarded the number three as a very bad number, the number nine being even worse as it contained three lots of threes so there was even more scope in it for the wolves! This was unfortunate, as I had to learn later that three was a holy number, but, in spite of all the religious teachings about the Trinity, I could never get the wolves out of my system!

My mother believed I was not very good with my hands and so thought I could be helped by working with a carpenter who had his workshop in the basement of our apartment block. I was duly introduced to this carpenter, who was a nice enough man, but of course he had to make a living and could not spend time teaching me carpentry.

He gave me a hammer and some bent nails and showed me how to hold the nail with one hand and hit it with the hammer, so as to straighten the nail. Then the nail could be used again. After a number of times of hitting my own hand with the hammer, I learned to be quite good at nail-straightening. It was not until several weeks later that my mother found out that I was not learning any carpentry but was straightening the carpenter's nails for him. At this point my visits to the carpenter abruptly ceased, once I had straightened possibly several thousand nails! This just shows how unintelligible the adult world is to a child. So many things are strange and hard to understand, so, if one is sent to a workshop to straighten nails, one just gets on with it, no questions asked. It is just one of those strange unintelligible things of which the world of adults seems to consist.

My mother had a sister, Edith, who taught in a school in Györ, about fifty kilometres' distance from Pápa. In order to visit her, we used to take the train to Györ, which was always a treat. Our Aunt Edith could also speak French, so we were happy to go and see her and there were always treats like poppy seed cakes and other goodies. I remember that on one of these visits I had found some treasures in her waste-paper basket and was in the process of sorting everything out, having put the various treasures in tidy piles, when it was time to go, because there was only one train that went at an appropriate time to take us back to Pápa. I was right in the middle of the most complex and exciting sorting activity, possibly learning a lesson in Logic provided by the environment, so in anguish I exclaimed, *'Mais j'ai encore mille choses à faire!'* and burst out crying. I was even more horrified when all my 'treasures', so tidily arranged, were simply swept up! It was a tearful journey home, with my mother trying to console me, trying to point out all the lovely red poppies growing along the track.

For years after that I always wondered why poppies grew only in places where you could never get at them and why you never saw them when you went out for a walk.

The three of us were able to live on the ten English pounds a month sent by our father from Wales as a regular payment to support my brother and myself. As the exchange rate between the Hungarian korona and the English pound became less favourable, it was more and more imperative for our mother to earn money in order for all of us to survive. This seemed not very possible in a small town like Pápa, so eventually we had to think in terms of moving to Budapest, where it was more probable that some work could be found. One such possibility appeared to be a job teaching dance in a private school, known as the Új Iskola (the New School). This had the advantage that, in the package offered, tuition for Gedeon and for myself was included. So it was not long before we found ourselves in Budapest, where our mother rented a room in a small villa in the midst of a very large garden which we both learned to love.

The garden was a paradise for children. There were a number of little copses of trees dotted about in it and paths criss-crossed in all sorts of directions. There was a little ruin at the far end of the garden, where Gedeon and I used to play fantastic games, imagining armies meeting in combat, fighting for possession of the ruin. There was a small fig tree right at the entrance of the ruin, which actually bore some figs, but unfortunately we picked them too soon and they always gave us collywobbles! The garden was so exciting to us that we decided to make a map of it. We measured all the paths, the distances between the copses and the length and the width of the entire garden by pacing out the distances. It took us several weeks of work, but eventually we came up with a fairly accurate map, using a definite scale, whose ratio I cannot remember now. I suppose this might have been our first introduction to geometry, in the original Greek meaning of the word, which means measurement of land.

All three of us, Gedeon, our mother and myself, shared one room. Gedeon and I slept in a large double bed, our mother in a single bed. She always rose at four o'clock in the morning in order to work on her system of human movement and to prepare appropriate lessons to give at the Új Iskola. This meant that she needed some quiet time quite early in the evening, so we were not allowed to talk after a certain hour. To both of us this seemed too early, as we could never go to sleep that early. So we invented a tactile alphabet, which consisted of touching each other's hands in certain places in certain ways to symbolise different letters and in this way we could carry on a conversation in bed, in the dark, without disturbing anyone. In fact, I do not know whether our mother ever found out that we had invented such a method of communicating silently in the dark! In the morning we all took the tram to a certain spot not very far from the Új Iskola and went to a small dairy to have a drink of milk and a roll, which was our breakfast. Then we walked up a small hill to the school, where we stayed for most of the day and our mother gave her dancing lessons to the various classes.

It took me some time to learn the Hungarian orthography, even though it is almost entirely phonetic. I remember using the letter J instead of the letters ZS for quite a long time for the sound which stands for J in French. We also had German lessons, during which we had to learn the now old-fashioned gothic writing. I remember that the German teacher thought I was not writing the gothic E very well and asked me to write a lot of gothic Es, so I could improve my writing. Then she forgot about me and when she came back, I had filled up a whole book with gothic Es. She duly praised my efforts but told me that I was really overdoing things. I did not understand this, since I thought I had done precisely what I had been asked to do! Another mystery of the adult world!

We had exams at the end of the year, most of them oral. I remember very clearly my first mathematics exam. I was asked to write a subtraction exercise on the board and then carry it out. I wrote one in which both the numbers had several digits but made very sure that, in the number 'below', every digit was smaller than the digit of the corresponding number above it, so that I would not have to 'borrow'. Naturally, I easily obtained a correct answer. While I was working it out, I heard the external examiner whisper to the school examiner that was it not rather smart of this child to make sure there would be no borrowing? He suggested that they should pass me without any further ado on account of this particularly evasive performance! This just shows you that adults cannot whisper about children who are around without these children getting the gist of it!

Our mother was still not sure if we were to settle in Hungary or whether we should eventually go back to France. So we continued to speak French at home, even though our Hungarian was quite firmly established. To this was added German, which we soon learned again as a result of its being a compulsory school subject, or possibly because of the year in the *Kinderheim* in Vienna and the few months of German immersion in Oberammergau. German was the only second language taught at the Új Iskola, as French and English were reckoned to be less important; the countries where these languages were spoken being much further away. So, by the time I was eight, I was fairly proficient in all three languages.

We soon started learning Latin as well. I thought it was great fun to learn a dead language that people had spoken hundreds of years ago. To make things more fun, our mother introduced us to a person we came to know *as Frici bácsi* (or Uncle Fritz), who tutored us in the various ins and outs of the rules of Latin grammar. He made us learn both the rules and the exceptions in the form of rhymes that I believe he had made up himself. Gedeon and I then would recite these verses as we walked along the streets of Budapest, to the astonishment of the passers-by who naturally could not make out what these crazy boys were saying!

One day, while Gedeon was in bed with a feverish cold, my mother told me to go across to the pharmacy and get some medication for him. She told me what the medicine was called but in Latin. This was the first time Latin came in useful! Long after she was out of the house, but as soon as I thought the pharmacy would be open, I went over and asked the pharmacist for some *syrop hypophosphorosus compositus*, which I reeled off with the greatest of ease, to the amazement of the pharmacist. I suppose you don't often see an eight year old boy come into a pharmacy and ask for medication in fluent Latin! I paid for the syrup, walked across the road to our lodgings and administered the *syrop hypophosphorosus compositus* to my brother. It worked. The next day he was better and ready for school!

So the three of us were getting used to life in Budapest. The arrangement between my parents was that, during the winter, we should remain with our mother but during the summer vacation, which lasted two or three months, we should visit our father. These visits to our father would take place in various different countries, where we had relatives and friends, such as in the parts of Czechoslovakia and Romania that had belonged to Hungary before the war or in Italy, where Sari's mother had a *pensione* and later on even in France and eventually in England.

This was to be a kind of to-and-fro life, not only geographically variable, but quite confusing from the point of view of the clash of philosophies. Our father being a

Marxist (even though a crypto-Marxist, as he had to keep on the right side of the authorities in Great Britain!) and our mother being a Catholic, the to-and-froing also involved changing philosophies twice a year! The procedure had a number of effects for we learned or perfected various languages, which was to come in very useful both for Gedeon and myself. In fact, Gedeon's knowledge of Russian saved his life during the 1956 Hungarian uprising! My knowledge of Italian made it possible for me to have a considerable influence on the development of methods of teaching mathematics in Italy. One result of the to-and-froing of the philosophies was, in my case at any rate, a development of tolerance of widely differing points of view, beliefs and philosophies. As a result, in the late forties I became a Quaker, I suppose as a kind of satisfactory halfway house between atheism and Catholicism.

Chapter Four

Some Holiday Experiences During Pre-Teen Years

Growth is always a difficult process. Birth, growth, maturation, senility and finally death have been invented by nature to ensure eternal life to the race if not to the individual. So going through these stages is bound to be full of painful adjustments, as an individual does not have thousands of years to solve his or her problems but must telescope the process into three score years and ten, of late slightly increased, but still a short span in which to solve all problems! As a result, many of us moan about our unfortunate childhood and blame everybody but ourselves for all the unfortunate things that happen to us, not realising that we all have to live with the package with which we are launched! In this section I should like to tell about some of the events that shaped my package, without making any judgements about whose fault was what but simply telling how the events unfolded.

We moved away from our romantic garden after a year or so and my mother boarded us with families, so that she could be free to pursue her work developing her system of dance, the theory and practice of orchestics (from the Greek *'orcheomai'* which, I believe, means 'I move'). Eventually, we were able to rent a whole apartment on the corner of Krisztina körút and Csaba utca, in which two large rooms were made into one big hall for the purposes of teaching the 'Art of Movement', one room was an office and one room was reserved for the family, namely for Gedeon, our mother and myself. There was not really a place for cooking, so we usually got our food from a local soup kitchen, something like a 'meals on wheels' system. After some years at the Új Iskola, we both moved to a Catholic secondary school run by monks, in the centre of the city, which meant either a forty minute walk each way or a ten minute tram ride, depending on the vagaries of our financial situation!

As I have already intimated, our summers tended to be more eventful than our winters, as we always had to travel outside Hungary in order to catch up with our father and Sari. I should like to describe some of the events that have stayed in my memory from our years of trans-European peregrinations, as these might have been the ones that had an influence over what I did later on in life. So here are some scenes, not necessarily in chronological order, which will give some idea of the kinds of memories that have stayed with me from those early years.

Picture a little train, moving along slowly on a narrow gauge railway track, winding its way up the tortuous valley of the River Aranyos in the depths of Transylvania, which had been transferred from Hungary to Romania a few years previously. There were three different classes on the train. The first-class carriage had reasonably comfortable seating, the second-class carriage had wooden seats painted green, presumably to look comfortable in spite of the hard wood! Then there were several third-class carriages, where the wooden seats had been left their natural colour. The first-class carriage was quite empty; the second-class carriage contained my father, Sari, Gedeon and myself, as well as my Uncle Laci and his wife and three children, all girls, ranging in age from about seven to eleven. The third-class carriages were crammed full of peasants, their children and their innumerable bundles. Suddenly we heard a big explosion and the screeching of brakes and the whole train came to an abrupt stop in the midst of what looked like a primeval forest. The conductor came running through all the carriages, telling us in Hungarian and Romanian to be calm, that the engine had simply broken down and that another engine would soon be sent to rescue us!

To cut a long story short, we waited in the forest for several hours while they were sorting out the engine problem, since someone had to walk to the nearest village, several kilometres away, to phone for help. All the passengers got off the train and the children had a great time playing in a mountain stream cascading down the mountainside. The peasants opened their bundles and their flasks of wine and sang songs to while the time away. After another wait further up the line, where the tracks had been washed away by a freak rainstorm, we finally arrived at the end of the line, eighty kilometres from Torda where the line began, which had taken us about twelve hours! Fortunately the two horse-drawn carriages that had been ordered for us were still waiting there, so the two families piled into them and we rumbled along in the pitch dark for several more kilometres, until we reached the village of Verespatak, where we were to spend the summer. For an eight year old boy, this was quite an adventurous start for a summer holiday too!

It turned out to be a wonderfully memorable holiday too.

'Jön a ló és megesz minket és kész!' 'The horse will come and eat us up and that's the end!' You have to imagine an eight year old boy running into the house panting, dragging a little cousin of about the same age behind him and uttering the above somewhat surprising sentence. You have to go back to the summer of 1925, to a small mountain

village in Transylvania known to Hungarians as Verespatak and to Romanians as Rossia Montana, where we were spending the summer. My Uncle Laci worked in Kolozsvár (known as Cluj to Romanians) as the editor of a left-wing periodical called *Korunk* (*Our Age*) and as I have already said, we spent the summer with him, his wife and three children. The little girl I was dragging behind me was one of Laci's children and thus my cousin.

I have to explain that Verespatak is, or at any rate, was then, a gold-mining community. Most of the men worked in the local gold mine, but there was also gold found in the local streams. The village was full of these little constructions that looked like mills and I do not quite remember how but I know they were somehow activated by horses pulling some receptacle up a stream. When the horses were not required for this work, they were simply let loose in the village, so there were a great many horses grazing everywhere, roaming freely. This was clearly something I was not used to and a horse being so much bigger than myself, I assumed that it would have aggressive intentions.

My reasoning went something like this: The horse is coming – consequence: it will eat us – consequence: the end. The second piece of reasoning is quite acceptable to any adult but the first piece seemed even more obvious to my little cousin and myself! So logically, or psychologically, the coming of the horse definitely implied the end of our existence!

Apart from the first day of terror occasioned by the wandering horses, our holiday in Transylvania appears to have been a very pleasant one. There was a mountain lake surrounded by a forest of pines and firs, about half an hour's walk from the village and we would walk over there just about every day for a bathe. My father tried to encourage us to swim and he designated certain trajectories, marked by some trees or rocks on the shore of the lake and swimming to each of these landmarks would be rewarded by some desirable outcome, such as a bar of chocolate or a bag of sweets. The longest trajectory was to swim right across the lake, the reward being a large cream cake! I remember getting as far as the bar of chocolate, but I am afraid that the cream cake remained forever out of my reach, as I had not learned to swim more than about ten metres by the end of the summer!

Among the many memorable events which took place that summer, one very interesting one was a visit to the gold mine. One thing which struck me was that the miners had to be stripped naked after a shift in the mines, I suppose to prevent them from taking any of the ore home with them! It also occurred to me that it was strange we were not stripped naked as well; after all, we might have been stealing some ore!

Even though most of the people in the village spoke Hungarian, some spoke Romanian and we began to get interested in learning something about this language. I am afraid all that remains with me to this day are the words *'mike'* for small and *'mare'* for big. We sometimes would go and tease the Romanian-speaking children by saying to them, *'Romania mike'* and they would shout back, *'Romania mare!'* which was not exactly an edifying conversation! It is probably from such encounters that racism eventually develops.

Now picture the north eastern corner of the Adriatic Sea, with mountains rising directly from the shore and many islands scattered over the bay lying between Istria and the Dalmatian coast. Here lies the little spa then known as Abbazia (now known as Opatija, as, after the Second World War, this part of Italy was transferred to Yugoslavia, but is now a part of Croatia) where Sari's mother had a *pensione*, where we spent several summers, the first one in 1926. Mussolini and his fascist cohorts had by then consolidated their grip on power, but it was a far cry from the later excesses perpetrated for the glory of fascism.

Gedeon and I soon made friends with some of the local children and so began to pick up quite a bit of Italian. Our father encouraged us to learn both Italian and English, so he determined a going rate for the learning of one hundred words in both languages. One hundred Italian words were worth two lire and one hundred English words were worth ten lire. In spite of the added value for learning English, we found Italian so much easier that we both learned several thousand Italian words within a few weeks of staying in Abbazia. Hearing the words used by others gave us some rudimentary idea of the grammar, so by the end of the summer we even spoke Italian amongst ourselves.

I remember making friends with a little girl who taught me to sing *Santa Lucia*. I still remember the words:

> *Sul mare luccica l'astro d'argento,*
> *Placida è l'onda, prospero il vento,*
> *Venite all'agile barchetta mia,*
> *Santa Lucia, Santa Lucia!*

There were several other verses, one of which began with '*O dolce Napoli, o suol beato*' but I do not remember any more!

She and I used to swim out to a little rocky island, only a few meters away from the shore. Sometimes we would pick some long sticks of bamboo and help each other to move from submerged rock to submerged rock, to get over to our island without getting our clothes wet. Then we would sit on our island, looking out to sea and sing *Santa Lucia* together and we both thought this was terribly romantic!

There were certain bathing establishments with cabins, deckchairs and all that sort of thing all along the shore. We would sometimes leave our clothes on our island, leaving only our pants on for decency and swim round the next point and into one of these bathing places. In this way we could go on all the swings and slides and did not have to pay an entrance fee! When we thought we had had enough, we would jump into the water, swim out to sea, round the point and go back to our island.

The fascist government apparently made a strict rule that on the bathing beaches women were allowed to show only a certain amount of thigh, which was determined as the number of centimetres from the ankle to the leg of the bathing costume being worn. The carabinieri, the state police, had a wonderful time checking the thighs of all the most attractive women and fining them if they showed more than the regulation number of centimetres of leg. Gedeon and I sometimes watched them and we tended to giggle while the carabinieri bent to their official toil! On some days, the policeman in question would come up to us very crossly and shout, '*Scappate, scappate!*' which means 'scram' in the vernacular. Just to annoy the policeman, we would pretend not to understand and he would then ask us if we could speak Italian. If we thought the policeman was not a very dangerous one, we would say, '*Si, si, Mussolini, maccheroni, uno due tre! Eia, eia, allallà!*' and dutifully put our hands up in the fascist salute. After a few more shouts of '*scappate*', we usually took to our heels, before things got too hot! Sometimes, to vary matters, we would sing a part of the fascist anthem *Giovinezza*. They could not really run us in for being so patriotic!

Our father was usually quite lenient with us, believing in the so-called modern theory of laissez-faire education. He did not believe in corporal punishment, so Gedeon and I thought we really had it made. That is, we did until one day the tables were turned and we both received what could only be termed a good hiding! The crime we had committed was playing the following game: we had to run across the road when a car was coming and get as near as possible to the car without getting killed. On looking back, I think it is small wonder that the invention of such a daredevil game shook my father's belief in 'modern' education!

I remember being extremely proud of having been beaten and telling our friends with whom we used to play, *'Mio padre è molto forte, fortissimo!'* ('My father is very strong, ever so strong!'). *'Come mai?'* ('How come?'), they enquired.

'Oggi mi ha picchiato, come se fosse stato un animale feroce!' ('Today he beat me just as though he'd been a wild beast!').

My prestige immediately went up by several notches with the local children!

Our father was not only keen that we should learn Italian and English (apart from learning how not to be killed by passing cars!), but wanted to improve our Latin and mathematics. I remember being given a Latin book, in which the left-hand side of each page was written in Latin and on the right was the translation.

Of course the translation was in Italian. I remember looking at the translations and thinking that it really was not worth putting the Italian version in the book, it was so much like the Latin version. But we dutifully spent a little time each week learning some more Latin, thereby also learning some more Italian. Our father thought I was too young to understand algebra, so he gave some simple algebra lessons to Gedeon, which consisted mainly of solving equations. I was usually chased out of the room during these lessons, as I might have been a disturbing influence. This made me all the more curious about what all this algebra was about. One day I managed to hide behind a curtain before the lesson started and listened very intently to mysterious happenings to a thing called X that was supposed to be unknown, but in the end, by some magic, it became known! I simply *had* to know how this magic worked!

One day when Gedeon did not seem to understand the equation to be solved, my father used some stones to represent the mysterious X and used some matches to represent the numbers. I do not exactly remember the actual equation, but it looked something like this:

$$SSSS\ MM = SSS\ MMMMM$$

the Ss representing the stones and the Ms the matches used. The problem was finding out how many matches a stone was worth so that the pile on the left was worth as many matches as the pile on the right. The magic resulted from the outcome that each stone was worth three matches, but, from behind the curtain, I could not work out how this magic worked, although a swift calculation showed me that, for the two piles to be worth the same number of matches, each stone had to be worth three matches, no more and no less.

There were lots of rocks, little ones, big ones, on our little rocky island. When I found myself alone with my little girlfriend I told her about my mathematical problem. She was not interested in solving it at first, but finally, to please me, she agreed to co-operate by finding a lot of big stones and a lot of small stones and putting them in two separate piles. Then I constructed the equation by making two piles, using our big stones instead of the stones my father had used and the small stones instead of the matches. My friend saw almost immediately that in the first pile there was one more big stone than in the second pile and in the second pile there were three more small stones than in the first pile, so, for the two piles to have the same value, each big stone had to have the value of three small stones. I saw the logic of this and promptly decided that it was good to have such a bright friend! We made up some more equations with our stones, but to our dismay we could only solve a very small number. The reason for this being, upon looking back, that our big stones would have had to have fractional values in terms of the little stones and neither of us knew much about such esoteric, mathematical creatures as fractions!

My father and I often went out into the mountains on long walks, so we got to know the hinterland of Abbazia very well. The highest peak around was called, quite logically, Monte Maggiore and we were both determined to climb it. Gedeon was not so keen on these trips, so he stayed at home with Sari and Sari's mother. It took us several hours to climb the peak, but finally we made the triumphal last steps to the summit and looked down over Istria to the west, to the Adriatic Sea with its many islands to the east, and to the mountains on the horizon towards the north. It was a great moment for a small boy: the achievement of conquering a peak with a person with whom it was helping to make a stronger bond.

We lingered a little on the summit and before we knew what was happening, a sudden thunderstorm blew up, seemingly from nowhere. We spied a village on the other side of the mountain, which we thought would be the nearest place of refuge from the storm. But it was too far. As we ran down the steep slope in the direction of the village, lightning seemed to be striking the ground all round us and my father said we had better lie down. He explained that on no account must we shelter under any tree; perhaps a small shrub would be all right, as those would not attract lightning. We stayed shivering under the shrub for what seemed like eternity and when the storm started to let up we continued our descent. It was nearly dark when we reached the village and there was no way we could get back to Abbazia; there were no telephones in the village so we could not contact anyone to let them know what had happened to us.

We found lodgings in a peasant's house and after a meagre repast, we were shown into a room with one bed and nothing else, with a door that would not shut. We got into bed without undressing and tried to sleep, but there were many strange noises. The owner of the house, came back from the local bar quite drunk, stamping about the house and swearing copiously. I still remember the swear words he used: they were very unflattering to the Madonna and I am sure they are unprintable, so I will not repeat them.

'Are you very strong, Dad?' I asked my father.

'Don't worry,' replied my father. 'When somebody is drunk, they are weakened by the drink. I could easily deal with him!'

I was earnestly hoping that my father was telling the truth, but eventually the house quietened down and we both slept. In the morning we paid the owner a few lire and took the bus to Fianona, from where we made a telephone call to Sari and just caught the boat for Abbazia. There was quite a crowd at the quayside to meet us, as the story of our adventures had spread far and wide in the little town and many, including Sari and Gedeon, were anxious to see us back safe and sound.

Another memorable thing we used to do on our Italian vacations was to go to Fiume. This town used to belong to Hungary before the First World War. It was the country's port and so was of considerable commercial importance. Many people still spoke Hungarian, this having been the official language before the war. At the end of the war the Italian writer D' Annunzio took it upon himself to march into Fiume with a small detachment of Italian soldiers and to plant the Italian flag there. Thus the borders of Italy were extended right to the small river that separates Fiume from Susac. During the Second World War the communist partisans, led by Tito, were not happy about the Italians so they in turn marched into Fiume and renamed it Rijeka, but they carried on right across Istria, occupying it right up to Trieste, which they renamed Trst. It must be very confusing to the local inhabitants, not knowing what country they belong to and what languages they should speak in public! Fortunately for Fiume, the word 'Fiume' means 'a river' in Italian and the word 'Rijeka' means 'a river' in Croat, so at least the meaning of the name of the town was not changed, even though the language was.

Our trips to Fiume were mainly shopping trips, but sometimes they were trips to the dentist. I remember that the dentist was very kind to me, even though the visits involved some painful experiences! He was Hungarian, having lived there during the pre-war Hungarian times. When the treatment was over, my father said we could return to Abbazia by hydroplane! I had never flown before, let alone been in a hydroplane! This was an experience to be relished. We were taken by small motorboat to the hydroplane, floating gently in the harbour. We all climbed in and the pilot started the motor, which was very loud. We taxied out of the harbour to get a clear run for take-off and then roared along at what seemed a terrific speed. You could not see a thing out of the window for the spray, which suddenly disappeared as we became airborne. It only took a few minutes to fly to Abbazia and the next thing we felt was a quick succession of very loud bumps as we landed on the small waves of Quarnero Bay. We were duly landed at the quay and I recall feeling that all the agony at the dentist was worth that wonderful experience of flying from Fiume to Abbazia in a hydroplane!

The following year Gedeon and I spent the summer in Czechoslovakia, again in a part of the world that had previously belonged to Hungary, so it did not seem as though we were very far away from home, as most people could speak Hungarian. Friends of my father's had a 'castle' near a little village called Körtvélyes (known now as Hrusov in Slovak). These friends had belonged to the Hungarian avant-garde literary circles in Budapest before the war and every summer a number of members of this intellectual elite were invited to spend a part or the whole of the summer in Körtvélyes, where they ran what they called a 'private university', where everything under the sun was discussed. Our mother took us as far as the border by train, where we were met by our father and we were then taken to the 'castle'. This was a single storey but very extensive building, set in the midst of a very large garden, which mainly consisted of large fir trees and a few fruit trees.

I remember our first meal, which took place round a very large circular table, adults and children alike sat randomly, or so it seemed, around the circumference of the table. Everyone seemed very polite. Servants were serving everybody and it was a kind of, 'Would you like some butter?', type of meal, with the adults conversing about things largely unintelligible to children, of whom there were six, all boys, including Gedeon and myself.

The next day only the adults dined in the main dining room and the children were served their food in the very large entrance hall next to the dining room. As soon as the food was put on the table by one of the servants, the other four boys instantly reached into the dish with their hands and grabbed all the food, so Gedeon and I remained hungry! In our case, this is what psychologists would call one trial learning, as the next day we both grabbed as effectively as the other boys and managed to get our share of food.

This was supposed to be 'modern education', I later discovered. There was another principle of this modernity, according to which children should be provided one environment in which they could do whatever they liked, so that they would not then resent the social restrictions prevalent in most other environments. The two Jászi boys, who were the children of the proprietor of the 'castle' had one room in which they could literally do what they liked. Gedeon and I were invited to see their room and we were surprised to be greeted by a huge pile of earth and stones, a broken set of floorboards and a big hole which the boys had dug in the room! When we asked them why they had made the hole, they simply said that they wanted to know what was underneath.

The private university sessions usually took place on the terrace at the back of the house, which was surrounded by very large fir trees, so it was well protected from the strong sun. We were strictly forbidden to be anywhere near the terrace while these sessions were in progress, which made us all the more curious to find out what was going on!

So Gedeon and I, while no one was looking, constructed a platform at a great height above the terrace, using the branches of one of the fir trees as a foundation and bits of wood we could find in the overgrown garden for making the platform. When this was ready, we climbed up the tree about ten minutes before the sessions were to begin and kept extremely quiet in our secret hideout, from which we could see and hear everything that went on down below on the terrace! Gedeon only stayed for one of the sessions, but I became a regular inhabitant of our lofty platform during the afternoon sessions of the university. I learned a great deal about how people differed in their political views, as well as a lot about philosophy. I had never heard of Plato or Aristotle before and I remember thinking how smart these ancient men must have been if they could entertain themselves simply through the process of thinking and reasoning. I was determined to find out more about all these things as soon as I possibly could.

The university was conducted either in Hungarian or in German, but this merely increased my interest in the proceedings. Sometimes the talk would turn to mathematics and in such cases it was my father who held forth most of the time. There was some talk about different kinds of algebra and since I had not realised that you could have several kinds, I pricked my ears up at this and wanted to know how this could be. After several sessions on the algebras, I thought I had gathered enough information to be able to make up my own, which I proceeded to do, making up the rules as I went along. The older Jászi boy was good at mathematics and I discussed my algebra with him. He pointed out that practically none of the rules of manipulation such as:

$$A \times B = B \times A$$

were true of my algebra. But I would not be deterred. I changed many of the rules of my game and after several more sessions with the Jászi boy, managed to construct several algebras in which most of the usual rules held good. I suppose this was the childish equivalent of a grown-up mathematician doing 'research' in mathematics!

I did not spend all my time in such intellectual pursuits! I was always very fond of nature and I would spend a lot of time studying flowers, shrubs and trees and watching birds. The large garden was mostly wild, not kept as a formal garden, so there was plenty of scope for this kind of relaxed activity. There was also a river not very far away from the castle, called the Ondova, where all the boys used to go for a swim. The children from the village of Hrusov also went there and when we saw that they all stripped naked to swim, we did likewise, seeing no reason why we should wet our clothes in the river! The river was quite deep and there was quite a fast current so it was fun to measure our strength against it and generally enjoy the cool water as a way of getting out of the hot sun.

Another form of entertainment was going for a car ride! Cars were somewhat of a rarity in those days and it happened that one member of the university, whose name was Kari, had an open car in which he would sometimes take two or three of us for a run. I remember looking at the speedometer registering sixty kilometres an hour and being extremely impressed! We invariably trailed a huge cloud of dust behind us, as asphalt was unknown in those parts, except in the larger towns such as Kosice. The bargain was that, after each car ride, the passengers would hose the car down. We all thought this was totally useless, as we knew very well that in the first few minutes of the next run the car would again be completely covered in dust!

Chapter Five

Pre-Teen Life in and Around Budapest

Many memorable things happened during the winter and spring months which stand out quite vividly in my mind's eye and so were probably events that had some influence on my later adult life.

My religious experiences certainly belonged to this type of happening. In Pápa we were introduced to the Roman Catholic faith: we went to confession and to Holy Communion; we went to Mass every Sunday, we even had rosaries and prayed Hail Marys, counting them on the rosary as good Catholics should. Our father did not approve of this, as he regarded all religion as antiquated superstition. There was a law in Hungary at that time that, in the case of mixed marriages, namely when one partner was a Catholic and the other a Protestant, the boys had to take the father's religion and the girls the mother's. Our father, being born a Calvinist, although an atheist now, thought Calvinism was much better than all the mythology that accompanied Catholicism. So, in order to obey the law, we had to go once a week to have lessons in religion from a Protestant pastor. Looking back, I think this pastor knew the situation and showing some sensitivity tried not to cause too much conflict for us between the Protestant and Catholic conception of things.

We seemed to spend most of the time learning to sing psalms, which I actually enjoyed, as both the texts and the tunes appeared to me very beautiful. While this was going on, my mother went tirelessly from government ministry to ministry to get legal permission for us to be raised as Catholics. When she finally obtained this permission, we had to tell the pastor that we would not be coming again as we were to be Catholics now. I still remember his kind but sad expression as he received this news! I also felt sad about it and about him, but I was only a child, so was powerless to be an active participant in this religious struggle going on between my parents.

Our religious problem was not only whether Protestantism was right or whether Catholicism was right: the very reason for having a religion, the existence of God and of eternal life, were all called into question during the summer holidays. I remember once writing to my father that in our last lesson we had proved the existence of God. He replied simply that this was a difficult question, I suppose so as not to arouse too much conflict. I also remember once discussing the existence of God with him in the holidays and asking him how he explained the miracles. His laconic answer was, 'There are no miracles!' Looking back, this seems to me just as much of a bigoted attitude as the one adopted by the so-called fundamentalists. It recalls Galileo's recantation of the movement of the Earth through space (to save his own neck) after which he quietly whispered to his friends as he got down from the rostrum, '*Eppur si muove!*' ('It's moving all the same!')It is amazing how we seem to arrange facts to suit our theories!

Possibly as a result of the attacks on my faith during the summers, but equally likely as a result of growing up and trying to think things out for myself, I eventually decided, round about the age of twelve, that all the Catholic mythology I had been taught was most likely fantasy, living only in the heads of priests and said to my mother that I no longer wanted to go to Mass or to take Communion. She was very sad and begged me to come, even if I found it hard to believe in everything. I replied that I could not be such a hypocrite and that was the end of the argument.

Of course we continued to have religious lessons at school, which were compulsory, but I did as little as possible, since I knew there was a rule in the school that nobody could be failed in religion. I suppose that, if they did so, they would have to admit defeat themselves, which they were not prepared to do. My rebellious attitude towards religion, in particular towards the Catholic religion, was to have a dramatic effect on my life, as I shall explain in a later section.

As the years rolled by, I became interested in many things. One of these was languages, possibly because I already knew four – French, Hungarian, German and Italian – and it seemed interesting to compare other languages with these. I remember on one birthday I received about five different grammars, among them a Romanian and a Slovak one, because my mother knew I wanted to know more about these languages. I was in seventh heaven! I already knew Latin quite well, I remember reading Lucretius's *De Rerum Natura* (which was *not* a prescribed text at school), particularly enjoying the second book in which he holds forth about the ultimate particle of matter – the atom!

Both Gedeon and I were interested in languages, so this remained something which encouraged a certain amount of interaction between us. At one point we decided that we should make up our own language. Before doing so, we had to think of a name to give our as yet unborn language. The solution came from a rather unexpected quarter. There was a kind of soap advertised on the trams and on many trams we could see writing in huge letters right across the side of the tram:

```
A   L   B   U   S
S   Z   A   P   P   A   N
```

meaning Albus Soap. To make it sound funny, we started to read it by taking a letter alternatively from the lower and the upper rows, so the advertisement then read:

```
S   A   Z   L   A   B   P   U   P   S   A   N
```

which we both jumped on as a very suitable name for our language. In due course of time this long name came to be shortened to:

```
S   A   Z   L   A
```

Now we had to invent the language!

We made up an extremely simple grammar; after all, why complicate things? Then we wrote a dictionary, with the kinds of words we thought we would be likely to want to use. After this we had TO write a Sazla primer. We did this as well and so after a few weeks of intensive work the Sazla language was born. All I remember of Sazla today is that '*ta*' was 'yes' and that '*ti*' was 'no'. My brother Gedeon, I believe, still guards our infantile efforts in linguistics. He told me that he managed to save the Sazla books throughout the Russian siege of Budapest, when our house in the Krisztina körút was on the front line for some weeks, with the Germans on one side of the street and the Russians on the other.

Gedeon and I had another common interest, the construction of imaginary countries. The imaginary world was called SENEID, which is DIENES written backwards and there were two countries in it. One was *called ITLOZ ország*, the other *ADIG ország*.

The word '*ország*' means 'country' and itloz is clearly zolti written backwards, which was the name I was known by; and adig is gida written backwards, which is the name everyone called my brother Gedeon.

We made an elaborate map of each country, with a road system as well as a railway network. Every town had a town plan in an atlas provided for each of the two countries, which included a city network of trams. There were schedules for the trains that ran along the imaginary railways tracks and there were even boat schedules so that we could travel from Itlozország to Adigország and back, taking trips in boats of regular shipping lines. We used to make trips, counting our journeys in 'real time', trying to make sure that the boat trips took place while we were at school or otherwise engaged. At one point I remember that Adigország was about to declare war on Itlozország, but peace was saved by frantic diplomatic efforts at the last minute. I suppose that we did not feel like redrawing the various maps and town plans, since much reconstruction would have to take place after a war!

Mathematics also continued to fascinate me. I recall reading René Baire's monograph '*Leçons sur les fonctions discontinues*' when I was twelve; the wonderful architecture of this work never ceased to amaze me and I really enjoyed his use of constructions as opposed to some authors who tended to use the *reductio ad absurdum*, which I always thought was a little suspect as a way of proving things. I could not but think that showing something could not possibly not exist was not as clever as actually showing that the something whose existence we were wondering about existed by constructing it! This book had a very profound effect on me. In fact, so profound that later, when I was preparing and writing my thesis for my doctorate in mathematics, part of it was about generalising some of Baire's results.

I also taught myself other aspects of mathematics, such as calculus and the theory of differential equations, which I thought were great fun. The curriculum we were doing at school I found trivial, for everything seemed rather obvious and easy, not much of a challenge, so during the mathematics lessons I often did not listen to what was going on but tried to amuse myself by solving various differential equations. One day I was doing just this, when I suddenly realised that there was dead silence in the classroom and that the teacher was standing behind me, watching what I was doing.

'You are not attending to what we are doing in class!' said the teacher rather severely, 'Come and see me after the lesson and bring that piece of paper,' he said, pointing to my scrawls, which were attempts at solving my latest equation.

At the end of the lesson I approached the inquisitorial bench of the teacher and placed my piece of paper on his desk. He asked me a few questions about calculus and about the particular equation I was trying to solve and when he was satisfied that I was not fooling he said, 'You were quite right not to listen. I tell you what; why don't you come up to my room after school and we can talk mathematics.'

I was extremely relieved and I could not wait for the time to come when we could discuss some 'real' mathematics! The teachers at my school were all monks and they lived in the same building where the classes were held, so it was easy to visit any of the teachers.

I will not bore the reader with the content of our discussions; let it suffice to say that, as a result of my frequent visits to the teacher's room, my interest in mathematics grew by leaps and bounds and we remained the best of friends until the time came when I decided to leave Hungary. In fact when I was once hospitalised for a lung infection, he came to see me at visiting time and we talked mathematics to our heart's content. He was very young and an excellent mathematician, having already published some original papers in *the Comptes Rendus de l'Académie des Sciences de Paris*. The other children in the hospital were quite convinced I was crazy to invite my mathematics teacher to hospital to talk mathematics, when I had the luck of missing school for several weeks! Of course, these children could not possibly realise how addictive the study of mathematics can become!

Many years later, in the mid-eighties, on one of my trips to Hungary I sought out my old friend, the mathematics teacher. Apparently he lived in a small village to which he had retired. I was not sure exactly where he lived or how to get to the village, so, when we were at a distance of about ten kilometres, I pulled up and asked a passer-by if they knew where Ferenczi Zoltán lived.

'Oh, you are one of his pupils, are you?' replied this person. 'A great many of them come to see him! Just follow the road until you get to a church, take a right turn there and you will find the cottage a little way up the hill!'

I duly followed these instructions and found my beloved teacher, who welcomed me and my wife with open arms. We talked about old times and I left him copies of some of my publications. A few years later, on another visit, I tracked him down in a priest's home in Székesfehérvár. I recall telling him how appreciative I still was of his contribution to my career, as, without his enthusiastic teaching, I would probably never have become a mathematician. Then he told me that, during the time of his own training, he used to go and listen to my father's lectures at the university and that it was my father who had given him the enthusiasm to take up mathematics seriously! So I had unearthed a chain of events which linked my interest in mathematics to that of my father's, the chain starting much earlier than I could have suspected. I even started giving private lessons in mathematics to some of my school friends. One of these was the son of a member of parliament and when we had finished our lessons and he had considerably improved in his performance, his father paid me what I considered to be a handsome fee, which I promptly gave to my mother, as I knew that our financial situation was not too rosy!

I was also very interested in finding out about what the whole country was like, so I joined a tourist organisation, so that I could stay in their various hostels, which were dotted over the country in some of the more scenic spots. I used to do these hikes quite on my own, sometimes staying the night in a peasant's house, sometimes in the Tourist houses that belonged to the organisation that I had joined. I had one problem: I was afraid of the dark. I was not sure what to do about it, but then I remembered reading about Pavlov's experiments with dogs, in which he explained how he could condition and de-condition dogs through a reward system. Surely I could de-condition myself of the fear of the dark? Pavlov knew that when the dog salivated, some reward had to be given, to make the link between what was to be learned and a pleasant outcome. What was to be the pleasant outcome for me? Since I liked the country, in particular I liked mountains, I thought a view from the top of a mountain might act as a satisfactory reward. So what I had to do was go for an all-night hike which finished on top of a mountain. I began doing this, as frequently as I could arrange it, starting out on a Saturday evening, walking through the forest during the night using map and compass and ending up on top of a mountain at dawn. Then I could watch the sunrise and then have a good sleep, before wending my way down the mountain again to the nearest railway station and getting the train back to Budapest. At first I did the hikes on moonlit nights, until I finally graduated to walks on nights of the new moon, when it would be pitch-dark.

For my 'graduation hike', I bought myself an acetylene lamp, took the boat from Budapest to Visegrád one Saturday evening and started on the long walk from Visegrád to Pomáz, via the Dobogókö, where I was to be at sunrise. Everything went like clockwork and when I reached Pomáz and took the train back to Budapest, I happily pronounced myself quite cured of the fear of the dark! I am not sure if Pavlov, or even Skinner, would have agreed with the psychological correctness of the treatment, but, as the proof of the pudding is in the eating, such approval does not seem to be necessary.

I slowly extended my range until I found myself walking in the Bükk, in the Mátra, in the Cserhát, in the Börzsöny, in the Pilis, in the Vértes and in the Bakonyerdö. I walked in all seasons, even in the dead of winter. The winters in the eastern part of Hungary are more severe than elsewhere and I sometimes wanted to challenge myself with tasks that involved a good deal of endurance. So, on a cold February evening, I took the night train to Miskolc, naturally taking third class, which only had wooden seats, but I managed to sleep in the luggage rack while the slow train trundled along, crossing the Hungarian planes. Slow trains cost only half as much as fast trains, so naturally a slow train it had to be! Arriving in Miskolc the next morning, I saw a different kind of country. Everything was under deep snow and the temperature seemed to hover around minus twenty degrees Celsius.

I then spent several days walking through the Bükk mountains, reaching a tourist house each time for the night. One day I found that I had somehow miscounted the distance involved, or had perhaps not reckoned with the depth of the snow, which meant much slower progress, so it was already dark under a starry sky when I was still about six or seven kilometres away from the tourist house I had to reach that evening. I was very tired and was feeling sleepy. Fortunately, I remembered being told that the last thing to do on a freezing cold night was sit down and go to sleep, for, as likely as not, you would never wake up – you would freeze. So I kept on making gigantic efforts to place one foot in front of the other, following the coloured signs on the trees which showed me I was on the right track! When I eventually saw a light twinkling through the trees, I was so relieved I nearly wept. I went into the tourist house and collapsed on the floor.

'What's the matter with you?' enquired one of the hikers who had obviously arrived earlier.

'I have just come from Tarkö, across the plateau, without skis and I am exhausted,' I replied.

'Let's have a look at your feet,' said the same man. 'I am a doctor; maybe I can help you.'

He tried to take my boots off, but he could not, for they were frozen to my feet. He boiled some water and put my boots in it to melt them, until he could, with difficulty, pull them off. Then he squeezed my feet and asked me if I could feel anything. I could not. So he brought a container of boiling water and another one which he filled with snow and he kept putting my feet from the boiling water into the snow and back into the boiling water and asking me if I could feel anything. For what seemed a very long time I could not, but eventually, when he stuck a needle in my foot, I said that I could just feel it.

'Good!' he said. 'I think you will not lose your feet!'

He went on with the treatment for about an hour, until all the feeling came back into both my feet. I was extremely relieved and could not thank him enough. He just warned me firmly not to go wandering about the forest alone, in the middle of the night and certainly not without any skis! I might not always find a doctor there at the end of the day who knew what to do!

There is a beautiful little corner of hills and woods on the west side of the River Ipoly, which puts it in Czechoslovak territory. I always wanted to explore it, but the problem was that you had to get a Czechoslovak visa and that cost money. Then one day I found out that if you had a 'certificate of poverty', you could get the visa *gratis*. So I went to the appropriate ministry and had no trouble getting the certificate, which I immediately took along to the Czechoslovak Consulate, where they stamped my visa in my passport *gratis*! So I was ready to explore my little corner of Czechoslovakia!

At the first opportunity I took a train to Esztergom and walked over the bridge to the Northern bank of the Danube (this bridge was bombed during the Second World War and has not been rebuilt since) and triumphantly entered the Republic of Czechoslovakia! I had some Czechoslovak koronas with me, so I was able to take a train to the nearest stop to my coveted piece of land.

It was a beautiful, quiet forest, growing on some rolling hills, full of all sorts of wild flowers. I walked and walked all over this little forest, soaking in the aroma of its ancient trees and enjoying the beauty of the quiet stillness. I did not meet a soul till I came to the other end of the forest many hours later, approaching the River Ipoly, the border between Hungary and Czechoslovakia. I had another look back at my special forest which had afforded me so much pleasure: the hills were gradually disappearing in the darkness that was falling. I reached the bridge, where I presented my passport.

'Are you sure you want to go back?' asked the border guard. 'You see, your visa will expire – it is only good for one trip.'

'No, I am going back to Hungary now,' I replied to the guard, 'but I must tell you, that forest and those hills are so very beautiful, you must be proud of them!'

He did not quite know what to say to this, but he gave me back my passport stamped with an exit mark and smiled at me as I left to cross the bridge. It was dark and so I sought shelter in the first Hungarian village which boasted just one *kocsma* or pub. There were many peasants sitting outside drinking wine and singing merry songs; they greeted me cordially and when I told them that I had just come over the border they were incredulous.

'*Oda át voltál,*' ('you've been over there?') they asked me and wanted to know how things were 'over there', as they called the territory beyond the Ipoly. Of course all I could tell them was that there was a beautiful forest and that the border guard was very friendly, so they returned to their wine and their singing.

One day I was walking through one of the Hungarian forests when I suddenly stopped short in my tracks: I had just seen two wolves standing quite still, a little way up the slope from the track I had been walking on. I recalled at this point that we had been told at school that animals could not stand a human stare! I had to think quickly and I decided to try. What did I have to lose? I also knew that the worst thing to do was to run. A human being could never outrun a wolf. So I stared at the wolves, remaining quite motionless. This mutual staring lasted perhaps a minute or two, after which the wolves turned their heads and started walking up the slope, away from my track. I waited until they had quite vanished amongst the trees, as any movement on my part might have been interpreted by the wolves as an impending attack. I have learned since that most people believe wolves are very peaceable animals and very rarely attack a human being, so it is quite possible that there was no need to use any kind of strategy; I could have just walked on and nothing would have happened. But of course we shall never know! My big mistake was to tell my mother about the wolves. She was horrified that I could have been in such deadly danger, or so she thought and instantly decided not to let me roam the forests alone! The problem was, with whom should I share my expeditions to the country?

I was very disturbed at the thought of having to give up my hikes, but, as it happened, I need not have been. My mother soon found a much older person, who was equally fond of the wild places in the country and was happy to join me on my trips. His name was Pápa Miklós and he was somewhere in his mid-twenties. So from this point on, we travelled together everywhere and became the best of friends.

Chapter Six

The Last Two Years in Hungary

In the year 1930 our parents decided that Gedeon and I were old enough to travel alone and we were allowed to travel back from Abbazia to Budapest, via Venice! We took the night boat from Abbazia to Venice, which skirted round the Istrian peninsula. We were travelling fourth class, which meant staying on deck or sharing a bare room without any furniture with a whole lot of peasants, their children and their variety of belongings. Naturally, we opted for the former.

We found a couple of deckchairs and were just about to doze off when the ticket controller wanted to see our tickets. When he noticed that we were fourth class passengers he practically shouted at us, saying, '*Niente Liegestuhl! Vierte Klasse nix Liegestuhl! Quarta classe, qui non potete rimanere! Nicht hier bleiben!*' (in a mixture of German and Italian he was saying, 'No deckchairs in fourth class, you cannot stay here!')

We tried our pretence of not understanding, but it didn't work.

'*Allora, rapporto al capitano! Rapporto al capitano!*' ('Well then, report to the captain!')

He repeated this several times and pulled us out of the deckchairs. When he saw that we had no intention of moving, he said, '*Dormire per terra,*' ('Sleep on the floor') pointing at the floor, putting his hands against his cheeks to mime sleep. We said something like, '*Molto bene,*' and dutifully lay on the floor. The rocking of the boat soon sent us to sleep.

In the early morning our boat anchored outside the Piazza San Marco. Literally hundreds of gondolas surrounded the boat, a trip on one of these time-honoured Venetian instruments of transport being the only way to get ashore! We felt very grand, paying our passage with our few lire as we were landed, with our luggage, outside San Marco cathedral. Having done all the usual touristy things, such as going through the cathedral, visiting the Doge's Palace and walking around the very narrow streets and bridges of this ancient city, we took the *vaporetto* to the railway station and embarked on the night train for Budapest.

Nothing untoward happened during this momentous, first, independent journey, so after this it was taken for granted that Gedeon and I could travel across Europe by ourselves.

The following summer, that of 1931, we were invited to spend the summer in London, where our father now had a post at Birkbeck College and a modest flat in Muswell Hill, in one of the 'desirable' areas of North London. So our mother put us on the Paris train in Budapest, but not before giving us some lectures about how formal and polite people were in England, telling us to behave likewise!

We spent five days in Paris, staying at a modest hotel, seeing the sights, walking in the Bois de Boulogne and in the Jardin de Luxembourg, bringing back memories from our earlier childhood. In fact we went to visit the building at Rue de Colisée trente-quatre; looking over the building, we were both amazed how small everything looked. The 'enormous' hall in which Duncan had organised his speeches and performances seemed no bigger than an ordinary church hall! Of course, we had not realised that size is relative and that we assess the size of objects in our surroundings relative to our own size and, as our own size increases, the size of objects relative to ourselves, decreases!

The time came to take the train from the Gare du Nord for Dieppe. We took the evening one, which meant we spent the early hours of the morning on the Channel ferry to Newhaven. We lay down on some suitable benches and had a good sleep and were woken up by the bellowing of foghorns as our boat was sliding into the harbour at Newhaven. We quickly got our belongings together and were soon ready to disembark.

We did not understand a single word of what the passport officer was saying, but nevertheless he stamped our passport with a permit to enter the United Kingdom, so we were allowed to pass through and on to the platform, where the train was waiting to take us to London. My first reaction to the train was, 'Oh, how small!' since, in England, platforms are constructed to the level of the entrances to the carriages, whereas on the Continent you have to climb up and into the trains!

Then I said to Gedeon, 'Look! Talk about being formal! See that word written on all the carriages? SMOKING. SMOKING. SMOKING – everywhere! We are not dressed properly to travel in an English train!'

Of course for us, knowing French and not knowing English, the word 'smoking' meant a dinner jacket, which we certainly did not have!

'Look over there,' said Gedeon to me. 'There is one that says NO SMOKING! You don't have to wear a dinner jacket in that one!'

So we boarded the train in a no smoking carriage, not realising till much later about our misunderstanding of the word smoking! The train ride seemed much smoother than what we had experienced on the Continent. There seemed to be very little noise from the train, although this was made up for by almost everybody coughing. I wondered if all the formality and politeness had given all these people TB!

We were happy to be met by our father at Victoria Station, who helped us with our luggage to the underground, which we duly admired. The oldest underground railway in the world, I believe, is the one in Budapest, which only has one carriage. There is now quite a network but in the thirties the only line was this very old one. There was a joke going around when Hitler, during the war, changed the Hungarian left-hand driving to right-hand driving, when people half seriously said to one another, 'We are all driving on the right now, but the underground is still driving on the left!' The Paris metro was a great improvement on the old Hungarian metro, but the English 'tube' was something else again! We greatly enjoyed the long slick trains which did indeed go in and out of the tunnels as though they were toothpaste being squeezed out of a tube! We also relished the experience of coming up the escalator at Archway Station, after which we took the bus to the terminus at the top of Muswell Hill.

Our introduction into life in England soon started by being introduced to my father's many friends and colleagues. One of these was Richard Cooke, one of his colleagues in the Mathematics Department at Birkbeck College. He was married to a very striking looking woman whose name was Rosalind and they had two children, a little girl of ten and a fairly new baby. Although my English was practically non-existent, I could carry on a reasonable discussion about matters mathematical, as the terminology was really quite international. So I had a repertoire I could rely on for using during our proposed visit.

It turned out that we were invited one afternoon to tea. It was explained to me that tea was not merely a time for drinking tea, although that activity was also an essential part of the procedure, but another inescapable component of the exercise was devoted to delicately consuming very thin slices of bread and butter adorned with even thinner slices of cucumber or cheese, to be followed by delicious biscuits, which were to be eaten while balancing a cup and saucer full of tea with milk in one hand, while the eatables had to be handled with the other hand. We need not have worried. The Cookes were not a conventional English family and they were looking forward to meeting these wild Hungarians from Budapest!

Richard Cooke had had polio as a child and so had to have a splint on his leg to help him walk, but he was nevertheless a very active person, being, for example, an excellent table tennis player. He also played the piano very well and was particularly fond of playing Chopin. But, first and foremost, he was a mathematician and so was pleased to be a friend of my father's and was looking forward to chatting with me, as he had been told of my interest in mathematics. His little girl had been told that two 'little' boys were coming to tea and so she got all her dolls ready, to show to her visitors. When we came in, she looked shyly at Gedeon and me, not knowing what to say. I recall thinking that she was a cute little girl, but I had neither the English vocabulary nor the temerity to start playing with a ten year old girl!

The talk between Richard and myself soon turned to such mathematical topics as the proof of Bolzano's theorem, which is about infinite but bounded sets having at least one point of condensation around which most of the elements of the set would 'congregate'. This was, of course, totally unintelligible gibberish to the poor little girl, who was absolutely shattered! I later made up for her disappointment, as, when I eventually moved to England a year later, we became the best of friends and then, when she was barely old enough, I married her! What is perhaps even more remarkable is that we are still married, after over half a century of many adventures which we shared together and with our children.

We were introduced to another family, who had children closer to our ages. They had two girls; one was fourteen, called Fifi and the other was sixteen, known by the name of Dandy. I do not think these were their real names, but that is how we came to know them. Naturally, Fifi became my friend and Dandy was obviously meant for Gedeon. We spent many happy hours together at the local swimming pool, which was called the lido (pronounced 'lie-dow'), as well as playing ping-pong. We even learned to play bridge, which had recently replaced the older card game known as whist and we outdid each other in bidding high, then doubling and even redoubling our respective bids, reaching astronomic scores, which fortunately were not to be paid in real money! There are always new ways to be invented for teenagers to impress members of the opposite sex and bidding high in bridge was probably one of the least noxious of these!

I was keen to explore the English countryside. My father gave me a few shillings and said one day I should go and take the train from Victoria to Dorking and explore the rather pleasant country around that little Surrey town. So I did just that. I remember being fascinated by the electric trains of the Southern Railway, which had no overhead wires, as electrified trains do on the Continent, but took the current from a third rail alongside the track. I could not help thinking that this was a dangerous way of powering an electric train and wondered how they would organise level crossings safely! My father was right: the country around Dorking is very beautiful; on my first trip I remember walking quite a long way through forested country to a high elevation with a little lookout tower on it, called Leith Hill. From the top of this tower I could feast my eyes on the rolling hills of Surrey, with lots of woods everywhere, interspersed with a field or two in the distance. I did not meet a soul during my whole walk through these woods and I almost felt I was back in a Hungarian forest, except that the vegetation looked somehow different. The green was a different kind of green, the trees had different shapes and certainly many of the wild flowers seemed unfamiliar to me.

I had much time to think while I allowed the feel of the country to soak into me as I walked silently through the woods. So this was England: Muswell Hill, Kentish Town; the woodlands of Surrey; strange electric trains; all bound up with a strange language so hard to follow and even harder to speak. Was my destiny bound up with this country? Or was I to live my life in the backwaters of Europe in a truncated little country, even though there was much that was beautiful left in it?

I wandered through the Surrey countryside with such thoughts coursing through my head. Sitting in the train on the way back to Victoria Station I tried to weigh up England, as a fourteen year old boy brought up on the continent of Europe could. I was not impressed with the girls here. They did not seem to me to be pretty; in fact the thought ran through my mind that they looked rather like horses, possibly with the exception of the shy little girl of the Cooke family. Fifi certainly was not pretty, but she was friendly and she was fun.

The sound of the language seemed to me ugly, somewhat like the barking of dogs, not musical like Italian or French. Even the writing of the language seemed to make it look barbarian. The sign WAY OUT did not seem to be right, I preferred EXIT which at least expressed the idea in Latin, but SORTIE would have been much better! The climate was not the best: there were a lot of grey, drizzly days, although there were some good sunny days as well. But there was one thing that impressed me: the people everywhere were friendly, even the policemen! I was engrossed in such thoughts when I realised that the train had come to a stop and the guard was shouting something that sounded like, 'All change!' I was a bit bewildered how one can 'change' when the journey was already at an end, but then this was England and you had to learn something new every day!

Gedeon and I were also introduced to the English theatre. I recall being taken to see *She Stoops to Conquer* and not understanding much of the language or of the plot. It was years later that I found out the meaning of the word 'stoop' and then wondered how one could conquer in such a position! We were also taken to see *Hamlet* at the Old Vic, which was much more fun as we knew the plot from the very good Hungarian renderings of Shakespeare and looked forward to all the gory details in act five! My father realised that we were possibly saturated with English and offered to take us to the Everyman Cinema in Hampstead where we could see some French films! The 'talkies' had just come in and René Claire had started producing his epoch making films. So we saw *Sous les toits de Paris* and *Le millionnaire*, both of which we thoroughly enjoyed. I still remember the tune of the former, in which the song ends with '*Sous les toits de Paris, c'est l'amour!*'

By the end of August we were ready to go back. We had booked ourselves along a somewhat circuitous route, via Paris, Basle, Lugano, Milan and from there to Budapest. Unfortunately, we did not notice that our passports had expired in the meantime. Fortunately for us, no passport official noticed it either until we arrived at Basel, on the French-Swiss border. The Swiss passport officer told us, smilingly, that we were travelling with expired passports, but nevertheless put in an entry permit all the same, marked, '*entrée exceptionelle*'.

We started to roll through the beautiful Swiss countryside, feeling relieved. We got off the train at Lucerne and took the boat that plied the Vierwaldstadtersee ('Lac des quatre cantons') to the southern end of the lake at Flüelen, where we took the next train south, bound for Milan. We arrived at the border town of Chiasso, where the passport officer refused to let us enter Italy with our expired passports. There was nothing for it but to leave the train with all our baggage! We argued and argued and finally persuaded the powers that be to give us a three-day pass through Italy. This was necessary as we would need to get a Yugoslav visa for the leg of the journey between Milan and Budapest. We phoned the Yugoslav Consulate in Milan and, being Friday already, they said they would not be open for business until Monday morning at 10.00 hours. We worked out that, under these circumstances, we could not make it out of Italy in three days and asked the powers that be to extend our 'exceptional permit' to four days. Argue we did, but the official in question was adamant. Three days, or nothing. So he wrote a great '*Annullato*' across our permits and we were back at square one.

We thought Chiasso was not much of a place to be stuck in, so we took the next local train back to Lugano and there found a modest little hotel. We phoned the Hungarian Consulate in Berne and they told us to send the passports and within a week we should have them back, renewed! So we packed up our passports and sent them to Berne.

How were we going to live for a week in Lugano, not exactly the cheapest place in the world? Our money supply was getting thinner with every telephone call. We made some rapid calculations and sent a telegram to our father in England and another to our mother in Budapest, hoping to receive some funds as a result. Having done what we could, we settled down to our forced holiday in one of the world's most beautiful spots.

We climbed all the mountains in sight, visited all the picturesque little villages along the shore of the lake, such as Morcote and lived mostly on fruit purchased at the local market and bread and cheese and suchlike, so we never needed to go to any restaurants. Our supplemental money arrived from London in a few days' time and the Hungarian consul was as good as his word, for, after a week we received our renewed Hungarian passports. We paid our bill at the hotel, looked up our trains, sent a wire to Hungary stating our time of arrival and duly arrived there as promised, welcomed by our mother. I did not know it then but this was my last 'return to Hungary' for about thirty years!

Gedeon and I went back to school in September. We were in the same class, as Gedeon had been placed in a sanatorium for retarded children for more than a year, which must have been a mistake: he later proceeded to gain two doctorates, one in law and one in philology and speaks over twenty languages fluently. But, nevertheless, there we were, nearly two years apart in age, attending the same class at the Piarista Gimnázium.

It took me a long time to adjust to the formal methods of the Piarista Gimnázium after the more flexible attitudes they had adopted towards learning at the Új Iskola, where Gedeon and I had attended previously. So my grades were not very good, just about average or below average. Then one day, to the astonishment of my teachers, I began to do well in Latin, physics, chemistry, Hungarian literature; in fact in all the subjects being taught. My grades shot up, but not until I had been repeatedly examined, mainly orally, by a number of my teachers, as they wanted to make sure that nothing 'fishy' was going on. This revival did not apply to mathematics, as I had always been good at that subject, so I did not need to improve my grades.

It should be remembered that we are talking about the inter-war years, during which Hungary was still smarting from military defeat at the end of the war and from the Draconian peace treaty which had taken away two-thirds of its pre-war territory. We still learned about the geography of Hungary as pre-war Hungary, the borders drawn along the chain of the Carpathian mountains. We were told that the parts now under the control of foreign troops were temporarily occupied and that one day soon they would be retaken and the one thousand year old Hungary would be re-established within its ancient borders. All this was, of course, wishful thinking but it was there in the atmosphere wherever you went and it was particularly noticeable in a school. There were pictures circulating with the map of 'old' Hungary, inside which was drawn the truncated Hungary and a crown of thorns covering the occupied territories; the caption underneath was: *Nem! Nem! Soha!* ('No! No! Never!') We were told that our Hungarian 'brothers' were living under cruel occupation and repression and that we should be ready to fight these enemies of our people at the earliest opportunity.

I seem to have been a kind of 'doubting Thomas' all my life and, turning these things over in my mind, I did wonder about the veracity of everything we were being told. So I determined to go and see for myself what was going on in the so-called occupied territories. There was a longish break in the school year at Easter, so I decided to go for a long hike across Czechoslovakia, namely the part of that country which had belonged to Hungary before the war. I secured my certificate of poverty and the necessary visa from the Czechoslovak Consulate and took a train to Selmecbánya (Banska Stiavnica in Slovak). From there I was meaning to walk in a Southerly direction to Ipolyság (Sahy), which town lay at the border between Hungary and Czechoslovakia. I wanted to take in, among other things, the peak known as Sitna on the way, as I could not imagine a trip without climbing a mountain. The weather was good, though there was still snow lying in the mountains, so I enjoyed my communion with nature during the day and finding out about village life in the evenings. I had no difficulty in finding somewhere to stay each night, since there was always a willing peasant household where I could get a bed, either quite *gratis* or for a very small sum.

'How are you faring under the Czechoslovak oppression?' I would ask, if I was staying with a Hungarian-speaking family.

'Oppression?' was the astonished reply. 'There is no oppression! We have our vote; our children can go to Hungarian schools; we have work! We never had it so good!'

This appeared to be the general reaction from the Hungarian-speaking population. I could hardly wait to tell my friends how our teachers were having us on about all this 'oppression' business!

I enjoyed my hike across the mountains and valleys of Slovakia. I duly climbed Mount Sitna, which was over one thousand metres high and covered with an ancient forest of firs. There was thick snow still lying on the ground and the forest was beautifully still, the silence only disturbed by the crackling of the crisp snow underneath my feet. I eventually reached Ipolyság, crossed the border into Hungary and took the train back to Budapest.

I told my school friends about my experiences in the 'occupied territories', some of whom were surprised, but some had long thought there must be something not quite right about the anti-Czech propaganda. But I had not reckoned with the school authorities. I was soon summoned to the principal's office.

'You are a traitor to your country!' the principal screamed at me. 'I hear you have been spreading lies about our poor unfortunate brothers in the occupied territories! If you do not immediately recant everything you have said, I shall be obliged to expel you from the school!'

'Mr Principal,' I replied. 'I have not been lying. I have seen how things are for myself. It is your teachers who have been lying to us. As a matter of fact, I do not wish to continue to attend your school, as I prefer to preserve my own integrity! Goodbye Mr Principal.'

So saying I walked out of the room and took the tram home. On the way home I turned over what he had said about my being a 'traitor'. Was I a traitor to my country because I wanted to find out what was really happening? I tentatively decided that being a traitor was relative: it all depended on your point of view, on your philosophy and perhaps, above all, on your previous experiences. I had seen many countries, spoke a number of languages, realised that there were different possible points of view on any particular question; and who was to say which one was the 'right' one? A traitor in one place was probably a hero in another.

When I arrived home, I told my mother what had happened. We discussed the situation and my mother eventually arranged for me to be examined at the end of the school year; thus I became a *magán tanuló* or private student. I continued to study at home from the various textbooks we had been following at school. My aunt Kató, who was a classics scholar, helped me with my Greek studies, as we had started reading *Antigone* by Sophocles, which, although very beautiful, was too difficult for me to study alone. I enjoyed reading Virgil, Ovid and Lucretius. Ovid's *Tristia* moved me quite a lot; this was written by Ovid while he was in exile on the shores of the Black Sea, where he expressed his loneliness and his passionate desire to be back in Rome, in some of the most beautiful lines of poetry ever written. For further emotional experiences, the poet, Catullus came in handy, as I was just beginning to be aware of the stirrings of romantic love and his verses spoke directly to my troubled heart.

My mother decided at one point to organise some dancing lessons for us in which boys and girls would learn to dance the usual dances then in vogue. Since I was just beginning to be interested in the opposite sex, I thought I would join this 'course'. All the boys and girls were in their teens or even younger and since practically no schools in Hungary at that point were co-educational, the 'course' proved very popular. I recall having great fun dancing with a little deaf girl who could lip-read very well. I could 'talk' to her without uttering a single sound and she would reply, in a somewhat raucous voice, as the techniques of teaching deaf people to speak were not all that well developed at that time. But holding a little girl about the waist and talking with her without making a sound, represented a double enjoyment for a dance! There was another girl of about fourteen, Ilonka, who swept me off my feet with her perfume, her delicate gait, slim figure and picture postcard pretty face! So this is how I learned the foxtrot, the waltz, the rumba and probably a few others whose names I have forgotten!

There was a skating rink just opposite the building in which we lived and during the winter, every evening, a band played, so that it was possible not only to skate but to dance on the ice. I took every opportunity to dance with Ilonka, but Gedeon and my cousin Gyuri were much more 'ladies' men' than I was and I never got even as far as kissing her! So I fell back on my cousin Aga, who was about twelve at the time and we became very friendly, although nothing particularly romantic resulted from the friendship. She introduced me to some of her school friends, so I soon found myself dancing with a girl on each side, talking sweet nothings to them and generally trying as well as I could to learn how to communicate with the opposite sex!

One memorable experience of that year was a walk from Miskolc in Eastern Hungary back to Budapest, which was the last big hike Pápa Miklós and I did together. The hiking route had been worked out by many previous walkers who had walked in all the regions of that part of Hungary. The route was marked by three horizontal stripes of white, blue and white, painted on trees along the way. From each tree so marked, you had to spy out the next one with the white and blue stripes. In this way a large forest could be crossed without any fear of getting lost!

We had a tent, so we did not have to arrange to stay in tourist houses and we did not have to trouble the local peasants with nightly lodging requirements. The length of the trip was somewhere between two hundred and three hundred kilometres; since we had plenty of time, we never had to do more than about twenty kilometres of walking on any particular day. In those days there were no commercial campsites; we just had to find a place each night where there was good drinking water nearby, preferably a spring and a fairly level piece of ground for pitching the tent. We carried basic foods with us in our knapsacks, but we were able to purchase fresh fruit, vegetables, milk, butter, cheese and suchlike from the peasants whose houses we occasionally passed. It was a very good 'give and take' exercise, each of us having to learn to respect the other's point of view and requirements.

We took turns in cooking the supper. One night we were camping about twenty metres from an abundant spring and it was my turn to cook. When I had finished cooking, we had our meal, but it was a very hot night and we were drinking a lot of water.

'Hey, don't drink up all the water!' I said to Miklós.

'Isn't there enough in the spring?' retorted my friend.

This particular piece of dialogue has remained in my mind after all these years, as I realised the utter selfishness it implied on my part. It was my turn to provide, so I would have to walk over to the spring and fill up the water jug and I just did not feel like doing that! And the spring was only a few metres away!

At another feast we were having, Miklós asked me why I was a vegetarian. I do not recall giving any satisfactory answer, but I do remember that we started discussing the pros and cons of meat-eating. One of Miklós's arguments was that, since nearly everybody ate meat, was it not anti-social not to eat meat?

This really touched me to the quick and I immediately said to Miklós, 'Cut me a piece of that salami! If being a vegetarian is anti-social, I cannot be a vegetarian!'

Miklós cut a large piece of salami for me, which I proceeded to eat with relish! And this was the end of my childhood vegetarianism, until the Second World War broke out. Finding myself in rationed wartime Britain, I became a vegetarian again, since the meat ration was so diminutive, it was not worth having and you got a lot more margarine and cheese if you had a vegetarian ration book!

We followed the white and blue stripes right across the Mátra, Cserhát and Börzsöny mountains, right to the banks of the Danube, where we were able to take a ferry across to Visegrád. The last lap of the hike took us across the beautifully forested hilly country of the Pilis and we ended up, tired but happy, on the outskirts of Budapest. We then 'broke the rules' and each took a tram to his own home. I did not know it then, but this was the end of our last trip together and, although we kept up the friendship by writing letters to each other at frequent intervals telling each other of our respective rambling achievements, we never saw each other again.

The time had come for my examinations. Most of the examinations were oral, on a one-to-one basis and I had successfully passed in geography, Latin, Greek, German and physics when it came to the examination in religion. Of course, I really had not prepared much, if anything, in that subject, as I had come to agree with my father that all that sort of thing was just the 'opiate of the people' and not worth spending time on.

The priest who was examining me was very kind and as I could not answer any of his questions, he asked me what I did know and would I tell him something about that. I said I had studied the Reformation and held forth for some minutes about Luther and Calvin. I was passed with 'sufficient', which was right next to 'insufficient', which was the name given to 'fail'.

The teacher of Hungarian literature was listening to all this and was really furious.

'All you know of the Catholic religion is the Reformation! That is scandalous! Let us see what you know about Hungarian literature!' he yelled at me in a very loud voice.

He rapidly asked me some questions, but did not wait for the answers and noted down that I did not know the answer. After about half a dozen of these questions, to which I was not given enough time to reply and so was judged ignorant as a consequence, the teacher yelled at me again, 'That's enough! I am afraid I must fail you!'

Naturally, I was somewhat upset, and realised that in September I would probably have to take the same examination and most probably my examiner would be the same teacher!

Next was my mathematics examination, which consisted of a friendly chat about all sorts of mathematics that the other teachers who were nearby could not follow and my friend Ferenczi passed me as 'excellent', which was the top grade. As it happened, this turned out to be my last visit to a Hungarian school until much later, when I came back to teach them how to teach!

It was now summer vacation time and so time to get ready for our journey to our father, with whom we were to spend the summer, as we had done for several years. The place was Haute Savoie and the name of the village where we had to go was Burdignin; at some distance away from this village was a little hamlet called Chez Nicoud, inhabited by Monsieur and Madame Sauthier, who worked their little farm there and eked out their existence by taking in paying guests for the summer.

Our father had sent us the money for the fare to Haute Savoie, which arrived in pounds. Gedeon and I found a way of changing these much sought-after pounds into Hungarian pengös at a much better rate than the official exchange rate, so we went to a travel agent and booked a very roundabout way to get to Burdignin. There was also a 'foreign university' at Perugia and there were extremely favourable fares from any point in Europe to Perugia. We discovered that one did not actually have to go to Perugia as long as the ticket was purchased. So we ordered two tickets to Perugia, via Fiume and Ancona, the section between Fiume and Ancona being, of course, by boat. Then we booked another leg from Ancona to Faenza, then to Florence, then to Genova, then to Nice, Marseille and finally to Geneva, from where we hoped to make it to the Haute Savoie village of Burdignin!

We enjoyed our second Adriatic crossing, this time travelling third instead of fourth class, so we were able to have a good night's rest without being referred to the captain for misbehaving! The train ride from Faenza to Florence is stupendous. The train chugs across the Appenines, over many viaducts, across innumerable deep gorges and a great many tunnels with breathtaking views all along. It is a pity that the Germans blew up this line during the Second World War. I believe they paid to have it remade at one point, but somehow, in true Italian fashion, the money found its way to other greedy pockets and the line is still in its wartime wrecked condition, probably never again to be restored.

We spent a few days in Florence, dutifully pacing through the Pitti and Uffizi galleries, as well as visiting the cathedral and the nearby baptistery with its beautiful Doors to Paradise, with many biblical scenes shaped in relief in what looked to us like gold!

I recall being told that the only thing to see in Genova was the cemetery, so we visited this rather extensive and ornate conglomeration of mementoes to the departed. Next stop was Marseille, where we walked around the Vieux Port, not realising how dangerous such walking was, especially after dark! But I suppose the local gangsters realised that two young boys were not likely to have much, so we were not harmed!

After another night trip we arrived in Geneva. Our tickets expired here, but so had most of our money. We had enough to take a tram to Annemasse, which was already in France and so in Haute Savoie. We put all our remaining francs together and decided that we had enough money to take the bus to Pont de Fillinges, after which we would have to think of ways of getting to Burdignin, which was up a long valley, past a little town called Boège. We decided that Gedeon would sit down on the roadside with all the luggage and that I would somehow get to Burdignin and get help. I had never hitch-hiked before, but, as it happened, I did not need to, because soon somebody driving a very old-looking car stopped for me. I said I was looking for my father, who would be somewhere in the region of Burdignin. Fortunately, he was going to Burdignin, so he dropped me in the middle of the village. I then enquired for Chez Nicoud.

'*Ah, c'est très bien, vous voulez voir Monsieur Sauthier?*' ('Oh, that's good. You want to see M. Sauthier?').

I was not sure of this Monsieur Sauthier, but I was pointed firmly in the direction of Chez Nicoud, which was on the edge of a large forest, reached by climbing what seemed to me a very steep hill! My father and Sari and a colleague from Wales were there sitting outside in the sun and they welcomed me but wondered what had happened to Gedeon! Having duly greeted Monsieur Sauthier, I explained to him that Gedeon was waiting, probably miserably, on the roadside near Pont de Fillinges!

Soon a transport was arranged and within the hour we were both sitting down to a good meal, served on a rustic table in front of the house. Thus ended our tortuous journey from Budapest to Burdignin and our holiday in the mountains of Haute Savoie began!

This holiday stands out in my memory as a very significant one, as well as a very pleasant one. I remember going on a great many long walks by myself as well as some two or three day hikes with my father. The longest one was a mammoth walk right into the town of Thonon, a pleasant little place on the shore of the Lake of Geneva. To get there I had to get over a col and then climb down into the valley of the Drance, which was more a gorge than a valley, vertical walls of rock rising out of the rapids, with the road winding along by the turbulent river, criss-crossing it many times as well as passing through brief tunnels. I did not have much time in Thonon to enjoy the lakeside, as I wanted to be back before dark, since the whole round trip exceeded forty kilometres. Nevertheless, I took a little time to soak in the atmosphere of Lac Léman, one of the largest lakes in Europe and gazed across to the Swiss side, where I could just discern what resembled Lilliput-like, little villages, no doubt in reality towns of some importance, such as Lausanne!

On one trip with my father, we set out to climb a certain particular mountain peak, whose name I have now forgotten. But to reach it, we had to walk to a mountain lake, which took us the best part of the day, walking through what must be one of the most beautiful parts of France. We reached the lake in the evening so we stayed the night in a little hotel built on the shore. The next day at dawn we were ready for the fray. Our path led through a large forest, to some alpine pastures, finally leading us to a very narrow, almost knife-edge ridge, at the end of which towered our goal, our mountain peak. We clambered along the ridge, being careful with every step we took, as on both sides there were drops of several hundred metres. At one point my father said that he did not want to go any further.

'Look, it's not very far now!' I said, pointing to the peak ahead of us.

'I tell you what,' said my father. 'Why don't I rest here and you climb the peak and then come back and we'll have our lunch?'

I agreed to this and went on by myself. I went along quite happily, hoping my father would be all right. What we had not reckoned with was that the ridge did not go straight up to the peak but had some up and down sections, so you had to descend a bit at a time before you could go up again. At the first one of these descents, my father and I lost sight of each other and my father thought I had fallen down the precipice which, of course, was not so, but he could not see the ups and downs of the ridge from where he was sitting! When I finally appeared on the summit, he saw me again and waved to me and of course I waved back.

Believe it or not, there was somebody else sitting on the summit! We introduced ourselves. Apparently he was a professor of philosophy at the Sorbonne! I did not realise at the time how funny that situation was! One talks about philosophers meditating on mountain tops, but one assumes that such imagery is unreal; it could never actually happen! But here I was on the top of a high peak in Haute

Savoie, overlooking the Dents du Midi, Mont Buet and Mont Blanc in the distance and here was a real philosopher meditating! My eavesdropped lessons in our treetop nest in Czechoslovakia actually came in useful, as I could carry on a reasonably intelligent conversation about Plato's and Aristotle's ideas. When I ran out of the little I knew about these ancient philosophers, I changed the subject and asked him some questions about the *Elan vital* of Henri Bergson, which he had expounded in his *Evolution Créatrice*, which I knew well, since my mother had translated it into Hungarian only a year or so previously and she and I had had many talks about how the universe 'endured'!

Then I said goodbye to my philosopher and worked my way back to where my father was waiting for me. He was getting extremely anxious and wondered why I had been so long! I excused myself by telling him about the philosopher, but he said that he would not be coming mountain-climbing with me any more as it was too much for his nerves! But we had our sandwiches and eventually managed to get down to our hotel safely, both of us ready for a good meal, beautifully cooked and served as only the French are able to do!

Gedeon never came on any of these trips as he was more interested in talking to people. He had long talks with Monsieur and Madame Sauthier. He also liked to study the local paper, which was printed in the local Savoyard dialect, which he then proceeded to study, getting all the information he could from the Sauthiers. At the end of the summer he was a reasonable Savoyard speaker!

My father and I also went on shorter walks, sometimes in the nearby forest and sometimes we would go down to the village of Burdignin or to Boège. On one of these trips we stopped at the little bistro at the centre of the village of Burdignin, taking our seats at one of the tables on the sidewalk outside.

I was quite surprised when *'le patron'* came up to me and said, *'Bonjour, monsieur! Qu'est-ce que vous allez boire?'* ('Good day, sir. What will you drink?')

On a previous trip by the bistro I had heard one of the locals ordering an *amer picon*. I had absolutely no idea what an *amer picon* was like, but, with my incipient masculinity welling up inside me, I said, *'Un amer picon, monsieur, s'il vous plaît!'*

My father ordered the same. He had enough sense not to ask me how I knew about things like an *amer picon*! We were two men together out for a stiff drink! Actually, the drink was not at all bad and we eventually treated ourselves to another one!

My father and I talked a lot about politics, social issues and the rising danger from the right. We talked about Hitler and what was likely to happen in Germany, and consequently, in Hungary. It seemed to me fairly inevitable that Hitler, representing a violent reaction to the injustices of the Versailles treaty, would get a mass following. My father at times gently hinted to me about possibly coming to live in England.

I thought about such possibilities during many of my walks alone in the nearby forest, trying to list and consider carefully, the pros and cons, practical as well as moral. My thoughts at that time ran approximately on the following lines.

IN FAVOUR

1. The menace from the right in Central Europe. In England one would be well out of it.
2. The developing bond between my father and myself. Life in England might be more emotionally satisfying for me.

3. I would not have to retake the examination in Hungarian Literature in September.

AGAINST

1. It would make my mother unhappy.
2. I would have to learn a new language.
3. I would live in a fog and never see the sun again!
4. I would have to learn to get on with Sari.

I finally decided that the pros outweighed the cons and wrote a letter to my mother asking for her permission to return to England with my father and Sari. The reply came by return of post, my mother very wisely saying that I was fifteen years old and hopefully wise enough to take such a decision and so I was allowed to go with her blessing!

In a sense this decision was the end of my childhood. Children do not make such momentous decisions about their lives. I was very much aware of this and it caused me a certain amount of heartache. The carefree days of childhood were over and the time for responsible behaviour had arrived. To express my feelings I wrote a poem, but, since I could not yet write it in English, I wrote in French. Here is what I wrote.

> *Le voile déchiré de l'enfance passée*
> *Me parle tout bas d'une voix délassée:*
> *Laisse tomber sur moi une larme timide,*
> *Un regret souriant à l'aurore défunte;*
> *Les désirs de la vie vont bientôt te saisir,*
> *Le péché de l'amour et l'ivresse du plaisir!*
> *Enfant moribond, penche vers mes seins*
> *Que je voie tes beaux yeux se fermer à jamais,*
> *Que je puisse déchirer mon amour et le tien,*
> *Que je puisse effacer le soleil qui paraît!*

Je gardai les roses de ton âme enfantine,
Je chéris tout doucement tes joues qui pleuraient,
J'allumai dans tes yeux la lunmière marine
Je remplis ton coeur de l'arôme des forêts.
Le jeu est fini et la fin est amère,
Je quitte doucement les pays d'ici-bas.
Je quitte ton coeur par la brume des mers,
Ma vie est passée, ne me regrette pas!

Chapter Seven

England

Alea iacta est; the die is cast. I was crossing my Rubicon. I had decided to put a full stop to my experiences up to that moment; everything to do with my life in Hungary was to be regarded as past and I was to throw in my lot with my father and so with his adopted country, which was to become my adopted country also. Looking back on it all now, it was really a surprisingly easy decision to make. I already spoke four languages so it seemed that to learn a fifth one was not such a big deal. Although I had some friends in Hungary, these were relatively few in number and I did not feel that they formed a really essential part of my life. The only real emotional hurdle was coming to terms with the hurt this would represent for my mother and this was greatly alleviated by her very wisely written letter in which she gave me full adult power of decision. And, finally, it would be wrong to say that I was not to some extent influenced by my impending examination in Hungarian literature, which, of course, I should not have to take if I decided to go to England. If indeed this fact was enough to tilt things towards England rather than towards Hungary, then I should be grateful to my irascible Hungarian teacher who had to make a point with me about the way I had conducted my examination in religious knowledge and so felt compelled, come what may, to fail me in his subject.

The language problem did worry me, as I remember asking my father if I could not go to school in Paris, as we had lots of friends there and I already knew French. A very definite no to that suggestion made me ponder English weather, which would have to become my weather; being used to hot sunny summers and lovely snowy winters, the weather prospect appeared to be a sacrifice I had to make in order to come to a country where people were free to talk about anything they liked and where the policemen were helpful!

Gedeon thought he had no choice but to go back to Hungary, for he thought it would be utterly cruel to our mother for us both to leave, so he never even asked her about it; in fact, I am not altogether sure he had wanted to leave Hungary. He was a much more sociable person than I ever was and he had lots of friends; he had also established himself as a dancer in our mother's school and often performed on the Hungarian stage. Also, being two years older would have made a difference. But, however all this may have been, I have a feeling that my brother Gedeon somehow never forgave me for leaving at that point, since, as a consequence, he and I were to have very different chances in life and very different experiences. So we never became adult brothers, for our experiences of life as well as our philosophies diverged, and as time went on, we had less and less in common to share with each other.

So the time came when Gedeon took the train back to Hungary and my father, Sari and I went to Paris and thence to England. It was lucky that I was not yet sixteen, since I could enter the country under my father's passport. On my sixteenth birthday, the following September, I simply obtained from the authorities what was then called an Alien's Registration Book and was allowed to settle in England, since the authorities did not know that I was not already living there!

One of my father's ex-colleagues from Wales, George Grube and his wife Gwenyth had rented a cottage in Kent, where they invited us to stay. Kent is a very pleasant part of England, a good place to spend one's first weeks after deciding to come and live in England. This was fine, but my father had to take some steps about sending me to school. He tried a number of public schools such as St Paul's, but, when they heard that I could not speak English, they said they were sorry but they had no facilities for teaching English as a second language. This was the reaction from all the schools he tried! As it happened, Gwenyth had been offered the job of teacher of Latin at Dartington Hall School in South Devon and she suggested that my father might try that school. She said it was a crazy kind of school, it was supposed to be very 'modern', but perhaps it might be all right for me. My father had a long telephone call with Bill Curry, the headmaster, who said he would come and see me in two weeks' time. My father said to me, 'You have two weeks to learn English! Go to it!'

I was somewhat taken aback, but I accepted the challenge. I looked around on the shelf and, among the many classics I found *David Copperfield*, which I was determined to read. I also found an English-French dictionary lying around, so, armed with these essential ingredients for learning English, I sat down to read. Fortunately, English has very little grammar compared with other languages and so, with the dictionary (which also had a list of the more usual irregular verbs), I was able to make headway with the book. I also read some of Julian Green's short stories and since these had been translated from the French, somehow the English seemed to follow easier!

In two weeks' time Mr Curry duly arrived and wanted to speak with me. We were introduced to each other formally. I clicked my heels and bowed in true Central European fashion and expected the interview to begin.

'Have you read anything in English?' asked Mr Curry.

'Yes, I am reading *David Copperfield*,' I answered in a very thick Hungarian accent. 'I have read about half of the book.'

'Tell me about it,' said Mr Curry.

In somewhat broken English, I tried to give him an idea of the plot to the point where I was in the story.

'How long have you been studying English?' was the next query.

'I started two weeks ago,' I replied.

At this point he indicated that the interview was at an end and went to talk with my father in another room. After he had departed, my father came towards me smiling, finding me already at work on *David Copperfield*, and informed me that I could indeed go to Dartington Hall School. So the continuation of my education was assured and I could look forward to some interesting times in one of the most avant-garde schools in the country.

We soon left the Grubes' cottage after this momentous interview and arrived at the Highgate apartment which my father had rented from an interesting family. After some frantic shopping, which included things like gumboots, to be worn for felling trees, I eventually found myself at Paddington Station, where I was to board the school train to Totnes, along with a whole lot of other boys and girls who were also about to embark on the same exercise. We had reserved seats, but I remember spending most of the time in the corridor, trying to talk to the other children. There was a girl of about fourteen who could speak Italian so I had a chat with her and I eventually found Matthew Huxley, Aldous Huxley's son, who could also speak Italian. Later I believe I thought that was 'chickening out' of the problem to be tackled and, in spite of my English being still very broken, I recall that I attempted to engage in a discussion with Michael Young (now Lord Young) about communism. I remember telling him that we could not possibly have had that sort of conversation in the corridor of a Hungarian train as almost certainly we would have ended up in jail.

We finally disembarked at Totnes Station, where cars were waiting to take us to the school. I was impressed by the luscious green of the trees, which made a long tunnel over a section of the road between Totnes and Dartington village. I already had visions of exploring this new country on foot at the earliest opportunity. The school building was absolutely new, having just been completed and I noticed the date 1932 prominently inscribed over the front entrance, which was that very year.

We were taken to our rooms, each child having his or her private room. My room was in the senior house, opposite the room where our house father, Oscar Oeser, a South African, had already been installed. There was no segregation of the sexes, boys and girls having their rooms along the same corridors and sharing the same shower rooms. In fact, during the two years of my stay at that school, I do not remember any rules relating to boys or girls as such being in force or even suggested. We were just people and gender was no more a deciding factor than left- or right-handedness or the colour of your eyes or hair.

There was much adapting to do, but it was not really a painful process. I would rather describe it as a learning process. First of all, I had to learn to speak, read and write English. Although there were two Italian-speaking children and I could speak to my French teacher in French and to the house mother in German (she was from Bavaria), there was no getting out of learning to communicate properly with the rest of the children and teachers!

I made friends with a boy of about my own age, David Robson. He came from Haltwhistle, a small Tyneside town and we often went for walks together. He had somewhat rough manners and he swore a lot but in many ways he was a boy after my own heart; I liked the fact that he could express himself without inhibitions, which had the salutary side effect of teaching me English, even though it might have been a very David Robsonian English at times! He would laugh raucously at my mistakes in English and kept correcting me. I tried to keep copious notes of all his corrections, which I afterwards put together into a kind of self-made grammar. I recall that one day, while we were sheltering in a ruin for we thought it was going to rain, I looked at the surrounding countryside and said, 'It rains there and it rains there,' pointing at a shower over some woods some distance away, 'But no rain here!'

David gave one of his raucous laughs and said, 'Ha! Ha! Ha! It rains there! That's not how you say it, you know! The correct thing to say is "it's raining", don't you even know that?'

I made a mental note, as I did not have my notebook with me. It was my first lesson in the use of the continuous present in English. I never made the same mistake again!

After several weeks of unconscious language instruction by David, we were sitting in the common room, and he was again chiding me for some error I had made, when he noticed that I was writing it all down in my notebook! He was horrified to realise that he had actually been teaching me English. He never corrected me again! Fortunately, I was fairly well advanced by then so that I could get the rest of my information about the mysteries of the English language from sources such as books and dictionaries, not to mention my English teacher.

I found the English mathematics very strange. There seemed to be an enormous amount of very involved arithmetic, coupled with some extremely elementary algebra, with a little sprinkling of geometry added for good measure. I saw no good reason for carrying out these complicated arithmetic problems, which in fact were not problems at all, but simply instructions to carry out certain given operations. We had an arithmetic exam, not an official one, just one to see how we were progressing. My score was zero. This was not entirely due to my zero competence in arithmetic, but to the fact that each question made use of the result of the previous question and, if you made a mistake in the first question, you naturally got everything wrong, which is what happened to me. Fortunately for me, everyone else did very badly and Oscar, who was our mathematics teacher as well as our house father, gave us all a lecture, emphasising the momentous fact that even Zed ploughed the exam and I was supposed to be good at mathematics!

I enjoyed the bit of geometry we had to do, especially the bits they called the riders. Every question consisted of a theorem to be proved and then a rider, which was a geometrical problem to be solved in which the theorem in question was to be used. I always thought it was a bit silly to have the theorem first, as that took away the challenge of the problem, as it essentially told you how you had to solve it.

I continued to study calculus from Vallée Poussin's *Calcul Infinitésimal*, which I had with me in Haute Savoie, so I brought it to school as mathematical amusement. I remember during one mathematics lesson Oscar was holding forth about calculating the value of *pi* by measuring half the circumference of a regular polygon with unit radius, giving the polygon more and more sides as we went along and so getting a better approximation every time.

I interrupted at one point and said, 'Why not use Wallis's formula?'

Oscar had never heard of Wallis's formula. So I thought I had better not make any more suggestions and embarrass the teacher.

Latin being one of my best subjects, I always enjoyed Latin classes. I soon realised that 'English Latin' was much stricter than 'Hungarian Latin'. In English Latin we could only use words and expressions that had been truly used in classical times and any other Latin was considered sloppy. So a lot of my Latin words that which had been okay in Hungary got underlined in red pencil by Gwenyth. I would then go to the library and look up these words in a voluminous Latin dictionary, which would give me quotations from Latin authors when the said words were in use. Most of the time Gwenyth was right, but sometimes I was able to find quotes from classical times for my words underlined in red!

Gwenyth used to invite members of her class to have an evening meal with her. She lived in a cottage at Tigley Cross, about three miles away from the school on the Plymouth road. We had to run down to Shinners Bridge to be in time to catch the Plymouth bus, which stopped outside her cottage. She would cook us a great meal and chat with us. These evenings were very important to me. I think she regarded her role there as partly someone in *loco parentis*, since she had been partly responsible for my being there. At the close of our evenings she would take us back to school in her car, so we would be back before bedtime, which was at 21.45 and usually strictly enforced by both house mother and house father.

Our French teacher was called Monroe. He spoke French very correctly, with quite a good accent, albeit a somewhat artificial one. During my first week at the school I chatted quite a lot with him, as I was not able to chat so easily in English as yet. I always remember my first written assignment, in which I made every possible spelling mistake, like spelling '*vous*' as '*vouz*', as of course that is how it sounded when you made the liaison! I can still see in my mind's eye his remarks at the bottom of my exercise, 'I judge from this that your spoken French is much better than your written French!'

On looking back, I must admire him for his tact. How thoughtful of him to point out that I could speak well instead of pointing out that I wrote badly!

It did not take long for me to learn how to write French and I did not have to learn any formal grammar, although I did have to put my mind to learning how to translate, both from English into French and from French into English. I never became very good at this particular skill and to this day I remain a very indifferent translator. It seems that being conscious of an expression in one language actually inhibits any corresponding expression in another language from lodging in my own mind. I suppose if you learn languages from scratch, so to speak, you are unaware of the links between the languages you know and so translating poses particular difficulties.

I tried one trick, using articles in Hungarian newspapers my mother kept sending me from Budapest, which involved translating an article from Hungarian into French and then translating the same article from Hungarian into English. Then I would compare the French and the English texts to see if they could be said to be translations of each other. They were of course, more rotations than translations, to use a mathematical metaphor! This trick did not seem to help very much, so I soon gave it up.

Learning English was, of course, my principal preoccupation. Since there was practically no English grammar, I was reduced to finding out how the language worked from first principles. I listened very carefully and tried to make some sense of how people put their words together to make sentences. I realised that the unit of communication was the sentence, so I set out to learn sentences, which then afterwards had to be put into categories, so that a sentence could be generated from another sentence simply by changing one of its elements. But even this method had its difficulties. For example, when I was trying to sort out the difference between 'some' and 'any', I came across 'transformations' such as:

I want some   *I don't want any*
I see some   *I don't see any,*

in which case I realised that most people would say,

I can see some

instead of

I see some,

and then negating, I would get,

I cannot see any,

and wondered what the difference was between,

I cannot see any

and

I don't see any.

I would ask Mr O'Malley, our English teacher, but I could glean no satisfactory answer, except some vague talk about what he called idiomatic usage, which I finally understood to mean a total lack of rules. It seemed at times that in order to learn to speak English like an English person, every sentence had to be learned separately!

There were, however, some rules I could formulate. It was not long before I discovered the difference between the use of the normal past tense and the perfect tense. I realised the difference between

> *I have learned a lot of English*

and

> *I learned a lot of English at school*,

was that in the former sentence, the time referred to was the past up to and including the present, whereas in the second sentence the time referred to a period of time in the past, but definitely not including the present. This became clear to me in sentences such as:

> *Have you ever been to France?*

and

> *Yes, I was there last year.*

It would be wrong to say, 'Yes, I have been there last year', as 'last year' is definitely over if we are already in this year and so the proper past tense would need to be used, not the perfect. Many foreigners do not ever learn this distinction, since in most languages the distinction is blurred. The distinction is made in Spanish in the same way it is made in English, but at this point in time I had no knowledge of Spanish and in Northern Italian, which is what I knew, the usage was similar to French usage, the perfect tense being used in both cases. Later, while I lived in Florence, I learned that it is best to use the *passato remoto* when telling a story, but, even in Florence, events in the very recent past (such as five minutes ago) would still be referred to by the use of the perfect tense.

It took me some time to get hold of the use of the plural. I certainly wondered why 'hair' was singular (as opposed to *cheveux* in French or *capelli* in Italian, which are always used in the plural) and yet the word 'trousers' was always used in the plural, in spite of the fact that we were dealing with a single object! I finally decided that all this was part of idiomatic usage and therefore there could be no rules governing the use of singulars and plurals!

I also missed the use of the subjunctive, which is not used in English except by language pedants. The difference between:

Non credo che abbia ragione

and

Non credo che hai ragione

(I don't think you are right) (I don't think you are right), is expressed in English through the use of varying the intonation. In the first sentence the speaker is trying to suggest that you are not right, but in the second he is telling you that you are not right! Such subtleties are never found in grammar books, since grammars are about structure and not about communication! How wrong we are to teach children grammar when we try to teach them a foreign language! Children do not want to know any structures – they want to communicate! And they will learn to do so with surprising rapidity in the appropriate language environment, whereas they never seem to learn to speak a second language, however extensively we try to feed them with grammar!

The vagaries of English spelling presented another formidable difficulty, although I found less difficulty than my schoolmates did in the case of words of Latin origin, known to rabid Anglo-Saxons as long words. I had to be careful to refer to Latin instead of Italian! For example the word '*comunicare*' has one M in Italian but the Latin '*communicare*' has two Ms and so the English word 'communicate' has two Ms.

I recall an English lesson when we were reading *Hamlet* and each child was allocated a part to read. I remember I was reading the part of the King in the fifth act and during the mess-up when the cups were being supplied with their ration of poison, the King declares, 'In this cup an union shall be thrown', which I read as, 'In this cup an onion shall be thrown' and there were roars of laughter! The joke had to be explained to me, as at that point I had not worked out how and when to allocate the letter U its various sonorous values. So I was not sure when the letter U had to be sounded as 'yoo' 'in 'use' or as 'u' in 'but', so I had given it the latter rather than the former value! Of course, instead of an onion, some poison had already been poured into the cup!

Hamlet was one of the prescribed Shakespeare plays we had to know inside out for our School Certificate Examination, which our group was to take the following June. So we used to practise giving each other 'context questions', which meant that a short sentence was taken out of the play and we had to place it in context. We tried to trip each other up with the most unlikely bits and pieces out of the play! I countered this offensive by first learning all Hamlet's soliloquies by heart, as well as some speeches uttered by other personages and I soon became the champion in 'context questions', and it was really hard to trip me up. Of course, I had an unfair advantage, namely that of *not* knowing English too well and so the Shakespearean English did not seem to me to be so strange as it seemed to the other children. Also since some of the words used in Elizabethan times had meanings rather closer to their meanings in Latin than in modern English, I sometimes had less difficulty in understanding the language than did my schoolmates.

On Tuesday afternoons we had 'outdoor work'. This was probably done for slightly left-wing, political reasons, to make us realise how 'the other half lived'! Be that as it may, every Tuesday afternoon all the seniors, about twenty of us at that point, would put on our 'working clothes', which included gumboots and marched out to 'work'. There were several work stations and most of the time we could choose what kind of work we had to do. There was 'pulling up mangles' (to this day I am not very sure what mangles are!), feeding the pigs and cleaning the sties and felling trees. My favourite activity was felling trees. We took axes and hacked away merrily at some quite well-grown trees, so each tree took about twenty minutes to fell. Then we had to drag the trees down to the bottom of the hill on which the wood stood and pile them on to a waiting truck. Loud shouts of *'Timber!'* echoed through the wood as we bent to our work, which usually lasted about two hours.

We all seemed to like the tree-felling, but we thought the work with the pigs was a bit much. So one Tuesday we thought we would go on strike. We locked ourselves in the loft for about an hour, after which one of us went to the adult responsible for outdoor work as the representative of the strikers. Our complaints were referred to the headmaster, who eventually agreed with us that it was reasonable that only those who really wanted to would go to do work with the pigs. As there appeared to be no volunteers for this work, this aspect of the outdoor work session simply lapsed into oblivion!

There were no lessons on Sunday, so I used Sunday for my exploratory trips into the country. Unfortunately, Sunday buses started rather late, which meant that it would be about midday before I reached the edge of Dartmoor, which soon became my favourite haunt. With the aid of an inch to a mile map and a compass, I criss-crossed the south-eastern section of the moors in all possible directions, becoming familiar with all the tors in the area, as well as with the many scenic valleys. The Dart Valley is one of the most beautiful sights in England. The river roars through many rapids, but there are also a large number of quiet pools, which I used for a quick swim, even in wintertime! The utter delight of gliding through the brown peaty water of the Dart, overlooked by huge cliffs on one side and densely wooded hillsides on the other, was something to which I really looked forwards during the week.

The section of the valley between New Bridge and Ashburton was not open to the public. Having had some experience of getting permission to walk through such areas in Hungary, I proceeded to find out who owned the land in question. I asked my English teacher, Mr O'Malley, whether he would like to explore the forbidden part of the Dart with me and he assented at once. So we wrote a joint letter and asked for a permit. By return of post we had the permit which was made out to 'Mr O'Malley and Friend' and I recall being very proud to be considered Mr O'Malley's friend! So the following Sunday we went to the moors together and walked downstream from New Bridge, following the tortuous path alongside the river. We soon reached the lodge, where the keeper duly checked our permit and waved us on. There was a sort of cart track all the way, leading up to a high point on the summit of a rocky promontory known as Lovers' Leap. Indeed, from this high point you could look down into one of the deep pools of the river several hundred feet below and I wondered if indeed some unhappy star-crossed lovers had leapt to their death from that point! This was a very memorable walk for me and, to make sure I would never forget it, I glued our permit into my diary of trips, which I had kept for several years, giving details of everywhere I had been, with the description of the conditions and, where possible, accompanied by photographs I had taken!

I think I was very lucky to have Oscar as house father. He really took his job of looking after us seniors very seriously. Most evenings he would invite us into his room, after we had had our showers and were ready for bed and usually five or six of us would sprawl on the floor while he played us classical music on his gramophone. Of course the records were the old 78s and he used fibre needles, which had to be cut with a special needle-cutter after every side. This gave us the chance to think about or to talk about the piece of music we had just heard, or Oscar would say a few words about the next part he was about to put on. In this way I got to know and love some of the Mozart violin concertos, Bach's Brandenburg concertos, as well as some of the best-known Beethoven quartets. It was good to slip into unconsciousness in my little room opposite his, with the strains of such lovely music still echoing in my ears.

I remember once being in bed with a fever. Oscar came in several times during the day to take my temperature. Of course, it was all in Fahrenheit degrees so I always had to make a rapid calculation to work out what the corresponding temperature would be in degrees Celsius. In fact, I amused myself in bed by making a number of such calculations and then memorising the results, so that next time my temperature was taken, I would know what it 'really was'.

It seemed to me that, after a month or two at Dartington, I had completely adjusted to my new life in my new country. I exchanged regular letters with my mother, Pápa Miklós and my cousin Aga, the daughter of the aunt who had instructed me in Greek, with whom I had been very friendly. I made some new friends, such as David Robson, as well as a number of others, with whom I often went for long walks round the estate while the official sporting events were taking place. My English had improved sufficiently for me to communicate with ease in that language; I had learned to write French properly, my Latin was progressing well and mathematics did not pose any problems, so I was well on the way to taking and hopefully passing my School Certificate Examination which we were all to take the following June. I was looking forward to going 'home', which was now our apartment in Highgate and to sharing my experiences of school with my father.

Chapter Eight

The Dartington Years

At the end of the autumn term, Oscar offered me a lift up to London in his car. This was a great thrill for me, as I had not then done a road journey across England. We duly packed the little car and set off in the morning for London. The going, by today's standards, was slow, as there were no motorways and hardly any bypasses, so we ambled along the country roads, averaging about twenty-five miles an hour! The journey seemed very different from a car journey on the Continent. There were so many bends in the roads that the constant turning made me dizzy. I asked Oscar why the roads were so twisty in England.

'Oh, that is because all these roads were originally just the cart tracks between smallholdings. With time these tracks were improved, until finally they became roads for traffic between towns. In France, Napoleon made the roads, so they are all straight and lined with trees! Ours are cradled inside hedges full of beautiful wild flowers!'

We followed the A30 through Honiton, Salisbury and Basingstoke and chatted while I drank in all the aspects of the English countryside which were new to me. The hedges, I thought, were a typically English invention, which at first irritated me, as I could not see what lay beyond. But after Oscar's remarks extolling the beauties of the English hedgerows, I started to observe these strange structures with a different eye. Unfortunately, on the outskirts of

London, in the region of Osterley, the car broke down and had to be towed to a garage. So I paid Oscar half the cost of the petrol used, which had been the agreement and took the underground train to Highgate.

There were three rooms and a kitchen and bathroom in our little apartment, which was on two floors. On one floor was my father's study, in which there was a desk, some armchairs and a divan where he slept. Next to this was the dining room, in which there was a table and some chairs, the table being transformable into a ping-pong table, and a bed had been placed in it for my use. On the next floor up, which was an attic, was Sari's room, which she eventually turned into a painting studio. My father was quite surprised at how well I spoke English, so he invited some colleagues from the university to meet me.

What he did not reckon with was the kind of English I had picked up from the boys and girls at Dartington, where free expression was a primary article of faith. Four-letter words were used with great frequency and I naturally assumed that this was a part of normal English usage.

We were sitting round the dinner table when I said, very loudly, 'Pass the fucking butter, will you?'

There was a deathly silence. Nobody spoke for several minutes and it was not until after the guests had departed that my father had a chance to give me some paternal warnings about the appropriate use of language.

Sari reminded me that there was a very good-looking girl downstairs, who was about thirteen and asked if I would like to meet her.

'Oh yes, certainly,' I replied. 'I shall probably fall in love with her!' I bragged.

'You might indeed do just that,' replied Sari.

She could not have been more right. This emotional catastrophe did indeed happen to me during the Easter holidays and although the beautiful girl, whose name was

Fiona, seemed to reciprocate my feelings, at thirteen she was clearly not ready for the passion of a sixteen year old! But more about this later.

I began to have some differences with Sari, although in no way was she the storybook stepmother! Her taste in music was restricted to certain classics. I recall one day, when I was listening on the radio to Schubert's Trout Quintet, she came in and said a word in Hungarian, '*Mütrágya!*' Translated into English, this means 'artificial manure'. I really felt put down, as though my own taste in music simply did not count and felt very sad that a work which I considered beautiful could be so cruelly denigrated. A would-be 'mother substitute' should have known better.

After a few outings to Sadler's Wells and the Old Vic to build up my Shakespearean education and one or two visits to the Queen's Hall to put me right in my musical culture, I again found myself on the school train.

What a difference a short four months can make! I could now speak English fluently and had even learned some of the subtleties of the language, so I could understand and make up jokes. The handling of jokes in a foreign language is like breaking the sound barrier. Being at home with jokes (and particularly with puns!) tends to assure the learner that he is now an accepted member of the club. So my journey to Totnes went off smoothly and I quickly settled into the routine of school life.

I made friends with a lot of children, some as young as twelve, others my own age or older. I remember being friendly with Matthew Huxley, who was a few years younger than myself, as well as with John and Kate Russell, Bertrand Russell's children, as these children, in spite of their tender age, were interested in discussing interesting social and political questions. David Robson remained a staunch if somewhat weird friend. Then there was Brenda, who was about fourteen then, with whom I used to go for a tour of the estate during official games times. We both enjoyed the walk along the bend of the River Dart, ending up in North Wood, where we revelled in discovering the profuse flora, which did not seem to die out even in the winter months. I was also on good terms with Michael Young and Michael Straight, with whom I played tennis quite often.

There was a philosophy session once a week in Mr Curry's room. I believe that took place after supper on Tuesdays. This was a kind of debating society but a very informal one and we discussed politics, religion, socialism, Marxism, capitalism and probably many other 'isms' in vogue at the time. There was no attempt on Mr Curry's part to brainwash us in any particular direction, although he made it clear that he himself was an atheist and so found it difficult to discuss the pros and cons of various religions.

To add to our political education, about every month a politician by the name of Gerald Heard came to talk to us. I remember at one of the sessions he was getting very excited about 'all those stupid little countries all over the place', which he thought ought to be united. When I chimed in and told him that I was from one of those stupid little countries myself, he was somewhat embarrassed, until I said that I entirely agreed it would be a lot better to have a federal Europe, with which he heartily agreed. This was a few years after a certain Monsieur Briant, prime minister of France, had asked all over Europe if the countries comprising the continent would like to federate and, as those who are old enough might remember, it was turned down by every single country in Europe.

During the early thirties there were very few modern co-educational boarding schools in England. There was Bedales, near Petersfield; there was Frencham Heights; and then there were the Quaker schools, such as Sidcot, Sibford and a number of others. But the most notorious one was

Summerhill, run by the educational wizard, A.S. Neill, who thought that all problems could be solved by applying the right psychology and that otherwise children should be left quite free to do what they liked. One day Neill came to visit the school. Mr Curry was showing him round the individual bedrooms each child had, the well-furnished common rooms, one for each house, when Neill exclaimed, 'I haven't seen one broken chair or window! Everywhere seems so tidy! Really, Mr Curry, you must use some repressive measures in your school!'

At this point he was just walking down the corridor outside the seniors' common room, in which David Robson and I were having a long-distance spitting competition. The door to the corridor happened to be open and David's spit, which was very long, landed directly on Neill's face! He carefully wiped off the spit with his handkerchief and said to Mr Curry, 'Well, it's not such a bad school, after all!'

There was no religious instruction at Dartington, but we did have a Sunday evening meeting. These meetings were held at the Great Hall, mostly in a little upstairs room called The Solar. Mr Curry invited all sorts of eminent people to talk to us and to anyone else who would care to listen. There were political talks, sometimes touching on the burning social questions of the times, on philosophical topics or on the fine arts, mostly on music and dance. In this way we were kept well informed about all the controversial issues that were exercising people's minds.

At Dartington, music was always regarded as an important part of education, not necessarily the actual playing of music, but most certainly the appreciation of it. There were regular concerts at the Great Hall, to which any of us could go. I remember that the Griller Quartet came down a number of times and played Mozart, Beethoven, Brahms, Schubert; they even played the Débussy Quartet, the only one he ever wrote, whose structure and tunes have stayed engraved in my memory ever since. Sometimes the Griller Quartet would come to the school building so that we could talk with them and ask them to play certain pieces. This was extremely valuable and brought music to us as a cultural reality.

There was also a dance school in operation, in one of the new buildings near the Great Hall, run by Margaret Barr. She taught what was then considered 'modern dance', which would probably be called modern ballet now. But she also offered an evening a week of country dancing and quite a few boys and girls from the school elected to learn this art, if indeed an art it is! In any case, we all enjoyed prancing about the floor, describing devious curves around each other while socialising with each other as well as with those from the rest of the estate who had chosen to take part in this activity. We learned some Scottish dances, which were very energetic and others of English origin such as 'Haste to the Wedding'. I much preferred these dances to the ballroom ones: the latter had many erotic overtones which sometimes bothered me, while the former was a truly group activity, in which all members of the group had to be equally skilled in order for the dance to be a successful social as well as an aesthetic experience.

During my two years at Dartington I kept up my exploration of the countryside in the county of Devon. There were not many corners of Dartmoor I had not visited and I began to be able to navigate across the moors by familiar landmarks, mostly tors, so that I hardly ever had to look at a map. I also eventually walked the whole coastal footpath between Bigbury on Sea and Brixham and on these walks, mostly by myself, I just immersed myself in the marine landscape of cliffs, beaches and sea, adorned by bracken, ferns, gorse and heather, always full of the cheerful sound of seabirds. At the end of my two years, the moors, the coast, the wooded coombs and picturesque little villages had soaked into my consciousness and become a very real part of me. Devon was to become a recurring geographical theme through my life right up to the present. As I am writing these lines in our Devon house, I can look through our northern window which overlooks the Dart Valley, with Dartmoor on the horizon, Buckland Beacon and Hay Tor's double peak making it geometrically interesting. I have come full circle from childhood to old age, with the beauty of Devon providing at least part of the connecting link.

During my first year at Dartington, we went swimming at Folly Island, where there was a large pool in the Dart. We swam all the year round and we never wore any swimming gear to do so. There was just one problem: on the other side of the river there was a railway line, which carried passenger trains between Totnes, Staverton, Buckfastleigh and Ashburton. There were complaints made to the Great Western Railway by some of the passengers who considered nude bathing, especially mixed nude bathing, indecent and immoral. Eventually a very British compromise was reached. The driver of the train was to blow his whistle several times while approaching Folly Island and any offending bathers were to either dive quickly into the river or hide in the woods until the train had passed.

Folly Island was a beautiful place in which to spend one's leisure hours. It was not really an island but a piece of woodland, the trees growing right up to the banks of the river. The river was at least two metres deep around this area, but there were parts nearby where you could paddle. On the other side of the river, where the railway line ran, there were steep banks, also thickly covered with trees. There was no human habitation in sight and it was possible to dream oneself away into a faraway land of utter bliss! At the edge of the wooded section of the bank there was a large area covered with grass and this is where we used to sit and gaze into the water and, when we felt like it, dive into it and pit our youthful strength against the current.

Before we could go there on our own, we all had to become 'senior swimmers', which meant passing quite a stringent test. We had to be able to swim one hundred yards upstream, tow an adult person across from one bank to the other, surface-dive to a depth of six feet and recover a heavy object, usually a large rock and we had to learn how to break away from a struggling person in the water, not to mention how to give artificial respiration! If there were three or more senior swimmers in the party, they could go swimming in the Dart by themselves; otherwise an adult had to accompany them. These rules were always strictly observed and we all co-operated as we agreed that they were sensible rules.

During the Easter holidays of 1933, the inevitable happened and I fell head over heals in love with Fiona, our landlady's daughter, who lived downstairs. It happened through the meetings of the Panther Club. Our landlady, whose name was Joy Baines, gathered together a number of 'interesting' people about once every two weeks and they would discuss anything and everything, in much the same way as the 'university' members at Körtvélyes in Czechoslovakia would go into the ins and outs of the problems exercising the minds of the intellectuals of the times. Joy always prepared some very delectable 'eatables', all based on her modern and 'way out' ideas of what then was known as food reform and which is now regarded by most health practitioners as sound common sense! During these discussions I would lie sprawling across the floor, so that people had to step over me if they wanted to move at all and Fiona would lie right next to me, quite innocently I suppose, while all the discussion was going on around us. I often missed the main points of the discussion because I could not think of anything else but the gorgeous little girl lying next to me.

We soon began to meet more frequently, as the logistics merely involved moving from one floor of a house to another floor and we found we had quite a lot in common. She went to King Alfred's School in Hampstead as a day child, which seemed to be run on lines similar to those at Dartington. I took her to the theatre a couple of times and we established a firm friendship as well as a mutual romantic attachment, which in those days never amounted to more than kissing and cuddling and, as far as I am aware, neither of us ever thought of going beyond these rather innocent but definitely amorous activities!

During the following school term we corresponded copiously, accumulating huge stacks of love letters. I wanted to write her some love poems, but my English did not as yet extend to such creative areas, so I wrote them in French. I found out later that she had no idea what was contained in the poems, which was probably just as well, as they cannot have been very excellent literature. Much later, she had them translated into English and we talked laughingly about 'old times' when we were both adults, still looking back with much pleasure to our times together as incipient lovers! At the end of the summer term I received a letter from Joy, inviting me to spend the summer holidays with them in Worthing and of course I jumped at it. We had several weeks together in Worthing. We went for long walks on the South Downs, we swam in the sea and of course we kissed and cuddled! At one point, while we were swimming in the sea, I think she pretended to be in difficulties in the water so that I could 'save' her, which I promptly did, towing her to the pier, where several bystanders helped me haul her out of the water and lay her down so I could apply artificial respiration. I really think she was having me on, but I played my part and I bathed in the glory of being a great hero, saving my girlfriend from drowning!

By September I believe Fiona was starting to cool off. I somehow felt she was not the same as before. We went to the theatre once more before the start of the autumn term, but nothing was said. I wrote to her from school, asking whether she still loved me or whether this was the end. Her reply was that it was best to be honest in all things and that indeed it was 'the end'. So that was my first experience of the ending of a romantic relationship and I hoped I could take it philosophically. Of course she was only a child, at the age of thirteen what did she know about love? So I eventually filed the experience away as something to learn from, although I was not altogether sure what, if anything, I had learned from the experience!

The summer term, apart from writing love letters to Fiona, was taken up by getting ready for my first examination in England, known then as the School Certificate. I was to take mathematics, additional mathematics, English, French, physics and Latin. You had to get 'credits' in five subjects – English, one language and mathematics being mandatory – in order to obtain exemption from what was then called matriculation (matric for short). The possession of matriculation was the key to entering university. I did manage to get credits in five subjects, but I only managed a pass in English, so one more year at school for me was decreed.

I did not think it was so bad a result, as, nine months previously, my English had been practically non-existent. I decided to retake the whole examination in December and I promised myself that I would definitely get a credit in English if it killed me! No more dalliances with pretty little girls for me, just knuckle under and work!

The year 1933 was a fateful year in Europe. It was the year when Hitler was voted into power. The voters had not realised that, though you can vote someone like that into power, you can never vote them out! We soon began to see some of the results of these events in the shape of new arrivals at Dartington. Anna Freud's three boys, Sigmund Freud's grandchildren, arrived and Anna Freud stayed around the estate for several months to make sure the boys settled into school life satisfactorily. There was another Freud grandchild, known as Yo, who was not Anna's daughter.

Yo and I became very friendly. I remember taking her for walks round the estate and showing her the delights of the Devon flora on the hedges, mainly in the shape of primroses, which I picked for her and put in her hair. I suddenly realised that I was about to slide down the same slippery path I had with Fiona; had I not learned my lesson? Yo came into my room after one of these walks and sat on the bed, not quite knowing what to expect, nor was I sure what to do or say next.

'I don't think we can be so friendly,' I started to say.

'Why not? Don't you like me?' said Yo, looking at me, very surprised and perplexed, with her big blue eyes.

'Yes, I do, Yo, but I've really got to work this term,' I replied, somewhat uncertainly. 'Being friendly, like we are getting, takes up a lot of time.'

At this she burst into bitter tears and lay on the bed for about an hour, sobbing. I do not know to this day how I could have been so cruel, but, although I tried very hard to comfort her, I stuck to my guns and any romantic development between us was over. I shall never know what would have happened if I had not acted like that. Getting friendly and involved with the Freud family might have directed my life in quite different directions. I might even have married Yo in the end! But, as they say, the ways of fate are inscrutable and all I can do about that sad episode now is ponder.

There was a Jewish boy, Rheinhardt, who came to Dartington at this point. His family had left Germany just in the nick of time and he was sent as a boarder to Dartington. We immediately made friends, partly because I could speak with him in German and I had remembered my first days and weeks when I could not express myself in English or understand fully what people were saying. We invented a secret way to communicate, which was simply to pronounce all English words as though they were Hungarian words. Hungarian writing is entirely phonetic, so Rheinhardt easily learned to transpose English words into their Hungarianised forms. We used some Hungarian words, such as *igen* for 'yes' and *nem* for 'no' and every final Y we pronounced as '*ipsilon*'. It was quite a long time before it was discovered how we communicated in this secret language. Many people seemed to think that we had what they called a key word, which was the word 'ipsilon', which naturally occurred quite often, but of course it was not a word at all! One day I said to Rheinhardt, while looking at rather an indifferent map of Dartmoor, '*Rowttain mop*' (meaning, of course, 'rotten map') to which David Robson immediately replied, 'It is rotten, isn't it?' and the cat was out of the bag! At this point I explained to David how we had produced our secret language and he joined us in mystifying the rest of the boys and girls for a while, until it finally came out how it was done. After that it was no longer fun, so we dropped it.

I kept up a good friendship with Rheinhardt. We often went for walks together and sometimes he came with me on my explorations of Dartmoor. One good result of this friendship was that I decided I had a stupid complex about the German language; probably something had trickled into my subconscious from the age-old hatred of the Habsburgs felt everywhere in Hungary. I thought, therefore, that the thing to do was to learn German properly. I went straight to the library and took out the famous book, *Im Westen nichts Neues* (*All quiet on the Western Front*) and, with the aid of a German-French dictionary which I kept on my shelf, I quickly read through the book, writing out any word I had to look up, putting the German word on one side of the page and the meaning in French on the other. When I had a little leisure, I would go through my word list to make sure I really knew all those words. I think I went on to reading other German books as well and, what with sometimes chatting in German with Rheinhardt, my German became really quite passable, Rheinhardt sometimes exclaiming that I was speaking just like a German!

There was a modern go-ahead type of school in Germany called the Odenwaldschule. Soon after Hitler came to power, he closed all such schools, as they would not preach the doctrine of Aryan superiority. Inevitably, there was an influx into Dartington from this school. There was one girl, Mona, a little younger than myself, who had arrived from the Odenwaldschule and we soon made friends. There were no romantic aspects to our friendship, although we went to country dancing together sometimes and she would tell me tales about the Odenwaldschule and I would talk to her about Hungary. What with the Freud grandchildren, Rheinhardt, Mona and myself, we began to set up a little German-speaking clique. It is strange how children and teenagers are always anxious to become special by establishing groups, held together by the strangest of links. Mona and Rheinhardt and I often would sit together at mealtimes and mystify the other boys and girls by speaking to each other in German!

In December I took my School Certificate Examination again and this time I also got five credits. I did manage to get one in English, although I only managed a pass in physics. So I really had two more terms in which to learn whatever I liked and really enjoy what amounted to the end of childhood, although perhaps my real childhood had already ended when I decided to come to England.

In the summer term all the children went on a school camp. At the end of my first year the senior camp was on the South Cornish coast, the tents being pitched along a deserted beach, to which there was no access except on foot. Camping is always a good way to socialise: we had to learn to put up with each other's oddities and we certainly learned how to do without creature comforts. The sea was reasonably warm, so we swam a lot. We also used to walk into Polperro along the cliff path and enjoy the local cream teas. It is strange to recall that, for the sum of one shilling, you could get a cream tea which included scones and jam and cream, as well as a whole lot of very creamy cakes, not to mention innumerable cups of tea!

At the end of the second year, three of us got permission to go for a camping tour on our own, exploring Exmoor. We took the train to Ilfracombe. When we reached Ilfracombe we walked along the seafront, enjoying the scenic beauty of the place, when I suddenly suggested, 'Why don't we go to the flicks?'

'Oh yes, why not?' responded my friends.

We soon found a suitable cinema. They were showing *Tarzan*, which suited all of us. Then we looked at the prices. The first three rows were sixpence, then came the nine-pennies and quite far back, but not quite at the very back, were the seats for eleven pence half-penny.

'That's about our standard, I suppose,' said Björn, a boy of Danish origin who was part of the small group.

'I suppose that's it,' we replied and bought our tickets.

Tarzan was great and we all enjoyed ourselves, but it was getting towards evening and we had to walk out of the town in order to find a place to pitch our two tents. We managed to climb up on to Exmoor plateau and we pitched our camp in the midst of heather and gorse.

We all had a great time, spending most of the week on the moor, nearly always within sight of the Bristol Channel and on one fine day we could see right across to the Welsh coast.

I remember saying to Björn, 'Look, I can see Wales!'

Björn scanned what he could see of the channel and said in a disappointed tone, 'I can't see any!'

'Oh, I don't mean whales, I mean the coast of Wales,' I retorted.

'Oh,' said Björn, even more disappointed. He saw no great point in looking across the Bristol Channel simply to see the other side!

For me, as a more recent arrival in the British Isles, it was much more important to be in physical and optical contact with the various parts of these islands I had not yet explored!

For my last two terms at Dartington I lived in one of the junior houses, as there were now too many seniors to fit into one house; in any case, I had made friends with a number of the younger children, so I did not mind not living with the other senior boys and girls. One result of this was that the juniors elected me as their representative on the School Council, which was the official body that made and revised all the school rules and dealt with suggestions and complaints. I remember bringing up Rosemary's case, who was a Roman Catholic and had been teased about wanting to go to Mass on a Sunday, as well as for insisting on wearing a bathing costume to go swimming. I told the council that, although I was myself an agnostic, I felt that it was wrong to exert pressure in the matter of what people chose to believe and the same applied to bathing costumes: nobody should be pressured into being in the nude if they felt it was not right for them. It was decided to convey the school's apologies to Rosemary and to make it clear to her that her religious beliefs were to be respected and so were her views about wearing swimwear for bathing. I conveyed this message from the council personally to Rosemary.

It took about half an hour to walk from the school to Folly Island, so we asked Mr Curry why we could not have a swimming pool. He pointed out that such things needed to be budgeted for and that it was just not possible to build a pool without some advance planning. But the council appointed a committee to examine the problem. The committee made extensive inquiries around the school and came back with the suggestion that the great majority of the children would be willing to supply their labour free to dig the pool, if a foreman could be hired and the necessary material purchased. This was put to Mr Curry, who agreed that the much reduced amount of money could indeed be found at short notice. A rota was put up and children signed up for work, mostly between six and eight in the morning but also during free periods in the day or in the evenings. To cut a long story short, the pool was dug in a few weeks. It was lined with concrete, a pipe was provided and a pump installed to pump the water up from the Bidwell Brook, and the pool was ready for use. There was one thing everybody forgot: the chlorination plant. By the end of the summer term we were all swimming in some very green water, but swim we did and we did not have to walk to Folly Island if we did not have the time.

My last morning at Dartington as a pupil was quite memorable. I was woken up at about six in the morning by a very distressed Monica, a girl of about eleven whose room was next to mine, who weepingly told me that there was a rat in her room.

'Oh, Monica, don't exaggerate! It's probably a little mouse and it will go away down some little hole,' I replied, trying to wake myself up and soothe her fears at the same time.

'No, Zed, do come! It really is a rat! Big like this!' she retorted, opening her arms wide and then wiping off her tears.

There was nothing for it but to comply. Monica dragged me by the hand and into her room.

'Look! There it is!' she cried.

There was indeed a large rat crouching in the corner of the room. I looked at the rat and then at Monica's frightened little eyes and decided to go to war with the rat.

I took my slipper off, approached the rat slowly and then gave it a big hard swipe with the slipper. But I did not reckon on the rat's urge to live! It retaliated by swinging round very rapidly and biting my hand. This, of course, made me even more determined to win the war with the rat, especially after I had glanced at Monica's terrified face. By then, she was curled up at the farthest corner of her bed! Fortunately, the rat stayed where it was, seemingly ready to attack again, so I pulled one of Monica's blankets off her bed, threw it over the rat and stepped on the part where I imagined the rat's head was. I repeated this several times, stepping harder and harder each time, before I dared remove the blanket. This latter move revealed a very dead rat and Monica's frightened face turned into a broad smile. She jumped off the bed, rushed up to me, gave me a big hug and said, 'Oh, thank you Zed, thank you! I don't know what I would have done without you!'

And this was the end of the 'heroic saga' of Monica and the rat, as well as the end of my stay at Dartington Hall School as a pupil. I did not know it then, but I was to come back quite often, eventually to show the teachers some of my inventions regarding improvements in the teaching of mathematics!

Chapter Nine

The Undergraduate Years

I now had my passport to university, which perforce had to be London University, as my father's budget would not be able to bear anything else. But there was a summer to fill in before the new life of a university student could begin.

Way back in my early childhood, I remember being fascinated by the map of the British Isles, particularly by the many little islands that seemed to bespatter the top left corner of the map. These islands lie to the north west of the mainland of Scotland. Although I had not read much about them, I decided to gratify my early childhood curiosity, which meant going to the Hebrides.

This was the first time I was to undertake long-distance hitch-hiking, but I soon learned the tricks of the trade. I eventually formulated some fundamental rules of the game. As far as I remember them now, this is what they were:

1. Map out the route and do not be deflected from it by kind offers to take you farther, if farther is off the route.
2. If possible, never be dropped in a town, but, if you are unfortunate, to be so dropped, take a bus, tram or trolley bus to the outskirts, as hitch-hiking in a built-up area is very unsatisfactory.
3. Always stand at the roadside to hail a car – never walk – and look as cheerful and smiling as possible, even

after an hour or so of trying to get a lift! No driver wants a miserable companion!
4. Never tell the driver how far you have been that day. If somebody generously offers you a lift from Carlisle to Glasgow, don't tell him you have just come from London: it devalues his generosity.
5. In conversation with your driver, be alert, amusing and enquire about his life and his business. If you tell any untruths, make sure they are funny but, above all, consistent!

At a much later date I once told my lift that I was acting as an agent for a friend in Czechoslovakia who made artificial pips out of trees felled in the Carpathians, which he shipped from Danzig to customers in England. The pips were for making artificial raspberry jam. Unfortunately, he asked me to his house for a meal and it became really difficult to keep to the 'consistency rule'!

I did get to Glasgow at the end of my first day of marathon hitch-hiking. I had joined the Youth Hostel Association, so now I had to find the Glasgow Youth Hostel. I asked a passer by where it was.

'Och, d'you nae ken where it is?' the person replied. 'It's nae harrrrd to find! Tak the frrrrst trrrrn to the recht, then you go left past the kirrrrk, then trrrn recht again.'

He went on like that for quite a while, until I managed to interrupt him, saying, 'Couldn't I take a tram there?'

'Och aye,' he said. 'It's a haepenny ride, but it's nae worrrrth it!'

I burst out laughing at this and the poor fellow could not make out why I was laughing.

He finally walked away mumbling, 'Sassanach! Sassanach!' which I later learned was a derogatory term for the English. He was too far away by now, even if I had known what he meant, to tell him that I was really a Hungarian. As likely as not, he might never have heard of Hungary or Hungarians so it was just as well!

Before long I found myself on the boat from Kyle of Lochalsh to Stornoway, bound for the Isle of Lewis. From Stornoway I almost immediately got a lift to the west side of the island. Just past the scattered crofting community known as Uig, I decided to pitch my tent. It was about eleven o'clock in the evening and the sun had just set. Uig Bay stretched out to the west and beyond it lay the vast Atlantic Ocean. There was not a tree in sight. Everywhere, the country looked as if piles of rocks had been thrown about at random, with many small lochans in between these rocky ramparts. I picked a good lochan, which had beautiful water lilies growing in it, at least that is what I thought these aquatic flowers were which grew in profusion in the shallow parts of the water.

I found a level spot and pitched the tent. I had a murder book to finish, so, having slipped into my sleeping bag, I thought I would read until it got too dark to see. I became so involved in the plot that I never noticed that instead of getting darker, soon it was beginning to get lighter. I suddenly realised that at the end of June at these latitudes there was really no night! So I put my book away and light or no, decided to go to sleep.

I thought I would explore the west coast of Lewis, so the next day I packed up and started walking in a southerly direction. I was walking over giant slabs of granite or treading over springy purple heather and I felt really great, so far away from everything and everyone, with just the sky and the mountainous terrain my constant companions. I breathed in the sea air blowing in from the Atlantic and, forgetting everything else, I seemed to be just happy with my lot. I came eventually to the end of a long inlet from the sea, a kind of fjord, with just one croft standing a little way off the shore. There were a number of small boats beached on the rocky shore and I could hear the sound of domestic animals as I approached the croft. It was already quite late so I thought I would ask the crofter's permission to put my tent up.

The crofter and his wife greeted me cheerfully.

'May I put my tent up here somewhere for the night?' I asked in as deferential a tone as I could muster.

'Och no!' said the crofter. 'You will no be sleepin oot! We have a bed forrr you inside!'

I was not about to argue with the crofter and before long I was sitting by a peat fire, drinking a hot cup of tea.

'Will you no take a piece?' asked the crofter's wife.

A 'piece' was of course a piece of bread. I assented, but almost immediately I heard the crofter's wife ask, 'Will you no take an egg?'

Before I could say Jack Robinson, I was not only having several 'pieces', with not just one egg but several, plus a good helping of fried fish, which the crofter had just caught and which I took to be mackerel.

After we all had had enough to eat, the children came crowding round and wanted to see what I was doing. I was looking at the map, trying to find the little inlet where I was being so hospitably treated. The children could not speak much English; they were talking in Gaelic very rapidly amongst themselves, no doubt about this strange person and this piece of paper with a lot of squiggles on it called a 'map' which I tried unsuccessfully to explain to them. The oldest child did understand a little English and finally I got through to her what a map was and how you used it.

To make conversation, I asked if there were many people like myself coming by the croft.

'Och, yes, any amount!' said the crofter. 'Last year there were thrrree, but you are the firrrst one this year!'

The only way to reach this croft was by trekking through the mountains, or by boat. To go by boat meant a very long trip, often quite hazardous, so if ever they needed anything, the crofters would walk to Amhuinsuidh. In fact the postman came once a week from there, as they told me and he had to stay the night as it was too far to come and go back the same day. They were expecting the postman the next day. I spent only one night at this croft, as I was anxious to do my exploring! They would not hear of being paid anything, they insisted that I had been their honoured guest.

So I moved on to another part of Lewis; this crofting community was on the border between Lewis and Harris. I did manage to put up my tent there and no sooner had I finished arranging everything than a whole crowd of children appeared from seemingly nowhere. They all chattered to me in Gaelic so I was determined to learn some of the language from them. We played the following game.

I pointed at something and looked enquiring, so they would say what it was in Gaelic. I would write everything down, using international phonetic script. I would then read the words back to them and, when they roared with laughter, I knew I had pronounced something badly, so I asked them to say it again. In this way I kept correcting my writing and began to acquire a little vocabulary. The children were also good at singing, so they taught me some of their songs, which I tried to play on my recorder. Having worked out the tune on the recorder, I was able to write down the tunes in musical notation and file them away for future reference. When I wanted them to go, I said to them, '*Oidhche bha...! Tha mi sgìth!*' ('Good night! I am tired!')

To this day I am not sure whether I have the spelling right, since I never really learned to read and write this language, although I got to a point where I could make out quite a lot of what was going on and express some simple ideas. But I can say with certainty that the spelling is very little to do with how a word is actually pronounced. The Gaels, in this case, have out-Englished the English!

I finally made it to Tarbert in Harris, where I again made friends with crofters and their families. I took a number of photographs and eventually posted copies to all the crofts where I had taken them, as a token of gratitude for all the hospitality I had received.

In one of the crofts where I stayed, the crofter had violent toothache. I asked why he did not go to the dentist. It turned out that they did not have the money for the bus fare to Stornoway, not to mention the dentist's fee! So this was where I thought I could repay Lewis and Harris! I placed a pound note in his hands and said, 'Now go to Stornoway and see to your tooth!'

Fortunately, the bus route between Rodil, Leverburgh, Tarbert and Stornoway ran right past this particular croft, so I made sure the poor man got on the bus. He came back with the evening bus, absolutely elated. He had had his tooth out and was having no more pain! I thought I had thus paid my way, but I had not reckoned with the crofters' idea of generosity!

'You have been so kind,' said the crofter as I was leaving. 'You must take fifteen yards of the tweed I've been making, so that you and your father can have a suit made!'

I tried to talk him out of it, but there was nothing for it but to accept. He tied this enormous weight of tweed on to my rucksack and, with tears in his eyes, waved me goodbye. He went on waving until I turned the corner of the road and he could not see me any more!

I had many more adventures on North Uist, Benbecula and South Uist. I learned more Gaelic and some of the songs, having attended many of their ceilidhs. I visited Eriskay of love-lilt fame, then Barra and Vatersay, until I could go no further south on these islands on the 'top left corner of the map'. So I was fully satiated, purified and saturated with the magic of the North and was ready to embark on my career as a university student.

I registered as a student at University College, Gower Street, for what was then called 'Intermediate'. This university year does not exist any more, having been replaced by 'A levels', which are taken at school. There were some choices to be made. Mathematics was a subject that you could take in the Arts Faculty or in the Science Faculty. If I wanted to be in the Science Faculty then I would need to take four Science subjects, which could have been pure mathematics, applied mathematics and two sciences. I did not know much physics, even less chemistry and biology was a closed book to me, so there was nothing for it but to go into the Faculty of Arts. This required at that time Latin as a compulsory subject, as well as one modern language. These, I thought, would present no problems and then I could add the two mathematics subjects. I did not want to take the very elementary mathematics offered for Intermediate, so I got permission to take the courses the first year students were taking. These clashed with French lectures, so that ruled out French. Italian was not offered, so I was landed with German. I thought this was quite a good idea, since, out of the five languages I knew, German was my worst language. I thought I might as well take it so that I could improve in it.

I enjoyed the mathematics, as University College was doing what was then called Syllabus B, while all the other colleges in London were doing Syllabus A, which was a glorified progression towards more and more intricate problem-solving, with little or no emphasis on getting students to understand the basic mathematical structures on which all the problem-solving depended. In fact, the Syllabus B idea was the main reason I had picked University College, instead of any one of the others; in the latter I thought I would be wasting my time. How arrogant can you get?

Latin presented no problems for Gwenyth had trained me well to read and write classical Latin and I also enjoyed my German studies. The German lecturer was very enthusiastic about poetry. We read Goethe, Heine and Schiller and I certainly improved my competence in that language to the extent I could enjoy a leisurely read of any average German book without having to resort to the dictionary. There was time left for social activities as well as cultural ones, such as going to plays and concerts. I joined the Socialist Society, which is as near as I ever got to joining a political party. I was still a Hungarian national and I did not want to blacken my book by joining the wrong party! I suppose this was still the result of my impressions of the political regime in Hungary!

During my Intermediate year, the Socialist Society did not do very much, but, as things got worse on the international as well as on the domestic front, there appeared to be more things to object to and students are proverbially good at vociferous objections to what they consider the wickedness of the world at large! One such activity was the welcoming of the hunger marchers. Hundreds of unemployed people had marched south from Jarrow to object loudly to the government's indifference to their lot. We put our pennies together and organised a big feast for the hunger marchers, cooking food for them, serving them and talking with them afterwards, assuring them that we were on their side! At about the same time the bus drivers went on strike. During the previous General Strike of the late twenties, students had volunteered to drive the buses and so helped to break the strike. So we marched proudly alongside the bus drivers and conductors, shouting the slogans, 'We support the busmen's fight! Students will not break the strike!' and felt we were on the side of social justice. I really forget whether the busmen won or lost the strike; what I remember is supporting the fight against what I considered to be an exploitation of the 'masses' by the 'capitalists'.

The Nazi minister Von Ribbentrop was invited to come to University College by the university to donate some books. We, of the Socialist Society, thought this was totally absurd, that we should have nothing to do with these barbarians. It should be remembered that Herr Hitler, as he was referred to then, along with Signor Mussolini, were considered to be bosom friends of the West and any anti-fascist or anti-Nazi opinions were branded communistic and dangerous. So we lined up where we knew Ribbentrop was going to come and, as he walked past us, we shouted loudly at him, 'Fascism kills culture!' He walked past us very quickly and disappeared inside the building. There were pictures of us the next day in the *Daily Herald* and in the *Daily Worker*. The right-wing press simply ignored the whole thing.

During the first autumn term I was to experience some dramatic events which were to shape the rest of my life. Mina, the daughter of a colleague of my father's, attempted suicide but was saved by the use of the stomach pump. Her father, Richard Cooke, also tried to kill himself, but was saved too. So another thirteen year old girl was fated to cross my path! There had been Fiona, then Yo; now Mina was on my horizon. I had not become more friendly with Yo because I thought I needed all my time to devote to my studies. Was it not the same now, only more so? I was now a university student and surely I had to prove myself.

My father explained to me that Mina was unhappy because of many family complications; would it not be nice if I could befriend her? So inevitably Mina came along and I was landed with the job of consoling her! She told me about some if not all the things that worried her; we played ping-pong; she came to the Panther Club meetings and took Fiona's place on the floor while we dutifully listened to all the discussions, but our relationship remained entirely platonic. But something undeniable drew us together and, possibly for reasons we shall never know, we became fast friends. I took on the responsibility of providing her cultural education. I first took her to the opera to see *Hansel and Gretel*, which she enjoyed very much. Then I graduated her to Mozart, in particular to the *Magic Flute*, which we have both enjoyed throughout our life ever since. One of my attempts was unsuccessful: I took her to Covent Garden to see *La Bohème*, but after the first act she felt it was too sad and wanted to be taken home, so we did just that, but not before taking her to Carwardine's in Southampton Row to have a coffee. This was an extremely grown-up thing to do with someone in her early teens and she duly appreciated it!

Mina's family situation deteriorated rapidly. There were fights; divorce was in the air; her mother seemed to use her as the nursemaid to look after her sister, ten years younger. It seemed that whenever I asked her mother if Mina could do this or that with me, the inevitable refrain was, 'Yes, but then who would look after Diana?' the idea not occurring to her that a mother, not a sister, was the person who should be responsible for a child.

So we discussed the possibility of Mina's going away to school. But there were financial problems. With the divorce pending and property being distributed, bought or sold, there was no time to think of a young girl's education. So I got in touch with some of my school friends from Dartington and tried to put a trust fund together, so we could send Mina to school. It would have been good to be able to send her to Sidcot, a Quaker school with a very good reputation, but the fees were much too high and there was no prospect of a scholarship. My thoughts turned to Summerhill, A.S.

Neill's 'crazy' school, where psychology was supposed to cure all the ills children suffered from in broken homes.

I hitch-hiked up to Leiston in Suffolk, where Neill's school was situated, looked it over and had a discussion with Neill. He offered us a considerable reduction in fees. Although the school looked a real shambles, I thought, no doubt mistakenly, that Neill's knowledge of psychology and his knack of handling difficult or unhappy children, would balance things out on the positive side.

So for the spring term Mina was sent to Summerhill. It is of interest to note that suddenly enough money was found by the parents to send Diana to Summerhill as well. Clearly, there was indeed 'no one to look after Diana'. At this point Mina was fourteen and Diana was four. It soon became obvious that it was a mistaken move. According to Neill himself, that particular period was probably the worst the school had ever known or would know thereafter.

I took a long weekend off to go and see how Mina was getting on. She had some horrendous tales to tell me. There was a girl of thirteen there who had been at a convent and had become some kind of a sex maniac, who wanted to rope in all the boys as well as control the girls in her gang, which became the 'in' thing! Mina was told that 'you had to be fucked six times to belong to the gang'. When she refused the exercise, she was branded a lesbian and treated as an outcast. One of the older boys tried to rape her but did not succeed and finally Mina asked Neill if she could have her own room with a lock on it, having explained the situation. This was granted and she had a bit more peace.

There was a jeweller in Leiston where these very young girls used to go, one or two of them exciting him sexually while a third one would look for things to steal. They stole no end of watches, jewels and gold chains and the jeweller just let it go on, no doubt since he enjoyed what went with it. The jeweller once said to Mina, when she happened to walk in there while the gang were doing their thing, 'You are chicken, you wouldn't dare to steal anything!'

So Mina took the opportunity while the jeweller was enjoying his fun, took an expensive clock off the shelf and ran off with it to the school. An hour later she was back with the clock.

'Here is your clock! And don't you tell me I am chicken!' Mina roared at him and walked out.

On one visit, I am sorry to say, the gang got me to go to one of their 'orgies', which, looking back, was not very serious, as nothing like real sex ever happened. If 'being fucked six times' meant such fooling about, the gang might not have been as bad as we had thought. On the same visit I also went to look in on some of the classes. Some were run in quite normal ways, but in one class I saw the teacher with a fourteen year old girl on his lap doing her best to stimulate him! I do not suppose too much learning took place during such a lesson, in any case not the learning of whatever the subject was meant to be taught!

During the summer term Mina thought that things had really got to a point when she had to take things into her own hands. She had a nice watch which I had given to her, for which she organised a raffle. She packed her suitcase, then went to the nursery section where Diana was, put all her things together and took her and the cases to the station. With the raffle money she bought tickets to London and took the first train to Liverpool Street, appearing at the door of the Putney house which Richard had just bought for her mother according to the terms of the divorce settlement. There was much telephoning. My father and I were informed, Neill was called and that was the end of Mina's career at Summerhill! Later Mina told me that she had to run away, as she knew that, if she had consulted Neill, he would probably have persuaded her to stay and she had decided that Summerhill was not for her under any circumstances!

Of course there was the problem now of where to send Mina to school. Summerhill was obviously a failure; Dartington or Sidcot were too expensive. What was left?

On one of my repeated visits to Dartington, which I was still fond of doing, I passed through the New Forest and discovered the Forest School, at Godshill, near Fordingbridge. It was run by Cuthbert Rutter, who was a Quaker and a number of teachers on the staff were also Quakers. I thought Quakers had a good press so I thought this little place might be the solution.

I took Mina down there, explained the situation and Cuthbert was able to agree on a fee which the 'trust' could afford. Mina took a long time 'interviewing' Cuthbert, the headmaster. He told me later that he had never been so thoroughly vetted by any parent as by this fourteen year old whipper-snapper! Even though the academic record of this school was not all that good, there was a dedicated staff, just about thirty children and, with the Quaker influence replacing the Freudian one, I thought at last we were on the way to solving our problem!

Our friendship continued to grow and I went down to the New Forest very often for weekends, usually hitch-hiking both ways, to see how Mina was progressing in the new situation. She had made friends with the children and with the staff, although one of the latter became very keen on her and the whole thing threatened to become more than a friendship.

One summer I was able to persuade her mother to let me take her on a hitch-hiking cum walking tour, with no particular geographical aim in view. At first we made for the south-west, as I was still feeling the need to keep up the link with Dartington. We ended up in Cornwall, where a farmer let us sleep in his barn.

There were some rats about and Mina cried out, 'Oh, do cuddle me, I am frightened!'

'I can't,' I replied quite sternly.

At the age of fifteen Mina still had not realised that for a boy four years older than she to be cuddling her in the night could only lead to one thing. We were happy in our platonic friendship and I, for one, was not about to spoil it because of some beastly rats! So we continued our holiday in the same platonic vein!

We got across eventually to Ireland and, having hitched a lift from Dublin to Killarney, we found ourselves in this most delightful part of the world, teeming with leprechauns and fairies in every wood! One night we were in the Caha Mountains and, just as we were wondering where we would sleep, we saw a little thatched cottage a little way ahead. We asked if we might stay and the farmers said we could stay in the hayloft. They were terrified when we put the flashlight on in the loft because they were convinced we would burn the place down! They had never seen an electric flashlight! The flashlight stayed in the house. They shared their meal with us, which we had difficulty in eating, as after each 'course' they gave the plate to the dog to lick, before the next was served! When you travel with very little means and have no funds for overnight stops, you meet all sorts of interesting situations and you have to take the good with the bad!

When we arrived in the city of Cork, we thought it might be an idea to hitch-hike across the sea, back to England. We spoke to the captain of a small cargo ship tied up alongside the quay; he asked us in and offered us some lunch and he said he would gladly take us there, but he was sailing for Lisbon in a few hours' time, which is not exactly where we had planned to go! He very kindly filled our bags with ships' biscuits, saying, 'You never know when these might come in useful!'

You really need to be very hungry to enjoy a ships' biscuit for they are as hard as rock! But before we managed to return to England, having run out of most of our money, we were indeed glad to have a ships' biscuit or two on the road!

We did make it back to England eventually and hitchhiked back to London. The vicissitudes of this trip really cemented our friendship, which became one that was to last a lifetime.

I had a number of romantic attachments during my undergraduate years and I often discussed these and other weighty problems with Mina. We saw each other often in the holidays and I frequently made the trip to Godshill to visit her.

I became involved in a love affair with a Dutch girl called Cor. I really thought this was it and that we were made for each other. How wrong can a person be? I took her down to my precious Devon and we had a glorious, romantic time during the Christmas holidays. When we got back to London, it appeared that she needed a job. It happened that Cuthbert needed somebody for the job of a house mother for the younger children and, on my recommendation, she got the job. Of course, Cor did not share my ideas of eternal love and flirted with all and sundry at the school. This got back to me via Mina and I was shattered. I thought that was the end and it was time to end it by committing suicide. There are several lakes on Hampstead Heath and, it being winter, I thought the cold water would soon do the trick if I were to jump! So I went to the lake and stood there leaning over the fence, looking down on the dark icy water.

Next to me there was another man, who seemed much older than myself, also contemplating the deep dark waters. After some minutes he spoke to me, 'Are you thinking what I am thinking?'

'Do you mean?' I replied, pointing down to the water.

'Yeah,' he replied.

He told me a sad story about his girl who did not love him any more. I told him my story. Then I said, 'I have a little friend – she is only a child at school, but we are very friendly. Here is her last letter to me.' I pulled out a much folded piece of paper: Mina's letter.

'Can I read it?' he said.

I replied by passing him the letter.

He read it and burst out laughing! 'How silly can you be!' he said to me. 'That's the girl who loves you!'

This was indeed news to me, but I realised that, in the true sense of the word, Mina did love me and I loved her. I thanked my would-be companion in suicide and asked him if he would like some coffee. At this point he must also have changed his mind about 'shuffling off this mortal coil' and we enjoyed several cups of coffee together before we parted. Ships that pass in the night, I thought, but what an important ship this one was!

It took one year of Intermediate and three years of undergraduate work to get to the final bachelor of arts examination in 1937. During the three years of undergraduate work, all the courses were in mathematics, some pure and some applied. I am sorry to say that I was not a very good student as there were too many distractions. There was the political work I have already referred to, there were my romantic relationships, there was my continuing desire to explore the country and there was my deepening friendship with Mina, which all took up time and energy and so subtracted from what was available for study!

At one point I asked my father if I could live away from home. There was a place called Youth House, in Campden Road, which was run on modern yet fairly sensible lines, like the places which nowadays are referred to as communes. You did not pay much to be there, but you had a room to yourself and you had your meals in the communal dining room; you had to provide so much work each week in order that the place could run at the extremely reasonable rates charged. I remember one occasion when Gedeon, as well as Gyuri, my cousin, came over from Budapest and stayed nearby, but came and practically lived at Youth House. Everyone said that the three Dieneses wiped the floor with all the girls – nobody else stood a chance! In any case, it was not a repeat of my experiences on our Budapest skating rink, as I believe I did manage to make as much impression on the girls as my brother and my cousin!

So romances developed and died, some more serious than others. One romance involved a Russian girl called Olga, which threatened to become quite serious, but somehow things conspired to make it go wrong. One thing on which I was keen with any girlfriend I contemplated marrying was the eventual raising of a large family! Olga was not keen on children and this put me off. She really did try and seduce me on several occasions. On one such occasion, while I was visiting her in her home, she took all her clothes off, lay languidly in my lap and said very sweetly, 'I am yours!'

But I did not take the hint, if a hint indeed it was!

Another time we went camping in the wildest part of the Dart Valley on Dartmoor and she provided the male apparatus and expected me to make love to her. I had a very bad sty in one of my eyes, which was a disadvantage and, in any case, I did not think one could do those things to order, so in the end I never made love to her!

I made some more trips to the Hebrides, really making friends with some of the crofters. On my second trip to Tarbert, Harris, when I was still very ignorant of the Gaelic language, I was put up in a small thatched croft with an earthen floor, with children sleeping all over the place. I thought it was a lovely family atmosphere and I longed for the time when I would have found my girl and we were looking after a growing family of children.

It happened to be a very bad summer, raining cats and dogs just about every day. As I went walking along the unmade-up road, crofters would pass me on the way to cutting the peats and they would all say, '*Tha i fliuch as fuar!*' so I would reply in the same vein, saying, '*Tha, tha i fliuch as fuar!*'

This went on for about three days, every day being wet and cold. At last a fine day dawned and, as I met the first crofter on the road, I said to him, '*Tha i fliuch as fuar*'; to which he gave a gentle laugh and said, '*Cha n'eil, tha i priagh an diu*' (I am really not sure of the spellings; meaning, 'Oh no, it's fine today!')

It began to dawn on me, putting two and two together, that the '*fliuch as fuar*' must have been referring to the weather and did not just mean 'Good morning' or some such!

After a few such blunders, I managed to learn some elementary Gaelic, but I never learned the spelling, as I continued to write down exactly what I heard using the international phonetic alphabet, which I had learned trying to perfect my pronunciation of the English language! I learned a number of songs, which I played back to the crofters on my recorder, and we enjoyed some private *ceilidhs* in many of the crofts I visited. Perhaps the one I remember best is the one on the south coast of Harris, between Leverburgh and Rodil, known as Borisdale. The matriarch of the croft was a Mrs McDonald, who ruled over all the doings of the croft with a firm but gentle hand. In the evening the children would sit with the grown-ups by the peat fire and as the little ones fell asleep, the older ones would carry them up to bed, until all the children were finally asleep, the oldest one being carried up by an adult!

They had a very strange system with the cows. One of the milking times was round about midnight and then they had to go over the hills to find the cows! Sometimes they would wander quite a long way off, but I never knew of any time when they were not found, duly milked on the spot and the milk brought back in buckets. Some of the morning's milk would have been left to stand during the day, then in the evening the cream would be taken off for making butter. One evening I was presented with the cream, which had been poured into a tin with the lid tightly closed, and told to shake it. I did not realise that I was to make the butter for the morning's breakfast! I duly shook the cream for about twenty minutes, when Mrs McDonald told me to stop. We opened the tin, and sure enough, it was full of lovely fresh butter!

I sometimes went fishing for mackerel with some of the older children. This was a simple procedure. One of us would row very fast across an area of water where we thought the fish would be and another would hold the net at the stern of the boat, which would gradually fill up with mackerel. When we hauled in enough fish for supper, we rowed back to the shore, beached the little boat and took the fish up to the croft.

I also took part in cutting and eventually in stacking the peat. This was the crofters' fuel and it was essential work. They used very large spades and cut huge pieces of peat out of the ground, which were left to dry for a time. When they were considered dry enough, they were stacked in little heaps, so as to leave room for the air to blow through the stack, whereby the peat could eventually become dry enough to burn. When it was dry enough, it would be carried to an outhouse by the side of the croft and piled in there for future use.

I kept up my friendship with the Macdonald croft for many years and, after I was married, I would take the family there for holidays. The children in the croft of course grew up, got married and themselves had children. I remember one of them had twin girls and we went on sending them Christmas presents until we left England for Australia! We spent many family holidays in Lewis, Harris, Barra, Skye and Mull while the children were young, so they could enjoy the vast expanses of golden sands, the lochs, the mountains and, above all, the people, who were always very kind to us.

In the year 1937 I finally took my bachelor of arts examination. Since I had not been a very diligent student, I did not expect to get a good degree, so I thought I would spend the summer on the Continent instead of waiting for the results in England. My father said he could give me what I would have cost in food at home and that I would have to manage on it! As it happened, I had been instrumental in helping a certain French family, the Azémas, to get one of their children, who was retarded, into a suitable school in England. This school was run by the Tomlinsons, whose little girl Rosemary had been to Dartington, for whom I had pleaded years before at the Dartington School Council. The Tomlinsons were happy to have the Azéma boy, who had settled very happily in the school and so, as an expression of their gratitude, the Azémas offered to have me at their summer villa outside Sainte Maxime on the Côte d'Azur. This appeared to be a solution to the economic problem.

So I hitched to Dover, used some money for the ferry and soon found myself in Haute Savoie, for which I retained some sentimental attachment on account of the great holiday I had had there, during which I had decided to come to England. I pitched my tent at the far end of the Lac d'Annecy, next to a farm. There was a little narrow gauge railway line there, which the farmer used for collecting his hay and other crops. He had several pieces of rolling stock and I saw one of the little boys playing with it. I asked the farmer if it was safe for him to do that. *'Non, Monsieur, il ne devrait pas jouer comme ça! C'est à la suite d'un accident qu'il est devenu muet!'*

I soon realised that the little boy could not talk. But he understood what was said to him, so I tried to distract him from the rolling stock and played with him on the beach, skimming stones, and showed him how to make whistling noises with blades of grass. I stayed there for several days and the little boy and I became fast friends. When I was leaving I told him that I would be back. I did eventually come back, after the war, with my wife and children and of course he was a grown-up boy, working at his father's farm, but he did recognise me as soon as he saw me and gave me a heartfelt embrace! I thought it was good to know that some things we do for people out of compassion are indeed remembered and appreciated!

I eventually crossed the Alps by the Col de la Seigne, descending into Courmayeur and eventually into Aosta. Carrying my tent and other heavy equipment, I did not feel like looking for a suitable campsite. I changed my remaining francs into lire at a bank, counted them up and wondered if I would have enough money. I found a small hotel where I could have a room for the sum of five lire, which I decided was within my budget and, after a simple meal of bread and cheese and milk in my room, I settled for the night. The next day, after I had bought some fresh supplies of food, I just had about three or four lire left! It was imperative to get to the Azémas quickly!

I managed to hitch a ride to Turin, where I found a place in which I could have a bed for one lire and eighty centesimi. It was in a slum district; there was another man in a half-drunken stupor in the other bed in the room and the noise of merriment came though the window. It was not possible to shut the window, try as I might, so I had to sleep with the incoherent grunts of my drunken fellow guest and the sound of accordions and singing going on outside a short distance away. So this was zero hour. I really had to get going!

The next day I walked and walked and walked and nobody gave me a lift. My opinion of the Italians sank lower and lower with every kilometre and with every passing car! I almost reached Cuneo by the evening, a distance of about 80 kilometres, when I spied a hay barn and, without asking any questions, collapsed inside among the dry hay and fell fast asleep!

The next day I was luckier and got a lift to Sospel, then another one to Menton. Fortunately, at the French border nobody asked me how much money I had! That day I reached Sainte Maxime, with all my food supplies consumed and not a cent in my pocket!

The Azémas welcomed me with open arms, fed me and showed me to my room, which was an enormous improvement on barns and doss houses! It had a large window overlooking the Esterel country, which was beautifully undulating and covered by a seemingly never- ending forest of pine trees. I feasted my eyes on the scene for a while before I washed myself and finally collapsed into bed, immediately moving into the unconscious mode!

Sainte Maxime and the Azémas were just what the doctor ordered! We went to the beach every day: the sea was warm for swimming and the Côte d'Azur continued obstinately to be azure. Apart from the retarded boy, who was not really that retarded, they had another boy of about nine, Jean-Pierre, with whom I instantly made friends. He showed me his 'secret places' in the woods nearby, which I truly appreciated as a sign of real friendship on the part of a child! He used to address me as '*tu*' and I was quite honoured that he did so, another sign of friendship. His father would say, '*Il ne faut pas tutoyer Monsieur Dienes! Ce n'est pas poli!*'

'*Mais Monsieur Dienes est mon ami!*' Jean Pierre would reply.

I tried to keep out of this kind of family argument, but, as it happened, the whole thing was quietly forgotten and I continued to be '*tu*' for Jean Pierre.

One day Monsieur Azéma asked me if I would like to visit Corsica. I said I would, very much, but I did not have any money. This appeared to be no problem for Monsieur Azéma and he pulled out some banknotes from his wallet, said that I could have the money for a short trip to Corsica and wished me luck. So this was the beginning of an interesting series of experiences. I had always loved islands and a big mountainous one like Corsica certainly caught my imagination!

I made a few enquiries about costs and decided to hitch-hike to Toulon, from which town there was a night boat to Île Rousse. I booked a fourth-class passage, as I wanted to be as careful as possible with the money.

I had studied the map of Corsica while staying with the Azémas and realised that the highest mountain was Monte Cinto, about 2,700 metres high and that the nearest town of any size to this mountain was Ponte Leccia. I also knew that there was a railway line connecting Île Rousse with Ponte Leccia and Bastia and that you could also make the connection from Ponte Leccia to Ajaccio, the capital. I talked to several people in the town of Île Rousse and, when I told them I was going to the interior, they invariably asked me if I had a good gun, for I would need it. There were a few bears left in the forest, I was told, which were liable to be none too friendly if approached, but the main reason for having a gun was the bandits who operated in the area. I had never in my life handled a gun, so I thought I would ignore that advice, and booked a ticket on the train to Ponte Leccia.

This train reminded me of a train I had ridden on many years before in Transylvania. It was narrow gauge, it had wooden seats and it puffed along extremely slowly along hair-raising bends. It stopped quite often for no apparent reason, when people got off the train and shared their wine and their gossip with some of the local inhabitants. I tried to join in these conferences and I received more advice about guns. The engine driver would give a long whistle and all the passengers would pile back into the train. The train then chugged along until it stopped again, for the scene to repeat itself once more. I tried to make out the dialect people were speaking. It seemed to me a cross between Italian and Spanish, which I had little difficulty in deciphering. I found that by distorting Italian in a certain way, it somehow became Corsican.

We finally made it to Ponte Leccia, where I mailed a letter to my father, as I thought there would not be any more places to do so if I were going to penetrate to the interior of the island. I remember referring to Ponte Leccia in my letter as 'an important railway junction'! In fact, I believe it is the only such junction on the island, or at any rate was at that time. I have no doubt that, by the time I am writing these lines, all the old trains will have joined the scrap heaps and proper roads have been built, on which tour buses take international tourists from one end of the island to the other!

There were no signs of tourists at Ponte Leccia, only some peasants going about their business, going from somewhere to somewhere else driving their horse-and-cart means of transport. As I started to stride out towards Monte Cinto on a narrow dusty track, one of these carts overtook me and then stopped. The driver offered to give me a ride, which I gratefully accepted, as my knapsack was really quite heavy. As soon as we got going, he pushed a full bottle of local wine into my hands and told me to drink. I drank. But what wine! It was smooth, 'full-bodied' as the wine merchants say and certainly strong! We took turns swigging wine out of the bottle until it was empty. I noticed a definite change in the velocity of our progress as a result of this episode. Somehow the heightened awareness occasioned by the strong wine had passed over to the two horses which were pulling us up the mountain road as fast as they could manage!

My driver was surprised that I could conduct a reasonable conversation with him in Corsican! He did not realise that I was really talking to him in Italian, using a few transformations I had learned on my train journey from Île Rousse. He showed me a pamphlet distributed by the Corsican branch of the Communist Party and said to me how he liked it that they had gone to the trouble of writing it all in the Corsican language. I had a quick glance at the pamphlet. It was the usual incitement to rebel against the injustices committed by the landowners and capitalists, with which at that time I heartily sympathised. It reminds me of a piece of political wisdom I have seen recently, according to which 'If you are not a socialist when you are young, you have no heart, but if you are still a socialist when you are old, you have no brain!' Obviously, I was young and had a heart, so was a socialist!

My communist friend eventually dropped me at what appeared to be the last human habitation before the forest. I thanked him for the lift, put my pack on my back and started to follow the valley, which, according to my map, would lead me to the foot of Monte Cinto. The track became more and more indistinct and practically disappeared, until I found myself climbing over huge boulders in order to make headway. I did glimpse a fair-sized bear, but, fortunately, the bear was stomping about on the far side of a big ravine through which the stream ran that I was following!

It was getting late and I thought it was time to look for a place to sleep. I had not brought my tent, as I had decided to travel light. I must say that I had not altogether succeeded in this aim, as my pack was anything but light! But it held a couple of days' food, a primus stove and accessories and as few clothes as I thought I could manage with. I eventually came to what looked like a cross between a cave and a roughly built stone hut. The cave seemed to have been extended by stone walls, with a rough roof made of sticks and branches. I even found some old coats inside, so I decided that this was just what the doctor ordered. I had a small snack, slipped into my sleeping bag and before many minutes I was fast asleep.

I was awakened by the noise of animated talk and by being prodded with the business end of a rifle.

I suddenly realised that these people, whoever they were, must be the inhabitants of the dwelling place I had decided to sleep in and they had taken a poor view of my intrusion.

'Give me your gun at once and your money,' commanded the man with the rifle, speaking in a broad Southern French.

'I have no gun,' I replied, 'and all the money I have is in this purse. I handed over the purse.

'Don't trifle with us,' said one of the others. 'Nobody travels here without a gun and without any money! Give it to us or we kill you!'

'Everything I have is in this knapsack,' I said. 'You are welcome to have a look'

They emptied the contents of the knapsack, as well as the contents of my purse on the floor. They were amazed to conclude that indeed there was not a gun to be found anywhere and that the money I had would not go very far!

'My name is Spada,' said the one who had been prodding me with the rifle. 'This is my friend Carlo and this one is Francesco.' He pointed to his two companions. *'Nous sommes des bandits d'honneur.* You have nothing, so you are our friend. Welcome to our humble abode!'

I was naturally much relieved.

Carlo went to the back of the cave and brought a large bottle of wine, which he opened. The bottle was passed round and, under the influence of this excellent Corsican invention, we soon became very friendly and exchanged stories. They sang some songs, which I tried to learn. I remember one of them started like this: *'La luna splende sovra Bastia'* ('The moon is shining over Bastia') but I have forgotten the rest of it. Some were romantic, some were bawdy and some were about the battles that had been fought over the centuries by the *bandits d'honneur* against the rich and powerful. I had never thought that one could have such an entertaining evening with three bandits!

In the morning I told them that I was going to climb Monte Cinto. They thought I was crazy to even think of doing this. They told me there was nothing there but a great many rocks and that I would probably fall down a precipice and get killed!

Finally Carlo said to me, 'See this little dagger? It has my vendetta carved on it! I want to give it to you. You cannot walk around in the mountains of Corsica without any weapons! We have had such a good time together and it will remind you of these times!'

I was very touched by Carlo's generosity. I thanked him profusely, after he had assured me that he could easily get another dagger and carve his vendetta on the new one!

I put my pack on my back, thanked the bandits for their hospitality and started on my climb, waving to them every now and them while I could still see them. They really thought I was mad to climb that mountain!

The bandits were right about the rocks. I soon left the trees behind and after a while I left even the grass behind till there were only the rocks left. The going became steeper and steeper and towards the end some elementary rock-climbing was needed before reaching the summit. The view was utterly fantastic, with ranges and ranges of mountains as far as the eye could see! For me, being on top of a mountain was always a solemn time. I felt far from the hustle and bustle of everyday living and much nearer to a kind of cosmic unity, which I allowed to pervade my being as I lay on the summit rocks gazing up at the sky.

I spent several days wandering about in the Corsican mountains, descending into valleys to obtain more provisions and generally sleeping at the higher altitudes. One of the most memorable events of my Corsican trip was something much more simple than drinking wine and singing with the local bandits. I was walking along the ridge of a mountain range, beginning to wonder where to spend the night, when I saw smoke in the distance. Smoke means fire and fire means someone who must have lit the fire. So I made my way towards the smoke and soon a small stone hut came into view, surrounded by a flock of bleating sheep. In front of the hut was a wooden bench and on the bench sat a man.

He was an oldish man, or at least so seemed to one as young as I was then. He had a plate of potatoes on his lap which he was about to eat.

When I approached him, he greeted me with the very simple sentence, '*On partage les pommes de terre.*'

Upon saying so, he went into the hut, brought another plate out, placed one half of his potatoes on this plate and handed it to me, motioning me to sit on the bench beside him. At this point we both proceeded to eat our meal of potatoes, which we finished in silence, disturbed only by the bleating of the sheep.

Looking back on most of my life, I think this was one of the most wonderful dinners I have ever had. This poor shepherd was just going to sit down to his somewhat meagre repast of potatoes, when he saw another human being. His very first reaction, even before saying *bonjour*, was to share what little he had. I thought that if the world were made up of more people like this shepherd, we would have a better planet on which to live!

I took some fresh fruit out of my pack, which my new friend was happy to share with me. When we got talking, he soon was keen to show me round. In one part of the hut he was making cheese, naturally from sheep's milk. In the middle of the hut was the fireplace. He soon lit a fire and I noticed that the smoke simply rose up and went through the roof because there was no actual chimney built for the hut. When we settled down for the night, he told me that he would keep the fire alight all night, as it tended to get rather cold by the morning and that I was not to worry about it since I must be tired after climbing up as far as his hut from the valley. So we settled down for the night, lying 'comfortably' on the floor, with our feet near the fire.

In the morning, before leaving, I gave him nearly all my provisions, particularly the fresh figs and oranges I had purchased in the valley. He had not eaten any such fresh fruit since he came up from the valley in the spring! We then parted. I walked over the ridge and out of sight of the hut, but the memory of the dinner and the night in the shepherd's hut remains with me to this day, as clearly as if it had happened yesterday.

I spent several more nights on the island, sometimes sleeping outside under the stars and once in a little inn, for I was wending my way back to the coast to take my boat back to the south coast of France. I sang and danced with the locals, played with the children and generally had a very relaxing time, until I finally reached my port of departure and once more sailed into the night, gazing back at the lights of Île Rousse getting ever dimmer as we sailed on.

The Azémas welcomed me back very cordially. I spent another week or so with them, reinforcing my friendship with the boys. Monsieur Azéma offered to take me back to Paris in his car, a suggestion I gratefully accepted. I do not remember much of the journey back. We stopped for the night about halfway to Paris and I remember that he introduced me to the wine known as Chablis! Another thing I remember is that, when we were driving along a very straight road downhill, Monsieur Azéma put his foot on the accelerator and said to me proudly, '*On va faire cent!*' Nowadays for a car to do one hundred kilometres an hour is no big deal. In those days, obviously, it was a rarity.

When I arrived home, I was greeted with the results of my degree examination. They were not very good. My undergraduate years were over, during which I had learned a lot of things, but clearly, on the whole, not those things the university authorities had intended that I should learn. It was time to rethink my future. Was it going to be an academic one? Could I go on to doing serious research in mathematics, or in anything else, with such an indifferent beginning?

Chapter Ten

Two Decisive Years

I had a discussion with my father about what to do next and he suggested that I should go directly into a Ph.D programme, which in England does not require any courses, only the writing of a thesis and an oral examination, which is the defence of the thesis. If I did well as a researcher, I was assured that it would wipe out the bad effects of my indifferent degree in mathematics. I agreed to do this and duly registered as a student.

It was only the end of August, I had about another month in which to enjoy the vacation, so I decided to go to the Hebrides with Olga, with whom I continued to be on friendly, even romantic terms, although I never felt that our relationship had the hallmarks of a life-long commitment. The girl I was going to spend my life with had to be one with whom I could sink my identity in order to become a member of a twosome and eventually of a family with many children and I did not feel that Olga was the right person. She was not keen on having a family and that was a minus from my point of view. But, still, we were good friends. We had many things in common such as the love of music, nature and poetry and I had become reasonably competent in Russian! So we packed our bags and started off on a hitch-hike towards the north.

We soon found ourselves beyond Inverness, beyond Dingwall and decided to walk through Glen Affric and reach Loch Duich and Kyle of Lochalsh, from which we intended to take the boat to Stornoway. The walk through Glen Affric is one of the scenic wonders of Scotland. At that time, it was almost totally deserted. There was only a rough track along the glen and there was no human habitation anywhere. We walked happily along, higher and higher into the upper reaches of this wonderful valley, until we found the pony shed. We had been told there was this pony shed where one could doss down for the night, as it was too far to do the whole glen in one day. We settled down in our little shed, having cooked a light meal on our primus stove and sat gazing at the rocky mountains that rose up majestically on each side of the glen, until the last rays of sunshine disappeared from the summits. We slept well and early the next morning we were ready for the second part of our journey to Loch Duich.

We reached Loch Duich before midday, where we joined the dusty road that led to Kyle of Lochalsh. As we were walking leisurely along this road, we came upon a 50 or 60 foot vertical cliff on our right side, the loch being on the left side of the road.

Olga had a look at the cliff and said to me challengingly, 'I bet you can't climb that cliff!'

What does a young man do in such a circumstance? Obviously, he sets out to climb the cliff.

I mentally mapped out the route up the cliff. I was a reasonable rock-climber, having climbed in North Wales, in the Cuilins in Skye as well as in the Alps, so I did not think I would have any problems meeting the challenge thrown at me by Olga. I was almost at the top, just looking for handholds while stopping on a narrow ledge, when Olga shouted up to me, saying she wanted to take a photo. So I stayed there, straddling the rock face, while she adjusted the camera and took the photo. At about this point one of my handholds gave way. I had no time to feel for another so I felt my whole body becoming light as I began to fall.

The next thing I remember I was lying on the floor at the bottom of the cliff, with Olga bending over me, tears in her eyes, trying to get me into a truck she had hailed, with the truck driver giving a hand. I was only half there, but I soon began to feel intense pain all over my body and it came to me that this was probably it: I had fallen down the cliff and was probably going to die soon. There is nothing like such a situation to make it clear to someone how unimportant one really is in the cosmic scheme of things!

There was a rail connection between Kyle of Lochalsh and Inverness, where the nearest reasonable hospital was and they must have found a phone somewhere to phone the station so they would keep the train till we arrived. I remember being hoisted into the guard's van; a doctor gave me a hefty morphia injection, after which I went out like a light. It was a five hour trip to Inverness, so fortunately someone thought of having a doctor meet the train at Dingwall with another dose of morphia. We finally arrived at Inverness, where there was an ambulance waiting at the station to take Olga and myself to the Royal Northern Infirmary.

I was placed in a bed in a ward near the door with a screen round me, given another morphia injection for the pain and left to sleep till the morning. By the morning the screen had been removed and I could see the other patients in the ward.

There was a boy of about sixteen across the ward from my bed, who said very cheerfully, 'Hello! You know they have put you in the deathbed, don't you? The last two were wheeled out from your bed to the mortuary!'

'Thanks a lot,' I replied, trying to control myself, what with all the pain coming back. 'I fell fifty feet, so you are probably right!'

As it happened, during the next several weeks, he and I became very friendly and he truly apologised for the way he had greeted me!

There was one problem: it was three days before I was to become twenty-one and they needed my parents' signature before they could do anything for me in the operating theatre. Olga phoned my father, who took the first train to Inverness and of course signed the appropriate form. I was wheeled into the operating room and duly put under with a general anaesthetic. I had compound fracture of the left arm, cracked bones in both my legs, broken ribs and severe concussion. They patched me up and told me that I would need to be in hospital for at least six weeks. That was a relief, anyhow: I was not at death's door and everything was not yet lost!

Olga took several weeks off work to be with me, so I had her company as well as my father's while I convalesced in the surgical ward of the Royal Northern Infirmary. We had the film developed that included the picture Olga had taken of me and I stuck it in my album of trips, marked *La roccia*.

My deathbed friend very much enjoyed looking through my journal of trips, I remember that he really praised one picture I had taken near Dartington, entitled *Foggy Wood near Rattery*. It gave a mysterious impression of trees disappearing into the distance in the thick mist; it was autumn, the leaves were hanging on by a thread and it seemed to express some of the sadness we were experiencing, confined for a long period of time to a hospital ward!

Instead of six weeks I was discharged after three weeks and we all took the night train back to London. My father had taken out an advance on his salary so we did not need to worry about finances and could have all possible good things to help me get better.

By this time Sari and my father had separated, so my father and I rented a small flat in Hampstead, with just two rooms, a kitchen and a bathroom. We soon started discussing what I was going to do for my Ph.D work. We decided that it was to be about the foundations of mathematics, so I started to read Hilbert's *Grundlagen der Matematik* and several volumes in the Borel Collection about ways mathematicians thought to resolve the antinomies that were turning up in the body politic of mathematics, as well as works by Brouwer and Heyting on the so-called 'intuitionist' approach to the foundations of mathematics. I returned to my favourite childhood gem, namely Réné Baire's *Leçons sur les founctions discontinues* and soon started to formulate ways in which his beautiful results could be generalised.

As I discovered later, mathematicians are inveterate generalisers. When two mathematicians meet at a congress and tell each other about theorems they recently proved, the inevitable sequel is always a 'Yes, but supposing that...', meaning that it would be nice to know what would happen under different, more general conditions. Would the result still be true? Or would something else, perhaps something even more interesting, be true if you changed the initial conditions of the theorem?

During the next two years I turned over many such ideas in my mind and in the end I finally came out with a thesis which treated a number of very fundamental questions in the foundations of mathematics. But it was not plain sailing. Human and emotional problems still kept entering into the equation and at several points during these two years I was not sure if I would ever make it to the end!

The foremost of these problems arose as a result of my long-standing friendship with Mina. She had left school by this time, probably as there was no more money available. She had a job at one point, looking after a little girl, but she soon decided that she ought to get her matric so she could have a career.

Robert McKenzie, the science teacher from Forest School, had had to leave the school because he had fallen in love with Mina and the head said that one of them had to leave. So he left. But not very long after this, Mina left as well. McKenzie suggested to Mina that he could coach her to get through matric. Rosalind, Mina's mother, had some huts in a field near Benenden in Kent, where they thought they could live and do this work. Of course, human nature being what it is, it became more and more difficult for MacKenzie to live in such close proximity with a sixteen year old young girl with whom he was in love, without very much wanting to make love to her. At a certain critical point, no doubt, Mina got scared and 'escaped'. Where was there to escape to? The obvious answer was my father and myself. So a very disturbed and untidy-looking Mina turned up one day at our Hampstead flat, weeping, begging to be taken in. I threw my arms round her and told her that of course she must come in and tell us everything that was worrying her.

So she told us her tales of woe and asked us what we thought she should do. We thought she could register at the local polytechnic rather than risk a 'fate worse than death' in a Benenden field! So this is what we did.

But of course our flat was very small. I was twenty-one and Mina was going to be seventeen in November. I give the reader one guess. Yes, our platonic friendship came to an abrupt end and I, at any rate, felt that Mina was indeed my girl: we had been friendly for years and now I was head over heels in love with her.

Our relationship had to be remoulded along new lines! I remember after a concert at the Queens Hall, during which they played the Beethoven violin concerto, while we were coming home, she said to me, 'I am just a daily tenant! I am not sure of anything!'

And as we were walking to the tube station, she started humming one of the tunes from the concerto, the start of the cantilena, starting C C A F G R C F, with the words, 'Oh, how much I love you, my dear!'

I was really confused and didn't know what to think. She told me later that she had been very frightened and was not sure whether she should be frightened of me or whether she should accept me. When we got home, I hugged her tight and kissed her and told her that she was all I would ever want.

We slept beside each other but did not make love. We both felt that things were hanging in the balance and wondered how things were going to turn out. It was one of those situations, now formalised in the theory of chaos, when something very small could make a very big eventual difference. Neither of us was sure how to manage such a 'something very small' to make things good, so we left it to fate!

We all eventually moved to a larger apartment in Cranley Gardens on Muswell Hill, where we each had a room and things seemed to quieten down. Mina decided that this was a new life and that she was no longer a child and was not going to be called Mina any more. She chose the name Tessa, by which name she has been known ever since. She frequented the polytechnic and she also kept house for my father and me. My father gave it out that he had adopted Tessa so she and I became brother and sister, and, in a sense, we tried to put the clock back, but it never really succeeded.

I went on with my studies of mathematics. I even started writing a paper, the central point of which was a 'type theorem', classifying sets into various types. I wrote up a paper, having discussed it with my father to make sure there were no mistakes in it and submitted it to the London Mathematical Society. I was shattered when I received the reaction. One of the reviewers very kindly pointed out that my 'type theorem' was not true, giving a few counter-examples to drive home the point. In mathematics you cannot argue with the truth: things are either so or not so. This was not so! This was a lesson to me not to rely on anybody else, not even on my father, who was a first-class mathematician; but *errare humanum est* and nobody is exempt, it seemed, not even my father and myself!

So I went into the problem more deeply and invented ways of 'puncturing' sets, thereby creating more and more complex sets. This led to a revised version of my paper, which I again sent in to the London Mathematical Society. To my great joy, it was accepted and soon published! My first original paper in mathematics! And before I even got my Ph.D. I resolved to incorporate the paper into my thesis, which I did, for which I eventually received my doctorate.

I had a Dutch friend whose name was Floor, who ran a kind of 'lunatic asylum' in a large house on Greenham Common, in Berkshire, near Newbury. They were not 'dangerous' lunatics; they just needed to be looked after. I often spent the weekend there, hitching down to Newbury and playing bridge with the inmates. There was an old cottage about two miles from the big house, also belonging to Floor, in the middle of a thick wood, accessible only by an extremely rutted, cart track and in wet weather, of course, totally isolated. Floor offered that place for me to stay in any time I felt like 'retreating from the world'.

The situation in Cranley Gardens was getting too much for me. I did not feel any less in love with Tessa, in spite of the fact that we were now 'brother and sister'! But somehow she was not ready for me! So one day in the spring I decided to pack my bags, mostly books on mathematics and avail myself of Floor's retreat. I gathered some food for the cottage, but, whenever I felt like not being alone, I would walk over to the 'lunatics' and share a meal with them. I did a lot of studying in my 'retreat', trying to work out how people like Borel and Brouwer, in their different ways, had tried to solve the problems of the foundations of mathematics. I had plenty of time for meditating on the meaning of life, watching the birds and the trees and generally trying to knock some sense into myself. Both my thesis and my soul were beginning to take shape under the peaceful influence of my isolated cottage life.

Soon the summer arrived and what could be better for a lovesick young man than to go gallivanting about the world?

I decided to make for the Hebrides. I made my way to West Loch Tarbert, on the Isle of Harris, and found a place where I could board for a very small sum. I thoroughly enjoyed the beauty of the Harris mountains. I climbed Clisham several times (the highest peak in Harris) and investigated the many lochs and rivers around, in which I swam to my heart's content. I said one day to my landlady that I was going to see Mary Rose Island, which was a small island in a loch in a deserted part of Harris.

'Och, no!' my landlady said. 'You canna go therre! The fairies will have yer!'

She told me then in some detail that anyone who went there heard fairy music and was never seen again! Once a young girl with an illegitimate child went there, not knowing about the fairies and let her child play on the shore while she went to pick berries. When she came back, the child was nowhere to be seen, but she could just see the trail in the water of a big black swan, swimming away towards the island very fast!

She was told afterwards that the fairies changed into swans and stole children! She went looking for her child, but she never found her. She came upon a little child she did not know, sitting crying on the shore of the loch! Of course, everyone knew this was a fairy child, left in exchange for the child that was taken! This young girl's child was called Mary Rose; so ever since then the island has been known as Mary Rose Island and nobody dares to go there any more for fear of the fairies! I thanked my landlady for the warning and, more curious than ever, set out to find Mary Rose Island!

I followed the directions given me by a crofter and in about two hours' tramping over the hills, I came upon this very loch.

There was a small island in the middle of it, with many bushes growing on it. The place was idyllic! There was no human habitation in sight; the Clisham mountain towered majestically into the western sky; the loch quietly glistened in the foreground, its brown peaty water only occasionally disturbed by a surfacing fish. I stripped and plunged into the water.

It did not take me long to swim to the island, where I climbed ashore on to the rocky bank and lay down to enjoy the rays of the hot midday sun beating down upon my naked body.

I dozed off for a while; maybe I dreamt of fairies, but no 'real' fairies ever came! The bushes were laden with bilberries, so I picked a good handful of these and ate them before I decided to swim back to shore.

My landlady had clearly given me up for lost when I walked in as dusk was already falling!

'I thank the Lorrrd you arrre safely back,' she said to me as I put my things down.

'I saw no fairies,' I told her.

She seemed disappointed!

Some time later, in order to perpetuate the legend and my daring swim to the fairy island, I wrote the following poem:

I saw the black swan, I saw the black swan
Glide on the loch late yesterday.
I still see its trail, I still see its trail
Far on the loch, fading away!
I'll not see my child, nor even the trail,
I'll not see my child, to love and to hold.
I'll not see her ever, for black is the swan
And still is the trail, for now and for ever!

Then I added the following Italian version:

Vidi il cigno nero sul lago,
Vidi il cigno sull'acqua strisciar,
Gli ultimi raggi del sole splendevan
Mentre miravo la notte calar.

Vedo la traccia fatta dal cigno,
Vedo la traccia nell'acqua sparir,
Guardo il lago per l'ultimo segno
Della speranza che voglio sentir.

O figlia mia, dal cigno rapita!
Non ti vedrò, ne la traccia laggiù!
O figlia mia, che non t'ho sentita
Quando chiamasti: 'Tirami su!'

Non c'è speranza, giammai to vedrò;
Nero è il cigno che via ti portò!
L'acqua è cheta, non c'è più la traccia,
La piccola fata piangendo m'abbraccia!

I was always very keen on islands and I knew that at the western end of West Loch Tarbert was a small hilly island called Scarp. I thought it would be fun to go and visit it. I did not have any idea how I was going to cross the sea to get there, but I thought that problem would solve itself, once I came within sight of my goal.

I walked over to the end of the road and there was indeed a jetty as well as a boat, with three burly fishermen about to board it. I hailed them and asked them if I could come with them if they were going to Scarp. As an answer, one of them shoved an oar into my hand and told me to jump in. The other three jumped in, pushed off the boat from the jetty and started to row furiously! Naturally, I tried to row as furiously as they did, but it was never very good, as we did not keep a very straight course! Each of us pulled an oar with two hands and clearly my hands were no match for these fishermen's hands and arms, which bulged with muscles! The sea was quite choppy, with waves several metres high. We beached the boat on the island about twenty minutes after the most strenuous rowing I had ever done before or have done since!

Getting used to such things is what people call 'being acculturated'! I do not suppose that the textbook writers who write about such things have ever had an oar shoved into their hands and been told to row across a very choppy, narrow bit of sea to reach an island!

One friend I had made in Floor's 'lunatic asylum' was an older woman called Joy. She was in her late twenties. She was quite keen on me when I visited Floor's place and she decided to come and see me in Harris. In fact, she arrived a day or two before I meant to leave to track towards more southern climes. After a couple of days' rest on the shores of West Loch Tarbert, we set out towards South Harris. We took the bus to Tarbert and then started walking in a southerly direction, stopping at each of the little lochs we passed for a swim! The two of us swimming alone in the warm little lochs certainly encouraged romantic ideas and we soon found ourselves involved with each other.

We eventually arrived at Borisdale, where we were warmly welcomed by the MacDonald clan, in particular by the matriarch Mrs MacDonald! She fed us well and, of course, we had to stay at least one night, if not more. She soon sized us up and, when it was time to go to bed, she showed us to a little room at the top of the stairs, which contained nothing else but one large bed! These crofters certainly had the basics of life worked out! We stayed for a few days with the MacDonalds and accompanied them on their fishing trips and helped with the peats; in fact, we did just about everything except milk the cows.

There was a boat once a week from Rodil to Lochmaddy (on North Uist), so on that day we packed our bags, said goodbye in as good Gaelic as we could muster and walked to Rodil to pick up the *Lochmore* to Lochmaddy, from where we wended our way down to Benbecula, which was quite a feat. In those days there was no military base there and certainly no bridge to join North Uist to Benbecula. There was only one time during the tidal cycle when you could cross. You had to keep two cairns (heaps of stones) in line as you were walking, until you reached the first cairn, after which you looked for two more cairns to keep in line and so on. Fortunately, somebody with a horse and cart who knew how to follow the cairns was about to cross when we wanted to go, so we went together. It was a four mile crossing, so one had to be sure that one would not be swept away by the incoming tide!

We got a ride in a fishing boat which landed us in South Uist, where we enjoyed the beautiful sandy western shore, miles and miles of golden sands stretching away as far as the eye could see. Between the sands and the rest of the island was the *machair*, which was a grassy area on which a wild profusion of flowers grew. I now understood what the Hebridean bards meant when they sang so emotionally about the *machair*.

We encountered one of these bards at a little South Uist village called Bornish; he taught us the latest song he had composed for his son, who had left Bornish to go to Cape Breton in Nova Scotia, on the east coast of Canada. I still have the words in my notebook, with the music, the words written down in my usual phonetic script! Sometimes when I feel nostalgic for the Hebrides, I pick up my recorder and play the old bard's song and think about his golden sands and his *machair*!

From Lochboisdale it was possible to take a boat once a week to Tobermory in Mull. The boat called at Castlebay in Barra as well as at Tiree. We left the boat at Tobermory and walked to Salen, a small village on the Sound of Mull, where there was a youth hostel. Mull is a very scenic island, much tamer than any of the islands of the Outer Hebrides. We climbed Ben More (which just means 'big mountain' in Gaelic), the highest peak in Mull and swam in Loch Ba, which stretches into the interior of the island at the foot of Ben More, surrounded by trees, which we were happy to see as there are practically no trees in the Outer Islands.

We drank in the natural beauty all round us, which helped to make us feel more together and more involved, although I must say that I never felt anything approaching the passionate love I had felt for Tessa. But I thought, Well, you cannot win them all and if Tessa was not romantically interested in me, that was her choice and I must find other avenues for my emotional life.

In the meantime Tessa was also doing some thinking. She had gone over to France with a boy called Boyce, whom she had met at the polytechnic, on the understanding that they were to maintain a platonic relationship and Boyce was to look after Tessa and see to it that nothing untoward happened to her. Although Boyce had agreed to this verbally, he must have had different ideas in his mind. This soon became apparent and Tessa informed him that that was it and they would have to separate.

Tessa walked along the road and came to a large stretch of water, very inviting for a swim. She was used to swimming in the nude so she stripped and jumped in the water. But Boyce was not very far away. He picked up her clothes and hid somewhere near, waiting to Tessa to come out of the water, which she did, but when she saw that her clothes were gone and she was there on the bank with nothing on, she jumped straight back into the water. When some people came along, she shouted from the water, '*Au secours! Au secours!*'

They came down to the water's edge. Tessa explained to them that someone had stolen her clothes so the lady in the party brought what she could find, which were some riding breeches and a blouse! In this way Boyce was foiled, except that he also had Tessa's passport!

To cut a long story short, these people took her down to the South of France, where Tessa earned some money posing as a model for members of the artistic community in Vieux Cagnes. After many vicissitudes, in which she escaped with her 'honour' by the skin of her teeth, eventually she arrived back in England, helped by a kind British consul.

Tessa used this time to ponder her future. She decided that I was really the man for her at the same time I was deciding how I could manage to live without her!

When I returned to London with Joy, Tessa was horrified that I could be so heartless to go off with another woman! Of course, she had no problem in seducing me so that was the end of the Joy connection!

My father, Tessa and I all thought that it was no use re-establishing the *ménage à trois*, so we rented a room for Tessa in Willow Road, on the edge of Hampstead Heath, quite near the pond where my companion in would-be suicide had told me that this was the girl who loved me! I visited her there quite regularly and with each visit our bond became stronger. In the end I think Tessa wanted to make sure of me and said that we should be married. There were two conditions, however: firstly, it would have to be done at once, or she might change her mind, and, secondly, we should have ten children.

I agreed to both conditions at once and we applied for a special licence to be married at Hampstead Town Hall the day after next! Since she was a minor, she had to have parental permission and Rosalind, her mother, immediately gave it. My father was not so sure about this marriage business, but I was over twenty-one so he could not really object.

At the marriage ceremony, when it came to the point of Tessa's saying, 'Will you take Zoltan Paul Dienes to be your lawful wedded husband' she hesitated for some minutes. After I had put my arms across her shoulder, saying, 'Come on, Tess!' she finally said the fateful words and we were pronounced man and wife. Fifty-nine years later we are still married and, in spite of forebodings from Sari and my father, neither of us has ever regretted it!

We stayed at Rosalind's house in Putney for our wedding night and then went for a short honeymoon. We hitch-hiked up to Leiston to see Neill, possibly to make it right with him about that running away business. Then we hitched back to Kent, to Rosalind's little hut in a field near Benenden, where we spent some time relaxing and enjoying each other in utter abandon. We had practically no money, so we looked around for mushrooms and nettles to make soup and gathered blackberries for our desserts. Some of the poorer village children used to come to see us, usually at the times when we were about to sit down to eat, so we just had to share what we had with them, which we gladly did. I had not forgotten my lesson in the Corsican mountains when I had been the guest of a poor shepherd!

One day we thought we would go to the pictures. We hitch-hiked to Hastings, where we saw a lot of people lining up to see a film, so we thought it must be a good one. To make the money for the entrance fee, I pulled out my recorder and played some tunes I knew quite well and Tessa danced before the waiting people. When we finished and passed the hat round, we had more than enough to see the performance!

On returning to London, we rented a room in Tessa's mother's house until such time as we could find a place to live. My father offered us two pounds a week, on which we would have to manage, so clearly no place in London would be possible. So we went cottage-hunting, hitch-hiking through the country around London.

Tessa's mother had been brought up in the little village of Long Crendon, in Buckinghamshire. We thought this would be a good place to look, as we could be local and people around would accept us. Rosalind's mother, Tessa's grandmother, still lived in Long Crendon and we went to see her. She welcomed us with home-made elderberry wine and goodies to eat and wished us luck in our hunt.

We finally found a little cottage, the middle one of a row of three, in a small village called Moreton, about a mile from Thame, which was about two miles from Long Crendon. When we explained that Tessa's grandmother lived in Long Crendon and her uncle in Thame, Mr Howse, the owner of the cottages, who lived in one of the end ones, agreed to let us his cottage for the princely sum of four shillings and sixpence a week. In the other end-cottage lived a family with three children, several of whom had TB, but we did not know this at the time. We had found our cottage and that was what was important!

The place was unfurnished, but the Caledonian market, in London, came to the rescue. We already had some furniture in our respective rooms in Cranley Gardens, to which we added some we bought for a few shillings at the Caledonian market, which we trucked down to our little cottage and moved in.

By this time, believe it or not, Tessa was already pregnant with our first child. This was very bad news for my father, who said that we must get rid of the child as it would interfere with my getting my degree. Of course, we were not going to do anything of the kind! To kill the first fruit of our young love and our long-standing friendship? Out of the question! When my mother got to know about the coming child and my father's threat to its life, she sent a telegram from Budapest saying, 'Your child must live! I shall send food and clothes!'

She was as good as her word. We began receiving food parcels of Hungarian goodies, as well as parcels of baby clothes, made by her various religious friends and lovingly put together in little parcels that kept arriving with exemplary regularity!

So the problem of the child was solved. It remained to solve the problem of my Ph.D! Although I spent quite a lot of time digging our vegetable garden, which provided us with just about everything, lettuces, cabbage, beans, peas, marrows, cucumbers, carrots, parsnips and so on, not to mention the apples that grew on our apple trees, I also had time to write my thesis, which by this time had taken on its more or less final shape. I had a little typewriter and I typed and typed and typed! I submitted the thesis at the end of May, in June I had the oral; so I raced against the birth of our first child by several weeks.

The months of expecting our first child were wonderful times. We were alone with each other, we had a little home, we had our garden, we even had our music! We had a very old gramophone and every other week we went up to London for me to discuss my mathematical work, and then we would go to the record shop and gaze at the records. The owner of the shop once asked us why we did not buy any records if we liked music that much. We told him that we had very little to live on and could only afford very little.

'I tell you what,' said the record shop owner. 'Take this Bach Brandenburg Concerto you have been looking at and just come and pay me when you can!'

'But it might be ages!' we said.

'Never mind,' he replied. 'When you come to pay for this one, you can take another work and pay for that when you can!'

We took the Brandenburg and greatly enjoyed playing it in our cottage over and over again. Several weeks later, having put our shillings together, we came to the record shop and proudly paid. The owner then let us have the Mozart G minor symphony. And so it went on during these glorious months of Tessa's first pregnancy.

'This is the best time of our lives,' I said to Tessa once. 'We shall never be happy like this again!'

In fact, this was probably true. We were two people loving each other, expecting the fruit of our love as nature intended; we had everything we needed, even if we had to draw water from the well and dig the ground so that we could have vegetables. We went to Thame market every Tuesday and to Aylesbury market every Saturday, having planned how we would spend every penny. Big Jaffa oranges were a penny each. We could get liver for a shilling, enough for both of us. Milk and eggs came from a nearby farm, also at a very reasonable cost. In fact, one week we were able to save a shilling so we could go to the cinema.

'Two seats in the sixpennies,' I said to the girl selling the tickets to the cinema, giving her the shilling. She gave me three pence change back! She obviously thought Tessa was a child and so would go for half-price! From the window she could not see that she was obviously expecting a baby!

We paid many visits to Long Crendon and Tessa's grandmother always received us with great warmth. It was a three mile walk each way from the cottage, but it was worth it for the kindness and friendliness with which we were greeted; a glass of home-made wine and some home-made biscuits were always on the menu!

On the 29th of June we hitch-hiked into Oxford, to the Radcliffe Infirmary, for our fortnightly prenatal visits. The doctors said that Tessa had better be kept in, as her blood pressure was getting too high. So I left her in the hospital and hitched to London, with the intention of buying a pram.

When I got back to the cottage the next day, there was a telegram waiting for me which said: 'BABY GIRL BORN THIS MORNING STOP BABY DOING WELL.'

I immediately knew that things were not going well, otherwise they would have said 'mother and baby doing well'! So I got on the road and hitched into Oxford to find out what was going on.

When I saw the nurse she said to me, 'Your wife has had eclamptic fits. She is sedated now – she probably won't know you!'

I rushed in to see Tessa. She was lying there in bed with her lips swollen, for she had bitten them during the fit. I spoke to her, I stroked her, tried to soothe her, but she hardly knew me.

Then I saw our baby! Our first baby! She was wonderful: a beautiful little baby girl, so delicate and frail; a whole new life beginning, starting out under our care and we had better make it good for her!

After some days Tessa recovered from the eclempsia, with no immediately noticeable effects. I loved to see her nursing our baby. It seemed to me the final fulfilment of our love for each other, this little creature taking her milk from my chosen loved one. But war clouds were surely gathering over Europe. Everything was not going to be milk and honey. We were to go through many vicissitudes, be confronted with many problems, suffer much unhappiness, but at that time we were so happy when we brought our first child back to our cottage.

A few days later a parcel arrived in the post. It was my Ph.D degree! The following day another parcel came. It was my naturalisation certificate. I had become a citizen of Great Britain.

War was declared on the third of September.

Chapter Eleven

Getting Used to Wartime

War had been declared by my adopted country. After marching into the Rhineland, into Austria, into Czechoslovakia and getting away with it, the West's bosom friend, Herr Hitler, thought he could march into any country and get away with it. Really, one could not blame him for thinking so! But apparently Poland was just one too many countries to invade and all of a sudden bosom friend Herr Hitler became the enemy Hitler and we were at war with the country whose leader he had become. Millions of children had already been evacuated into the country, all dutifully carrying their little gas masks. The world waited.

I walked into Thame and presented myself at the recruiting office. I said I was volunteering for service. I found myself almost at once on the pillion of a motorbike, on the way to the office they had just set up for dealing with new recruits. I was soon talking to an officer, who took down my particulars.

'You have a doctor's degree in mathematics. Yes, I am sure we can use you. Let me just find the form for applying for a commission,' he said, looking among a pile of files in a filing cabinet. He took it for granted that I was not about to become a private, as that would be a waste of manpower. When I told him that I had only recently been naturalised, his face darkened.

'Oh, that's a different story. At this point in time only persons who were born British can be given commissions in the armed forces,' he informed me.

There was another officer there, who overheard this conversation. He took over and said to me, 'The drill for you people is, I believe, to go before a special selection board, which will decide what can be done in such cases. The board for this area meets in Oxford. I will give you their telephone number and you can make an appointment.'

In a few days' time I was sitting facing a number of Oxford professors and a couple of army officers in a room in one of the Oxford colleges.

The person who appeared to be the chairman of the board said to me, 'We are aware of your case. We have examined it but we have not yet come to a decision. We shall let you know what we can do for you in the next few days. Goodbye Dr Dienes. We do appreciate your desire to serve your new country.'

I greatly savoured the 'doctor' preceding my name! I hitch-hiked back to Moreton.

I did not have to wait long. In a few days' time I received a registered letter from the selection board which merely said: 'This is to inform you that you have been reserved for research until further notice.'

Thus ended my brief career in Britain's armed forces! Clearly, the members of the board thought it would be a waste to send me out as a private: they obviously thought that, since I was capable enough to do research in pure mathematics, I would be able to do more useful things at home than going over the top in battles. I must say that I really could not help being of that opinion myself!

Upon the birth of our first child, whom Tessa and I named Corin Ruth, my father increased our allowance to three pounds a week. With this bonanza we thought we should move away from our romantic cottage, as it was hard to be hygienic with a new baby under such conditions. True, we had a lot of help. Parcels continued to arrive from Hungary, as Hungary was not yet a belligerent; Björn's parents asked us to come and stay in their beautiful country house for a couple of weeks and it seemed that more than one person had had 'guidance from God' to come and help us at the cottage.

But these helpful episodes came to an end and we really had to go cottage-hunting once more. We used a carpenter's bag for carrying Corin and in this way we found it easy to get lifts on the road! Our first port of call was Floor's 'lunatic asylum', where Floor and his wife welcomed us warmly. We could even leave Corin there for some hours while we engaged in our hunts. The problem of finding a place had been made much more difficult by the outbreak of war, as now everyone was looking for a safe place away from London.

Eventually, we came upon a little thatched bungalow in a small village called Eastbury, which lies between Lambourne and Newbury. The rent was only two pounds a month, which we thought we could afford, even though it was twice as much as the rent of our cottage in Moreton. But I had had a fifty per cent raise in my allowance.

The cottage was called Wavy Elm because it was made of wood in the form of waves. We renamed the cottage Thatched Croft sometimes referring to it as The Palace. The 'cottage' was not really a cottage: it was a wooden bungalow with a thatched roof. It had three rooms in it, an entrance hall and a proper lavatory! The middle room, which was just through the entrance hall, was the kitchen. There was a bathtub, covered with a wooden board provided for culinary activities. The water for the bath was heated in a tank right next to it, under which was an oil stove for heating water. The other two rooms were on each side of the kitchen. There was electric power and we had to use electric fires for heating, since there were no fireplaces. But power was only a penny a kilowatt per hour, so we were sure that our budget could stand it! The bungalow stood on a grassy patch of ground, but there was no garden; nor were there any vegetables growing. We would have to start everything from scratch out there! But, of course, we thought it the lap of luxury to have running water, electric heating, a bathtub for which you could heat water without lighting a primus stove and, when heated, lo and behold, the water came out of a tap, directly into the tub!

Our landlord was the person who ran the village store, which was also the post office. The store was very near, so we did not have to go far for shopping. We gave notice to Mr Howse and moved to our new luxury bungalow, with a certain amount of nostalgia for we both felt for the place where we had been so happy expecting our first child. There was a little branch line from Newbury to Lambourne, which had a number of stops, one of them being the village of Eastbury. So if people wanted to visit us, we were quite accessible.

While solving all these personal problems about where to live, we also had to think about getting a job. We could not expect my father to continue to subsidise us indefinitely! I had two degrees, so I ought to be able to get a job teaching mathematics somewhere. Of course I did not have a teaching diploma, but this was not strictly required for teaching in a secondary school. I looked through the adverts in the *Times Educational Supplement* each week and sent off a lot of applications, but I never seemed to get a reply, let alone an interview for a job!

It was after Christmas, during the first days of January, that I was actually offered a teaching job in a secondary school in Lytham St Anne! There was one small problem: we were having one of the worst winters in living memory, with trains and buses being buried under snowdrifts and I wondered how we would ever get to Lancashire. I had to go alone, obviously, and when things got a little better Tessa and Corin could follow. I recall that I had to take thirteen different trains, wending my way across England by a tortuous route, avoiding the buried trains and closed sections of the railway system. I arrived finally in Blackpool, as the line to Lytham was still closed. It was late at night and there was about a foot of snow lying everywhere and there appeared to be no transport. It was a six mile hike from Blackpool station to the King Edward School in Lytham! Naturally, I set out to cover these six miles as fast as my legs would carry me. Well after midnight I found myself knocking at the headmaster's door. He could not believe his eyes when he saw me presenting myself for the job he had offered.

'However did you get here?' was all he said.

I stayed the night with the headmaster and in the morning I was introduced to the school, in particular to the other mathematics teachers and to some of the classes. I was then given my weekly timetable and allowed to depart to look for digs.

I found a nice landlady in a nearby street called Lake Road and sent a message to Tessa that, as soon as the trains started running, she could come and join me. Within a few days they arrived and we settled into our digs, our landlady being very kind and helpful with Tessa and the baby. After all, she was only eighteen and she needed some help and encouragement from someone older and more experienced.

I cannot say that I was a great success teaching mathematics to eleven to fifteen year olds. I had no idea how to keep discipline and the boys soon realised that they could do whatever they liked with me. This was quite a shock to me, as I could not think what else I could do to put food on the table for my little family! I told the Headmaster that I did not think I could go on doing the job and he accepted my resignation.

I managed to get a job in a preparatory school near High Wycombe so we moved down south again for the summer term. This work was much easier. I had to teach younger children, some as young as nine and I soon got the knack of interesting them not only in mathematics but in other subjects as well, such as physics, geography and even French.

This was the year 1940 and, as we all know, towards the summer the war was beginning to move from the phoney war stage to some real fighting. France was soon defeated by the advancing Germans, the Dunkirk evacuation took place and a number of us were convinced that nothing could now stop Hitler from crossing the Channel. The war was as good as over and we would have to brace ourselves to become occupied. Fortunately, Mr Churchill had other ideas. After his rousing speech about fighting on the beaches, we all pulled up our socks and got ready to resist the world's most powerful dictator in any way we could.

After some differences with the head of this little school, I resigned that teaching post as well, so I had to look around for yet another job. While the dogfights were going on in the sky between British and German airmen, I was busy scanning the papers for jobs. I finally found one at Highgate School. Most of Highgate School had been evacuated to the country, but a core staff remained to look after those pupils who elected to remain in London in spite of the war. I was to join this core staff and teach mathematics and physics and anything else the head thought I could teach! The bombing of London had started in earnest, so we decided that Tessa and Corin would stay in the bungalow at Eastbury, which we still had and that I would stay for the week at Cranley Gardens with my father and visit Tessa and Corin in Eastbury at weekends. My timetable included lessons on Saturday, as the authorities of Highgate School thought that Saturday school would keep most Jews from applying to come to their school so the school would not need to invent reasons for not accepting them! This meant that I could only have twenty-four hours a week with the family, as I needed to get back on Sunday night. There were no dawn trains from Lambourne to Newbury!

I thought that I should learn from my failure in discipline at Lytham school and start off being very strict. The school had a system of penal marks. A pupil got a penal mark for misbehaving and when these accumulated to three within one week he was sent to the headmaster to be caned. So when I walked into my first lesson I was determined to use this weapon. Naturally, the first thing to happen was that one boy threw his ruler noisily on the floor. This was the gauntlet.

'What is your name?' I said to the boy in question.

'Hoffnung,' the boy said.

'Well, Hoffnung, that'll be three penal marks for you. Go at once to the headmaster and tell him I gave you three penal marks,' I said.

'Three, sir?' groaned Hoffnung.

'Yes, Hoffnung, three! Off you go, double quick,' I replied.

There was deathly silence in the classroom. Hoffnung slowly walked out of the classroom. I said nothing. We waited. In a few minutes' times a very much chastised Hoffnung came back, holding his posterior. I had no further trouble in that class or in any other class, as the news spread like wildfire that this new teacher really meant business! I do not believe that I ever had to give out another penal mark during the year I taught at Highgate. Having established my authority, I was able to make my lessons interesting, as the boys knew what would happen if they started any funny business. I taught German, I taught geography, I taught physics. It was no use telling the head that I could not teach a subject.

When he asked me to teach geography and I objected, saying that it was not fair on the boys as I really did not know any geography, he said, 'Look, Dr Dienes, you have travelled in many countries – just tell the boys about them!'

So I told the pupils about the bandits in Corsica, sometimes inventing gory details, then I held forth about the mountains in France or the minority problems in Romania and Czechoslovakia and I suppose that they probably learned more useful 'geography' that way than if I had followed some dull textbook.

In one German lesson I taught them to sing the Lorelei song:

> *Ich weiss nicht wass soll es bedeuten,*
> *Dass ich so traurig bin,*
> *Ein Märchen aus alten Zeiten*
> *Dass kommt mir nicht aus dem Sinn!*

In explaining what the text meant, I thought they would learn more grammar than if they had to recite some grammatical rule out of a book.

In mathematics I always tried to give the boys problems that had to do with real life situations which interested them. For probability, we went to the horse races (mentally, of course) and worked out odds for bets or looked at our chances of winning at the roulette tables! For trigonometry, we measured the height of the tree they could see outside the classroom. They were impressed that they could do so without even leaving the classroom, merely by measuring some angles of elevation and the distance between two points in the classroom! In physics, I told them that people who tried to climb Mount Everest had trouble making tea because the water boiled at much lower temperatures so the tea leaves just floated on top of the water! They did not believe this, so I set up an experiment with a pump and a Bunsen burner. The pump was used to extract air from a receptacle in which there was some water and a thermometer immersed in the water. As the air became thinner, they noticed that the water began to boil at much lower temperatures. Then they began to believe me about the Everest climbers!

But, whatever I did, there was one boy who could never really work up an interest in these things. This was Hoffnung! He was more oriented towards art than science.

One day I saw him scribbling on a piece of paper, when he ought to have been listening, so I said, 'Hoffnung, bring me that piece of paper!'

Hoffnung complied and brought out a very interesting caricature of myself.

I said to him, 'Leave this picture on my desk and come and see me after the lesson.'

'Yes sir,' replied Hoffnung, with a slight tremor in his voice.

At the end of the lesson Hoffnung came up to the desk.

'This is a very clever picture, Hoffnung, may I keep it?' I said to him.

'Oh, yes, please do, sir,' said Hoffnung with relief.

'I think you don't enjoy coming to school. Is that so, Hoffnung?' I asked.

There was an embarrassed silence. He finally said, 'No, sir, I don't.'

'Then why do you come?' I said.

'My Dad says I must,' replied Hoffnung.

'Wouldn't you like to go to an art school instead?' I put in tentatively.

'Oh, yes sir, yes indeed I would,' replied Hoffnung enthusiastically.

'Well,' I said. 'Tell your father that your teacher thinks you are wasting your time here and that you would do much better going to an art school or college.'

The end result of this conversation was that Hoffnung did go to an art school and very soon became a well-known artist, making most interesting collages. Quite possibly, but for the caricature that he had drawn of me and our ensuing conversation, he might have been wasted learning useless mathematics!

During all these educational happenings the bombing went on unabated. We got used to the raids to such an extent that we began to ignore them and go about our daily business even during air raids, which were very frequent. I remember one air raid when bombs came whistling down over the West End. I was walking in Piccadilly Circus, when I felt a pat on my back and heard a voice saying, 'Do you remember those three penal marks, sir?'

It was pitch dark, for it was evening and the blackout was strictly observed, but I recognised Hoffnung's voice.

'Yes, I do. I am sorry I had to do that. I hope it didn't hurt too much,' I replied.

'No, that's all right, sir. It was all worth it to go to art school. I am getting on very well there,' replied Hoffnung.

We went our separate ways and we never saw each other again. He had a brief but illustrious career and died very young. I am not sure how he died, but I know that Hoffnung did not live very long. I am glad that I had been able to make his brief life a happier one than it might have been otherwise!

I had one student whom I taught on a one-to-one basis, as he wanted to take the 'distinction papers' set by Cambridge University in mathematics. He was very keen and very intelligent. He caught on to things very quickly, sometimes more quickly than I could myself! We had great times going through past distinction papers, discussing various ways of solving the problems. At the end of the school year he took his exam and got his distinction and I was very happy for him. The headmaster congratulated me. He should have congratulated the boy! He probably would have got his distinction without any help, he was so bright.

During the night bombing raids, all public buildings were watched by 'fire-watchers'. The staff and the older boys made up a rota and every so often, if it was one's turn, one had to spend the night on or about the roof of the school and try to put out any incendiary bombs that happened to fall. I was on duty on the night of what came to be called 'the fire of London'. Hundreds of planes were raining bombs all over the city and, as far as the eye could see, there were huge fires blazing. By some miracle no bombs fell on or even near Highgate School, so all we had to do was keep on the lookout and gaze at the fireworks. It seemed as though not a building would be left standing in the city by the morning. Next time I went into the city, I was amazed at how many buildings were still standing and how many were not even damaged!

My father's job was at Birkbeck College, which was an evening college, for people to be able to take a degree while they had a job to do during the day. It was clearly not possible to continue to run classes in the evenings because of the raids that seemed to happen just about every night. So the evening college was turned into a weekend college, lectures being held during the day on Saturdays and Sundays. We arranged for my father to stay at our Eastbury place during the week, so he would be out of the bombing, while I was at Cranley Gardens and at weekends he would come to Cranley Gardens and I would go to Eastbury.

When the bombing got really bad, I managed to get myself billeted at Reading, from where I could get to Highgate every day and take the train back in the afternoon. The trains were always packed, so it occurred to me to get into the restaurant car at Paddington, which they always had on the 4.15 train bound for Wales. For the princely sum of one shilling and sixpence, I could always have a comfortable seat and a pot of tea with scone and cake. I do not know to this day why none of the other passengers who were packed into the corridors like sardines ever thought of my simple trick!

One morning I was waiting on the platform at Reading station for my usual train to London, when a train pulled in from the west. I thought that I might get to London earlier on that train so I got on.

The ticket collector who saw me get on came up to me and said in a loud voice, 'This train does not stop at Reading.'

'In that case I did not get on,' I replied logically. I showed the ticket collector my season ticket and we both had a good laugh.

On Saturdays I would take an earlier train out of London from Paddington and arrive at Newbury just in time for my connection to the branch line. Almost every Saturday there were mothers with children getting off at Newbury station, who had been bombed out and had nowhere to go. I would sometimes bring them to Eastbury, where they would doss down on the floor until we could find somewhere for them to stay, which was quite difficult, as all the households were full of evacuees. These bombed-out people were usually poor working-class people who often had very different ideas of hygiene from ours and sometimes it really did try one's feelings of humanity to attempt to accommodate them even for one night! But there was a war on and we all thought we had to do what we could for each other.

My educational work went on with the background of the world collapsing all around us. This did not stop me from thinking about what I was doing and what my future was likely to be. I did not really agree with the way school discipline had to be kept; I did not like the punishment system, since it was always the same pupils who were punished so clearly it was not an effective deterrent from whatever we wanted them to be deterred! These thoughts kept going through my mind as I sipped my tea on the 4.15 to Reading, while the train tried to make its way along the bombed tracks, occasioning long delays at times.

I was still in touch with Dartington, but of course there was never time to go and visit any more. Towards the end of the summer term, Mr Curry offered me a teaching post at Dartington.

I talked about it to the headmaster of Highgate School.

He said to me, 'Don't be a fool, Dr Dienes! I can see your career mapped out clearly before you. You are a thundering good teacher! Not many young men like you could handle those fifth forms you teach! You will get a teaching post in one of the crack public schools and most likely finish up being headmaster at Winchester or even Eton or Harrow!'

'I don't know that I could identify myself with their educational practices,' I replied.

'That's nonsense,' he retorted. 'You are very young – you'll have more sense as you get older!'

But I was not impressed. I really could not see myself going along with the establishment! So I contacted some educational agencies in the hope of finding someone to replace me, so as not to let the school down. There were many replies, in particular from a man who had a Ph.D, a certain Dr Deutsch. He appeared to have all the qualifications I had. The head was not impressed. Dr Deutsch was a Jew. I frankly thought such a thing should not enter into the equation! Like it or not, I wrote a letter of resignation, stating that I had done my bit in finding a substitute and accepted the teaching job at Dartington.

The job at Dartington was an interesting one. Between nine in the morning and three in the afternoon, I was to teach the six to seven year olds and between half past four and six in the afternoon I had the children who had already taken School Certificate and were going on to higher things in mathematics and physics. Mr Curry called it a giant pincer movement, using a military metaphor. The job included the responsibility of being house father to children between twelve and fourteen, so we were given a room in one of the houses and Corin was given a separate suite in another part of the building, opening on to a large terrace, where all the baby matters could be dealt with. Food for all three of us would be provided in term-time, which we could have either in the dining room with all the other children or in Corin's 'private suite'. It seemed a good arrangement. Tessa was able to concentrate on Corin, having no other responsibilities and I could concentrate on my work.

I had never taught any six year olds before, nor had I been to a teachers' college to learn about any necessary or desirable techniques, so I had to start from square one and invent everything for myself. I wonder to this day how Mr Curry thought I could do justice to the job! There were thirteen children in my class and I had two rooms in which to teach them. I turned one room into a classroom and the other into a workshop. I made up instruction cards with pictures for teaching them arithmetic and got them to paint or draw pictures of stories that they invented and then showed them how to write down what was happening on the pictures. We used a lot of wood and cardboard for making models. At one point we made a whole village, each child making a house or a shop or a church, sticking the cardboard together with tape or glue. We used the village for making up more stories. I had studied a simple Nature book, so I would recognise the usual trees, shrubs and flowers and we went for many walks into North Wood, which was just at the back of the building and studied the trees and flowers, identifying them using my book.

There was a child whose name was Wanda, who had been brought up on a farm in Cornwall far away from any other habitation. Wanda had never played with any other children. On account of family problems, she had been placed at Dartington as a boarder and, of course, she was totally lost. Fortunately, we were having a warm and sunny autumn, so Wanda could run about outside, climb trees, at which she was expert, swing on the swings and no doubt do all the things she had been used to doing at the farm. Not having been to teachers' college, I did not know that you had to have the children in for classes even if they did not want to come, so I left Wanda to her own devices. She would sometimes come and put her little face right against the lattice window of our classroom. We smiled and waved and she smiled and waved back.

One day Wanda marched into the classroom and loudly announced, 'Zed! I want to learn to read!'

'That's very good, Wanda. Why don't you sit here next to Jamie? He has written a story and he will show you how to read it,' I replied.

Wanda duly sat down and, after Jamie had read one page of his story to her, she came to me and said a little sadly, 'I haven't learned to read! I'm going out to play again!'

I suppose she did not realise that learning to read would take time.

The playing went on for some time, until one day Wanda came in again and asked me what she could do.

'Look, Wanda, here is some paper, some paints and some paint brushes. Why don't you paint some pictures and make up your own story?' I said to her.

Wanda was pleased to do so. She stayed in all that day and painted a whole series of highly dramatic pictures, using very bright colours. The other children crowded round to admire her work and Wanda was in seventh heaven!

'Why don't we put them all up on the wall?' said Janet, a five year old, also there because of some family break-up. 'And you can tell us how they go and tell us the story!'

This was duly done as a collective effort. As she told the story, going through the pictures, Michael wrote down what Wanda was saying at the bottom of each picture, sometimes in a simplified and somewhat garbled version!

'I want to read my story now!' said Wanda.

The other children helped her read the story, sometimes correcting words that had been spelled wrongly! By the end of the year all the children, including Wanda and Janet, learned to read and write and do simple arithmetic, so it appeared as though I had not missed much in not having frequented teachers' college!

In the summer term I would take the children down to Folly Island, so most of them learned to swim as well. North Wood provided their textbook on biology, together with the dairy farm next door, which we sometimes visited. To widen their horizons, I used to read to them from the *Good Master*, which was about Hungary. We had many talks about Hungary. They thought it great fun that I could speak their strange language and I had to keep telling them how to say this, that and the other in Hungarian. Jamie was also interested in Spanish, since his father had been killed in action in the Spanish Civil War. His drawings were always about battles and sometimes he would ask me how to write things in Spanish underneath his battle scenes.

A very sad thing happened right at the beginning of our stay at Dartington. Tessa had a miscarriage. She was taken into hospital in Torquay, where the nurses were not particularly kind to her, as they thought she had brought on the miscarriage herself!

Actually, although this was sad, the problem was soon remedied when Tessa became pregnant with our second child, who was eventually born at the school during the summer term! The baby was a month premature. Labour might have been brought on by a twelve mile long walk on Dartmoor we did one day, as I was keen to show Tessa and Corin the beauties of that part of Devon! The baby threatened to be a *placenta previa*, but the doctor arrived just in time to hold back the placenta and allow our son to be born. We shall never know, but the birth might have given him some brain damage, which could have been responsible for some of his somewhat unusual behaviour in later life.

The well-known Irish playwright Sean O'Casey lived in Totnes at that time. One of his children was in my group and the other one, a little girl, Shevaun, of about Corin's age, had become a friend of Corin's and they used to play together on the terrace outside the nursery. Eileen O'Casey, Sean's wife, became friendly with Tessa and came to help with Nigel, our new-born, doing the washing or looking after Shevaun and Corin. She was so taken with Nigel that she offered to adopt him! Naturally, we were not going to let her do this, but we were happy that she liked the new baby so much, in spite of the fact that he had rather a curious look at birth, being a month premature!

During the year, the war was taking its inevitable course: more cities were bombed; more people were killed or made homeless; some stray bombs even landed near the school. We used to go to the air raid shelter when there was an alert, though we need not have bothered! The shelter was built right next to the chemistry laboratory and sometimes people were working with Bunsen burners even during an alert! If any bomb had dropped anywhere nearby, the whole laboratory would have gone up and us with it!

Possibly under the influence of the war, or however it might have been, I had started to think about our future again. Was I really going to spend my life teaching little children? I thoroughly enjoyed my time with the children – it was emotionally very satisfying work – but you did not need a doctorate in mathematics to do so! True, I was also teaching older children some higher mathematics, but I still had my doubts. I somehow got to know that there was a job going at Southampton University College in the Mathematics Department. I applied for it and to my great surprise I was offered the job.

So this was the end of my short but varied career in school teaching and the beginning of a more 'scientific' career. I discussed the projected move with Tessa and we finally decided to leave Dartington and face the real world! This meant another house-hunt! Tessa took a poor view of moving into Southampton, as it was being bombed rather intensively, so we concentrated our search in the Winchester area, from where I could commute with relative ease. We found eventually a converted chicken hut, the rent being thirty-five shillings a week! There was water laid on and electric light and an old smoky stove for heating and cooking. We decided that with two young ones it was better to have the inconveniences of the converted chicken hut than the bombing in Southampton, so we moved into our new home.

In September I started my work as a university lecturer in mathematics. This was a definite turning point and, although I did have some nostalgic feelings about my little charges I had taught at Dartington, I found myself much more in my element in a university mathematics department!

Chapter Twelve
The Peripatetic University Years

I spent two years teaching at the University College of Southampton, which was at the time an external college of the University of London, so London degrees were awarded by the college. They were formative years, during which I found out more about myself as a person as well as a great deal about how to teach mathematics to university students. From our chicken hut home there was a two mile walk to the bus stop, from where I took the bus to the outskirts of Southampton, where the college was situated, so each working day involved four miles of walking, rain or shine. My father had given me a waterproof cloak to use on rainy days for my obligatory walk and I became a well known sight in the neighbourhood, striding across the downs in my flowing gown. In fact, believe it or not, some people thought I must be a spy, as who else would be doing these secret walks every day?

There might have been something in the rumours, as by that time we had had dealings with spies, naturally unbeknown to ourselves. Across the road from our Eastbury bungalow lived a family of Russians, who appeared to do nothing else but make interesting-looking clothes. They were in due course arrested as spies and it transpired that they had sent messages encoded in the patterns they wove into the clothes they made! Apart from this brush with a spy ring, Tessa's mother, Rosalind, had let one of her rooms in Putney to someone who later turned out to be a spy, who was arrested and duly shot at dawn, as they say. So, assuming the police were aware of the above, they might have started the rumours themselves to ensnare me, assuming I was indeed a spy!

Since my first degree in mathematics was not a very good one and I now had to teach students much mathematics of which I was not very sure, obviously I had to quickly knuckle under and learn it! So I collected all the exam papers that had been set for the external BSc degree by London University for the past ten years or so, cut out the questions and categorised them into types. I put all the ones of the same type in an envelope, marked with a coding system. I then began to train myself to solve these problems. I wrote out the solutions, sometimes more than one solution for each problem and so made up a curriculum for my courses. In the honours course there was only one student, a girl. I still used the blackboard to show her the development of various mathematical structures, including any proofs, while she sat at a desk all by herself in the lecture room. She was a good student, so these sessions tended to be rather pleasant as well as mathematically instructive for both of us!

I was also able to return to my original mathematical work, which was a further development of the work prepared for my thesis. I was working on enumerable procedures, following on the lines initiated by Emile Borel, which eventually resulted in an interesting set of results. I did not publish these until after the war, when Emile Borel himself presented my paper to the Académie des Sciences de Paris, so I was following in my Hungarian mathematics teacher's footsteps!

I made friends with my colleagues, who often invited us to come and stay with them in Southampton, though Tessa was not keen to accept such invitations, because of the constant bombing of Southampton during the night time.

One night, while we were guests of the probability lecturer, Dr Pedoe, there was an unusually noisy air raid; Dr Pedoe was getting visibly more nervous every minute.

After a near miss, he got up and said, 'I am going out for a walk!'

'Don't be crazy,' said his wife.

He could not hear for he was already outside, striding along the road as fast as his legs could carry him. Apparently he was in quite a daze, but when he reached a bomb crater, he said to himself, well, one has fallen here, it is not likely that another will fall in exactly the same spot, so he lay down in the crater and went to sleep. It just shows how weak is one's mathematical training when it comes to applications in real life! The probability of a bomb falling anywhere was the same everywhere, independently of whether one had already fallen or not. Or so it would say in any textbook on probability, assuming it was written in wartime! But gut feeling is very strong in supporting the opposite view, especially when confronted with the reality of a bomb crater! But this was not the end of the story.

As the next day was beginning to dawn, Dr Pedoe was rudely woken up.

'Excuse us, sir, do you mind moving? We are the bomb disposal squad and you are lying on an unexploded bomb!' said one of the members of the squad.

Dr Pedoe told the story to his probability class the next day, to drive home the validity of the equal probability theory!

Tessa and I got very tired of the smoke exuding from our chicken hut stove, which tended to envelop us just about every day. We asked the landlord to do something about it – we even threatened to withhold the rent – but he would not be moved. To avoid the smoke, we had to do the moving! There was a small village called Pitt on the road between Winchester and Romsay and miracle of miracles, we heard that there was a cottage to let. It was an old thatched cottage, probably more than three hundred years old. The water had to be hauled up from a 150 foot well and there was no electricity. But the rent was only six shillings a week and there was a vegetable garden and a very pretty front garden with hollyhocks growing near the gate and roses round the cottage door, so we fell in love with it and immediately gave notice to our chicken house landlord and moved into our romantic cottage.

The move lengthened my daily walk from a two mile one to a three mile one and I had to make sure that I had drawn enough water for the day for Tessa and the children before I left. But the rent was much less and the outlook was much more pleasant. Corin also made friends with some of the village children and on Saturdays or Sundays I would take her with her friends on long walks through the nearby woods, sometimes getting as far as Romsay, where I would give the children some treat like an ice cream. There was also an anti-aircraft gun emplacement a little way up from the cottage, which was a bit of a bother at night when enemy planes flew over and the gunners tried to bring them down. You could not hear yourself think while the guns were blazing away, but, fortunately, the barrage never lasted very long. I do not know if they ever brought any enemy planes down, but, before we left, they managed to bring down part of our roof.

The soldiers who worked the guns would often pass the gate to our cottage while Corin was happily swinging to and fro on it. She obviously looked very cute, because invariably she came in with some chewing gum or other goodies given to her by the passing military!

During this time my father still kept on Cranley Gardens, but he was not too happy being in London because of the continuing air raids. During one of the raids, a bomb fell in our back garden but fortunately did not explode. There were a great many of these unexploded bombs around, so it took two weeks before the bomb disposal squad got around to removing it, so, what with one thing and another, we thought it would be better for him to move out. We contacted a person in Letchworth who was willing to have evacuees from London and my father moved in there. During the Christmas holidays we went to stay in Letchworth with our two children. The place was crowded to capacity, as there were other evacuees sharing the same house.

To add to the confusion, the children got whooping cough when Tessa was already pregnant with our third child! All this meant that Tessa and the children could not come back to Pitt cottage, so I had to go back alone.

I stayed at the cottage until the roof fell in, after which I found a billet in Winchester with a family with four children on a low-rental housing estate, from which I commuted to Southampton. They had one of those indoor shelters, which was a very large and long steel table, placed in the biggest room. When there was an alert, the children all scrambled under the table and the rest of us just hoped for the best. Thankfully, the worst never came: we all survived the raids. This was 1944 and everything was hotting up in preparation for D-Day. But Hitler was not taking things lying down. As a response to the Allied 'area bombing' of large German cities, the Nazis invented the pilotless bomb, nicknamed 'doodlebugs', which were sent over with a certain amount of fuel, calculated to run out over a target area, at which point the bomb would simply fall down and explode.

Our baby was due in February, so we booked Tessa in to one of the hospitals in Hitchin, only a few minutes away from Letchworth by car or ambulance. Our third child was born during an all-night raid, Tessa could hear the doodlebugs going all through her labour, but the baby was duly born in the morning and soon brought back to Letchworth. I was given a little time off from work to see mother and child! It was wonderful to have three children! We were nearly a third of the way towards our contracted number of ten!

Then a thought ran through my mind about the ways words sounded to parents as they had children. After the birth of the first child, the word 'child' certainly acquires a totally different meaning. We can talk about our child and the word has a very different ring! After the second child is born, the parents can savour the word 'children', which will have acquired a very different ring!

'Will you see if the children are asleep?' A simple request, but how charged with meaning and emotion and fulfilment! When the third child is born, it becomes possible to talk about 'all the children' and really to savour the meaning of the word 'all', which one could not use with a family of only two children. Then I wondered what the next stage would be, when (not if!) the fourth child had made his or her appearance on the scene. It came to me that it would be the word 'some' that would have the flavour of newness: 'Will you ask some of the children to come and help?' Such a sentence would not be possible with three, but would become possible with four children. At this point I could not go on thinking what would happen when the fifth child arrived, as Nigel, our second, was having one of his whooping cough fits and Tessa was giving a feed to our new one, Jancis, so I had to go to the rescue and hold Nigel until he stopped coughing!

During these tumultuous times I still had time to look through advertisements in the *Times Educational Supplement*, because I was wondering whether to move on to higher things. There was a job in mathematics at Sheffield University. I applied for it and I was offered it in due course. I accepted and gave in my resignation at Southampton.

At this point 'university family allowances' had been introduced and with three children we thought we would be getting more than enough to move to Sheffield and to survive there. There was no possibility of renting, as the government had introduced legislation which made it virtually impossible to get rid of tenants, so most owners of rented property would rather sell than relet. This meant saving enough money to put down so we could take out a mortgage! We had only enough money to handle day-to-day expenses, so we were stuck. This is where friends come in. One such friend, by the name Pat, offered to lend us one hundred pounds! To us, such a sum seemed an absolute fortune! But life evolves along strange and unexpected lines, which, in this case, meant thinking in terms of owning property!

We soon found a 'modern semi' on the outskirts of Sheffield, in a suburb called Totley, which was up for sale for eight hundred pounds. With a ten per cent deposit, we could just manage to buy the house with a seven hundred and twenty pound mortgage! We were able to get some 'furniture units', as we had had some bomb damage; the damage from the collapsed roof at Pitt was counted as bomb damage. With these units we were able to buy some furniture at controlled prices on the 'never-never'. We moved eventually into No. 4 Mickley Lane, in the suburb of Totley. Our would-be neighbour had whitened our steps and had a pot of tea ready for us. Yorkshire was a friendly and hospitable place. The neighbours had a little girl about Corin's age and the two of them soon became the best of friends. We soon settled into our new home and I could start my work at a 'real university' and be a 'real property owner' as well!

The work at the university was not very different from that in Southampton. However, there was one course on partial differential equations which I had to give to engineering students. I recall thinking that the way the textbook presented the subject matter could not possibly have been of any use to engineers: in partial differential equations, instead of arriving at a solution containing 'arbitrary constants' which had to be filled in from knowledge of the particular application at hand, one arrived at solutions containing 'arbitrary functions'! How useless if one wanted to build a bridge! But that is how I was told to teach it and that is what the students got, useful or useless as it might have been! I thought even the usual courses on ordinary differential equations were rather useless from the practical point of view. So I tried to make up situations, which, although sometimes somewhat unlikely, had some element of reality, not just a series of dy/dx interspersed! I remember that one of the examples was about a bull charging across a field. Here was the field: the bull begins to charge; I run for the gate. The bull veers, so it is always pointing towards me as I run. I just make it to the gate. The speed of the bull is a constant u feet per second, my speed is a constant v feet per second. The length of the field is a feet, the width is b feet. What relationships must obtain between a, b, u and v so that I indeed escape from the bull? Already at this stage in my educational career I was trying to think of ways to make learning fun for my students, although I was greatly hemmed in by syllabuses, textbooks and professors of mathematics, who are usually notoriously conservative and formalistic!

Totley was very close to the Derbyshire moors and dales. We would sometimes take the children on the train to Grindleford, only one stop away and then climb up on to the moors, walking across some beautiful woodland. We would spend the day out of doors and either walk back to

Totley across the moors or take the train back. I had a great wish to pass on to my children my love of nature and this was the beginning of their education in this direction.

During the first year at Sheffield, we sent Corin to the Dore and Totley High School, where she made reasonable progress, but we eventually thought that the education she was receiving was much too formal, by far too much unnecessary discipline, as we then thought. When she was naughty, she had to 'stand on the gratings' for lengthy periods. We never found out what and where these 'gratings' were! So we sent her to a more modern school in Hathersage, which was the stop beyond Grindleford on the train line. She took the train every day to school and she came back the same way. She very much enjoyed the school, which was run on Montessori lines, as well as the train trips.

One day I was walking along with Corin, passing a small railway station at which a train happened to be standing.

Corin said to me, 'Dad, can you do everything?'

'Yes, of course,' I lied.

'Well, then, can you make that train go?' asked Corin.

'Yes, of course I can – it's nothing!' I replied.

The guard was just about to lift his green flag to let the train move out of the station.

I started saying the magic words: 'Abracadabra hocus pocus, train *go*!'

At the precise moment I pronounced the word 'go', the train moved, accelerated and soon disappeared round the bend.

'Dad! You really made that train go!' said Corin to me, full of admiration.

Of course she did not know about the guard and the green flag and I did not let on until much later how that particular piece of magic had been performed!

The time arrived when Pat would have liked some repayment of our loan. Of course, with three children, school fees et cetera, we had no funds with which to repay any of it. But there was one possibility: house prices were increasing. We could sell our house and buy another, maybe for less money, or at any rate with less down payment. I went to a lawyer and asked him how we could do this. He suggested that we went to house auctions. I could write a cheque for the deposit at the auction and he promised to cover it up to a certain amount.

One day Tessa came home enthusing about a house which was part of a row, much nearer the centre of Sheffield, from which I could walk to the university. The auction was the next day. I never saw the house but took part in the bidding, and, by a miracle, became the highest bidder and so bought the house! We soon sold Mickley Lane and, with the difference, paid Pat, the lawyer and still had some money left in the bank! I was learning the elements of how to be a spiv!

At the end of our first year at Sheffield, we thought we deserved a holiday. We rented a little cottage on the Isle of Arran, a mountainous island in the Firth of Clyde. For the first time, we took sleepers on the night train to Glasgow, then a boat to Brodick and we installed ourselves in our cottage!

Soon after we arrived, we heard the news that the war in Europe was over. It was VE Day! At last we could look forward to better days, or so we thought! We invited Pat to share our cottage for part of the summer, as a sign of our gratitude for his generosity in helping us buy our first house! The children enjoyed the golden sands, except that Jancis filled her mouth with it on the first day as she thought the whole beach was made of brown sugar! What a disappointment for an eighteen month old little girl! Of course, she spat it all out in disgust and we helped her wash her mouth out. Pat and Corin and I climbed Goat Fell, the highest peak on the island. It was Corin's first strenuous mountain climb, but she acquitted herself very well, considering she was only seven years old!

Corin and I joined the Youth Hostel Association. In England, children had to be nine years old before they could come to hostels, but the age limit was five in Scotland, so on several occasions I took Corin to Scotland youth hostelling. At Eastertime, during our second year in Sheffield, Corin and I toured the Trossachs. This is a very accessible and very beautiful part of Scotland and our route included the ascent of Ben Lomond and a leisurely journey walking in a southerly direction on the 'bonnie bonnie banks of Loch Lomond'. But I had miscounted the date of the beginning of the next term and arrived back in Sheffield a week late! This was a catastrophe and, having realised it was entirely my own fault, I offered to resign immediately. This was not accepted since they needed someone to give my lectures!

But my days at Sheffield were clearly numbered and I had to look for another job. I went to an interview for a post as a mathematics lecturer at Chelmsford Technical College. There were only two of us on the shortlist being interviewed. One was Caleb Gattegno, the person who was madly popularising the coloured rods introduced by the Belgian teacher Cuisenaire. To Gattegno's chagrin, I was offered the job. So the continuation of my working life was assured, even though it was to continue in what was then considered an academically inferior institution.

Before the end of the year we did another 'spiv job'. We sold our house and bought another, moved into it for two weeks and then let it for the three summer months, furnished and then, with the money, we were going to have a holiday in the Hebrides with the children.

There is a small island called Canna between the islands of Skye and Rhum. Next to it there is an even smaller island called Sanday. Canna is about six miles long and about half a mile wide; Sanday is about one-quarter that size. There is a small footbridge between the two islands, which can be used nearly all the time except at high tide. On Canna there is one dairy farm and beside it is the laird's 'castle'. We rented the laird's castle for twenty-five pounds a month! We thought it was a real bargain! There was one boat a week from Mallaig that called on the inner islands, one of them being Canna, on the way out to Barra and South Uist.

One Monday morning, after spending a night on the train, we found ourselves on this boat. There were six of us, as my father had decided to come and share the holiday with us. It was a beautiful day when we disembarked on our island. We were met by the dairy farmer, who had the only truck on the island; he conveyed us about a mile up a dirt track to the castle. This was a beautifully appointed, old-fashioned residence, with a huge garden in the back, in which all sorts of vegetables were growing, ready for our use. There were some redcurrants, raspberries and all sorts of gorgeous goodies in profusion. There was no shop on the island, but the dairy farm had milk, cheese and butter and most of our food could be gathered from the garden at the back of the house. For anything else, we had to go to the post office and call the grocer at Mallaig by telephone and they would put a parcel of food on the Monday boat; then we would pay the purser, who would give the money to the grocer when the boat returned to Mallaig a few days later.

We made ourselves at home in the castle and started our summer holiday on our 'desert island' in a happy frame of mind. From our front window we could see the mountains of Rhum and, if we climbed the hill at the back of the house, we could gaze at the 'far Cuilins' over in the Isle of Skye, with their characteristic rocky ridges cutting into the northern sky. The Isle of Sanday was also in front of us, with the menacing mountains of Rhum towering some way behind. Sanday appeared to be a small hilly promontory, with a few crofts scattered about along its shores. There were about twenty-five people living on Sanday, five of whom were children. This meant that the government was obliged to send a teacher to teach these children. This, in turn, meant that Corin could go on attending school and make friends with the crofters' children.

I made a pact with the teacher by agreeing to teach her some French, in return for which she would teach me some Gaelic. So, at the end of school, I would sometimes go over to Sanday so we could have our lessons, while Corin and the children played, after which Corin and I returned, sometimes walking over the mud flats, or, if the tide were too high, by the footbridge. If the tide was right up, I would have to carry Corin part of the way to and from the bridge! Corin preferred to walk over the mudflats, for this was much shorter. As none of the children wore shoes, Corin did not like to wear them either: children do like to be like their peers! Unfortunately, this caused her foot to become infected and we were not quite sure what to do.

One evening we saw a trawler coming close to shore, so I hailed it and asked if they would bring back some disinfectant from Mallaig, since that is where they were heading. But we need not have worried! We had invited my mother to come from Budapest to share our Hebridean holiday with us and she arrived one Monday morning on the weekly boat from Mallaig. By some miracle, she was destined to solve the problem. The children came on board while the sailors were loading and unloading the cargo and we all hugged each other, happy to see each other again after so many years! My mother produced some Hungarian peasant clothes for the children, which they instantly donned. They rushed up and down the deck, to the great amusement of the few tourists on board. When they heard my mother and me converse in Hungarian, I heard one of them exclaim, 'Listen to this! They are talking in Gaelic! And look at all these children! They must be the local children wearing their national costume!'

I did not like to disappoint them by telling them that the Gaelic was really Hungarian and that neither the national costumes nor the children were local, so I let it pass. The captain and the crew were, of course, speaking to each other in Gaelic, but how could you expect some mere Sassenachs to know the difference!

When my mother saw that Corin had a bandaged foot, on arriving at the house, she started to devise the remedy. She put some milk to sour in the sun and, when it was solid, poured the resulting substance into a linen sack she found in the butler's pantry and let it drip for a few hours. Soon the sour milk had become cream cheese, which she proceeded to place round the infected part of Corin's foot, bandaging it up tightly so as not to lose any of the precious substance! After two applications of this treatment and long before the trawlermen came back with the disinfectant, Corin's foot was entirely cured and healthy-looking. I suppose that the treatment was a precursor of penicillin treatment, which was being discovered at about that time.

We explored the island. Corin and Nigel were old enough to come with me, but Jancis stayed with Tessa. On one occasion Jancis came in from the garden with her face almost entirely covered in something red.

'Have you been taking the raspberries from the garden?' asked Tessa.

'No!' lied Jancis.

She was only two and it must have been her first lie. The art of lying is not easily learned: she had not realised that the tell-tale red on her face should have been washed off first!

I once persuaded Tessa to come and clamber about on the rocks on the island, as I had the impossible idea that I might teach her to rock-climb!

She was so terrified by her very first attempt to clamber over a rock face which was not even very steep that she said, 'I never want to do this again!' And she never did. Her career as a rock-climber had started and ended there and then on that forbidding piece of rock on the Isle of Canna!

Our time in Canna soon came sadly to an end and my father went back to London. The rest of us, including my mother, had arranged to stay in Compton MacKenzie's house on the northern shore of the Isle of Barra. The Monday boat took us directly to Castlebay, from where the McNeills used to rule Barra with an iron hand. We were taken to Eoligarry Sands, a large bay, on the shore of which lay this beautiful, large residence.

While we were staying in Barra I got a message from a Professor Newman, who was professor of mathematics at Manchester University, in which he asked me to come to Manchester for an interview, with a view to a job in his department. I told him by phone that I had already accepted a post in Chelmsford, but he thought he could fix that, as there was enough time before the beginning of the term.

It would have taken for ever to wait for the boat so I decided to fly. There was an air service between Glasgow and Barra, but the times of the flights depended on the tides, as the plane had to land on the Eoligarry Sands! These sands were covered with water at high tide, so only at low tide could a plane land and take-off. We all trooped out to see the plane land, they lowered the steps and then I climbed in. In a surprisingly short time, flying over Tiree, Coll, Mull and the Mull of Kintyre, after drinking in the geography of these islands from the air, I felt the plane land at Renfrew. I did not have much money, so I hitched the rest of the way.

In Manchester I had a number of interesting talks with Professor Newman, including a discussion of my paper, which had just been presented in Paris by Borel. At first he was not unduly impressed and, frankly, did not believe in my theorems! After a two hour discussion, I finally convinced him that it was all correct, as it really ought to have been, seeing one of the world's greatest mathematicians had presented it to the Paris Academy on my behalf, so he offered me the job, saying that he would make himself responsible for the situation created at Chelmsford.

Now I had to buy a house and arrange for the removal of our belongings from Sheffield. We had money in the bank, as we had sold our third house in Sheffield, so I soon found a suitable house in the part of Manchester known as Whalley Range, in a road called Wood Road. I recall paying one thousand and two hundred pounds for a three-storey house which I thought would do us fine. I also arranged for the removal of our furniture, setting a date for its arrival, which we would have to make coincide with our arrival from Barra. I hitched back to Renfrew, took the little plane over to Barra and found myself once more in the bosom of my family in Barra. Corin had gone back to school, which was only about a mile away from our house, so we could settle down to a few more weeks of relaxed living, Barra-style!

In the meantime Tessa and my mother were getting very friendly. They were able to share some of their spiritual experiences, which had the effect of establishing a bond between them. Tessa had been having exciting mystical experiences, during which she came to 'know' certain things she did not know before, and, upon discussing these with my mother, she discovered that it was all in the Bible, Jesus having said it all before!

Our marriage had not been all plain sailing up to that point. Marrying someone of seventeen from an unhappy home does not provide the best prognosis for success. We had had our ups and downs and it seemed as if the time in Barra was doing us both good. Tessa was wondering what to give me for my thirtieth birthday, which was coming up soon. 'For your birthday, I will give you me!' she said a day or two before my birthday. I hugged and kissed her instantly and told her that nothing else could possibly have been a better gift!

I had a marvellous birthday party: we were all together; things were looking up; my mother was with us; and I could look forward to an interesting time in the Mathematics Department at Manchester University.

We arrived in Manchester at the same time as the furniture. You can imagine the pandemonium! Corin and Nigel would have to be got ready for school; furniture needed arranging in the house; I needed to become familiar with my working schedule at the university; we needed to make up a budget, as now there were six of us and we had to be careful. The result of the budget discussions was that we decided we needed to let some of the rooms. We had no problem getting tenants after we had placed a notice at the International Club, which said, 'Rooms to let at reasonable rents. Only black people or Jews may apply. Phone Chorlton 1070'.

We soon had a Nigerian student whose name was Tai, a Jewish girl and another girl, who begged to have a room and who loudly excused herself for not being either black or Jewish! Her name was Sheila and she had the room next to Tai's, on the 'attic' floor.

Tai and the children soon made friends, Tai was very patient with them, in spite of the fact that they all liked stroking his very fuzzy hair! The tenants cooked in their own rooms on primus stoves, or out on the landings. We were too simple to think of things like fire hazards. We enjoyed the three students being able to live their lives and organise themselves as they saw fit! The result of our libertarian attitude was that in the end Tai married Sheila and, when he had finished his studies, they went back to Nigeria and started the Mayflower School, which later played an important role in the Nigerian educational system. They eventually had four children, calling one of them after our little Corin! Tai became a well-known figure in Nigerian politics, known to all for standing up for the people's rights and liberties!

Corin and Nigel were sent to the local elementary school, while Jancis stayed at home. It was a bit of a shock for Corin, who had been used to a private school in Sheffield and lately to the island schools, where she was very popular because she could speak, read and write English so well, not like the other children, whose language of communication was Gaelic. For Nigel it was very new, as it was his first school and he was not even five, but they both adjusted quite soon and made friends with the local children.

My schedule at the university was a very light one. I had only three lectures a week: Mondays, Wednesdays and Fridays, from twelve to one. The rest of the time was supposed to be devoted to research! How much of the 'research time' was, in fact, devoted to family duties, I leave the reader to guess!

We made friends with a number of faculty members and their families and the children made friends at school, so we soon developed a very busy social life. I found that one of the boys I had been friendly with at Körtvélyes in Czechoslovakia, the older one of the Jánossy boys, was working on cosmic rays in the Physics Department. He and his wife had four children, so we re-established our boyhood friendship on a family basis. Later he got a professorship at Dublin University, where they provided him with a private lab for the study of cosmic rays. He sometimes invited us to his Dublin house, where they lived in relative luxury, compared with the house they had had in Manchester on a low-rental council estate.

The Christmas pantomime season was upon us. The children were now old enough to take part in this age-old form of English entertainment, so I would take them, together with a host of their friends, to the theatre and we all enjoyed laughing and singing and watching the antics of the pantomime actors together. The children all remember one particular pantomime, *Babes in the woods*, in which some bandits 'steal' the children and are going to do away with them.

The bandits would tie up the children on the stage and then face the audience and shout to them, 'Shall we murder them now?'

'No! No! No!' the children would shout back at the 'bandits'.

'Shall we murder them later?' asked the bandits.

More frantic shouts of 'No!' would come from all the children in the audience, sometimes with several running up to the stage to 'save the babes'.

'All right then,' the bandits would say to them. 'We'll let them go, shall we?'

They would then let the rescue party untie the babes, to tumultuous applause from the children in the audience. There were many such pantomimes and I really enjoyed taking all those youngsters, watching them getting worked up and taking part in the proceedings.

We were in Manchester for two years. My mother stayed with us for one year, after which she went to stay in a nunnery. The ups and downs and upheavals of family life – children's problems, marital problems – were all too much for her. This was a pity, as Tessa had just been accepted at the university as a mature student and was going to study social sciences while my mother took care of the children. But we could not afford the day care costs or the equivalent babysitting costs, thus Tessa's career at the university was prematurely aborted.

We spent the summer of 1947 on the Isle of Lewis and my brother Gedeon came to see us with his new bride, Maya, taking in Lewis as part of their honeymoon. We took over a whole croft on the west coast of Lewis; in fact, it was at the very spot where, years before, I had pitched my tent for the first time, looking out over Uig Bay towards the west. I felt a kind of 'coming back feeling' when I tramped over the same hills and swam in the same lochs with the children, where I had enjoyed myself more than fifteen years previously. The university vacation was three months long, so the children still had to go to school for a part of it. The school was about a mile away and the schoolteacher was happy to have Corin and Nigel come to school.

'Och, it will be great to hear English spoken in the playground,' she said to me.

I was actually hoping that the children would pick up some Gaelic, but most of the children knew enough English to communicate, so this desire of mine never really materialised.

There was a bull that grazed, quite loose, in the hills surrounding our croft and the school.

I asked one of the crofters, 'Is it not dangerous to have a bull loose like that?'

'They never toss anyone until they are three years old,' replied the crofter.

'How old is this bull then?' I asked.

'Och, I believe it is just aboot thrrree yearrrs old!'

Needless to say, I did not tell the children any of this, but decided to accompany them to school in the morning and meet them at the school to walk home with them!

Gedeon and Maya brought us a number of Hungarian goodies; some were eatable and some wearable. My brother was then in the Hungarian diplomatic service, so he had 'smuggled' them all into the country in the diplomatic bag, which customs are not supposed to open! We duly thanked them and they sped on their way, as they wanted to see as much as possible on their honeymoon.

The children grew and developed well during our two years in Manchester, in spite of more marital difficulties, which Tessa and I had trouble in resolving. We consulted a Jungian psychologist but it was a disaster. The psychologist tried to pit Tessa against my mother and at the same time violated her. When his wife found out, she committed suicide. I also had one or two sessions with him, but I did not feel he had much to offer.

I tried to escape sometimes from our problems by taking Corin and later Corin and Nigel, on youth hostelling holidays in Scotland and Ireland, but of course this was no way to address the problems we were having. I found out later that my mother secretly baptised the children, to save them from hellfire!

It was perhaps time to look for other climes and I started looking again in the time-honoured *Times Educational Supplement* for a job! My job at Manchester was not permanent, so I had to look for a better job in any case. I applied for and obtained a permanent lectureship at the University College of Leicester. The professor of mathematics there was a certain Professor Goodstein, who had been a student of my father's and who had leanings towards the 'finitist' conception of mathematics. This meant that we had overlapping interests in the field of the philosophy of the foundations of mathematics, so we struck a deal for a permanent lectureship at Leicester. This was the end of my peripatetic or wandering life as a university lecturer. I had spent two years at Southampton, two years at Sheffield and two years at Manchester. I was to spend more than twelve years at Leicester before deciding to branch out further and move out to Australia in a new field!

Before moving, we did another 'spiv job' in Manchester, buying and selling a house in Didsbury, this time selling it before we had really completed the purchase, so we had practically no legal bills and a net profit of five hundred pounds. We sold our Wood Road house for two thousand pounds and paid off the mortgage so we were ready to move 'upmarket' in our new surroundings in Leicester, in the hope of improving not only our economic well-being but our ways of living with each other as well!

Chapter Thirteen

Leicester

We bought ourselves what we considered a very nice house on the corner of Queens Road and Northcote Road, in a 'desirable' residential area, the price exceeding three thousand pounds! There were two reception rooms and four bedrooms, apart from the kitchen and the bathroom, and there was a pleasant garden. The mortgage repayments did somewhat stretch our budget, so again we had to think in terms of letting rooms or having paying guests. But these were quite small problems in relation to the milestones that were to be passed during our twelve year stay in Leicester. Of course, we did not know what these accomplishments were going to be when we started out on our Leicester life but this is as it should be. The future should unfold, constructively, out of the past and the present, allowing life to flow and for the participants in these life dramas to grow and develop, slowly but surely.

The milestones of our Leicester years were: the birth of our fourth child; the birth of our fifth child; joining the Religious Society of Friends (the Quakers); making strides in research into the problems of mathematics learning; publishing my paper on degrees of rigour; publishing my paper on abstract perception; publishing my books, *Concept Formation and Personality* and *Building Up Mathematics*; getting a degree in psychology; the invention of the manipulative materials known as the Multi-base blocks and the Algebra Experience Materials; publishing my paper, 'The Growth of Mathematical Concepts in Children Through Experience', a theoretical basis for the manipulative materials.

I was given a schedule of about ten to twelve lectures a week, which, with the preparations, was quite a full-time job, leaving a little less time for family and recreation. The college was only a quarter of an hour's walk away from our house, which helped in organising how my time should be spent. All the courses I was assigned were familiar subject matter, so really there was less preparation than I might have expected. My original notes from Southampton for solving problems set in past examinations came in useful again, as Leicester was also an external college of the University of London and had not at that point become an independent university.

I was given, among other courses, the introductory course for freshers on Number Systems. I always started by saying to the students, 'I should like you to forget, or at any rate ignore, everything you have ever learned at school in mathematics. I should like to start from square one and build from there. In this way you are all in the same boat!' I knew that the average teaching of mathematical concepts in schools took place at an abysmally low standard, so it was much safer to assume no knowledge at all and start from zero. Many of the students were greatly relieved when they heard this; others, the 'better ones', were somewhat offended. They soon found out that the truly mathematical way of looking at mathematics was very different from being crammed with complicated problem-solving strategies, applicable only to a certain restricted class of problems!

I never prescribed any textbooks. When asked what books they could read, I told students that this was really an individual matter, different people preferring different approaches and directed them to the library, where they could browse and make their own choices. I also never gave the same course two years running, although it would appear under the same name in the schedule. Notes taken for any other lecturer's courses could be sold at the end of the year to students the following year; notes taken for my courses had no value at all, as the course was never the same two years running. I sometimes suggested my old favourite Vallée Poussin, telling the students that this author did not just cover the ground but also emphasised the aesthetic and poetic aspects of the mathematical structures treated. But few of them could read French and even less frequent were the students I had managed to interest in the beauty of the subject! It was a few years later that I embarked on such projects and my subjects then would not be university students but children in elementary schools.

Tessa and I had to look around for tenants so we could balance the budget. Our first candidate was another new member of the mathematics department, whose name was Roy. He was a very meticulous kind of a person and spoke very meticulously, pronouncing every word carefully and correctly. We showed him one of the rooms and offered board for twenty-five pounds a month.

'I will have to think about it carefully,' said Roy. 'You see, when I first arrived, I came upon a place that said 'Roy's 'Ome' so I thought it was meant for me! The lady who leased me a room was not very sure whether she wanted me as a tenant. I found out why last night. The place was raided by the police and many of the tenants were arrested, including the landlady. Apparently, the place she had called 'Roy's 'Ome' was a brothel. So I am not sure where to go now. Can you give me some time to think about this?'

'We are in a hurry to get things organised,' retorted Tessa. 'We really ought to have your response to our offer now so if it is no, we can ask someone else.'

'I will then walk round the block and think about it,' said Roy and, so saying, walked out of the door. Five minutes later he came back.

'I will come,' he said quite laconically.

We found two other tenants, or rather boarders and in this way our family was truly squashed into a small space, but the budget had become feasible. Of course, the work had multiplied by several hundred per cent, part of which naturally fell on me and on our oldest child Corin, who was about nine or ten years old at this time. This caused some difficulties, because I had so many things to do in the morning, what with helping get the breakfast and seeing that the children were ready for school in time. So I was sometimes late for my lectures, or arrived dressed in strange ways. One of my students once asked why I wore odd socks! I looked at my socks and they were, indeed, odd! Clearly, there had not been time allocated for categorising my socks according to colour, so I had often ignored that particular attribute of my attire.

But the strangest thing which once occurred is this: I walked straight into the lecture room with my coat on, without going to my room to get my notes and disrobe, since, as usual, I was a little late. I went straight to the blackboard and started holding forth on some aspect of spherical trigonometry. At the end of the lecture, I realised that I had to do something about my coat, so I took it off and hung it beside the blackboard, then walked out. It was not until the evening, when I went to look for my coat in my room, that I remembered what I had done, so I went back to the lecture room where I had given the lecture on spherical trigonometry, took my coat off the hook and walked home. If somebody had told me something like that, I would have had difficulty in believing it. Yet it is supposed to be the prerogative of 'absent-minded professors' to do just such things. Sure, I was absent-minded but I was not yet a professor. That honour had yet to be conferred on me.

We had a number of interesting holidays while we lived in Leicester. The first one took us to the west of Ireland, to Achill Island in County Mayo to be precise. We rented a cottage in a small village called Dugort, where we looked after ourselves. Dugort is built on the shores of a very large sandy bay and one of the things I used to like doing was swimming the width of the whole bay, from one end of the sands to the other. The children went to school in the local Protestant school, where only very few children went, since most of the Irish are of the Catholic faith. The children thus got some truly personal attention from the teacher, who was the only teacher, but, since there were no more than about fifteen pupils, he was able to do a good job. He organised some parties and many people from the village would join some of the Irish dances which the teacher had taught to all the pupils.

We made friends with some of the people living in the village, particularly one family consisting of a retired major, his wife and their three year old little girl called Sorrel. There was a number of older, single people living in the village.

I was passing the house of one of these, the old lady sitting on the bench outside called to me, 'Oh, doctor, I have such pains in me legs! Would you give me something to take the pain away?'

She clearly did not make any distinction between a doctor of philosophy and a doctor of medicine!

I could not think of anything to say, but finally I said to her, 'You pray to the Good Lord, my dear and your pains will be gone by the morning!'

'I will do that, doctor! I will, to be sure,' replied the old lady.

The next day, as I was passing her house, she called to me in a very cheerful voice, 'Oh, doctor, I did pray and pray last night and, as you promised, all my pains have gone away now! I am so grateful to you, doctor!'

The previous day had been rather a damp day, but this was followed by a good, dry, sunny day! I did not have the heart to suggest to the old lady that the weather might have had more to do with her pain going away than any intervention on the part of the Almighty!

Another older lady once asked me to come in and have a chat with her. I saw no harm in it and I even accepted a drink, which she fetched with great ceremony from a hidden corner of her large kitchen. I did not realise it at the time, but what she offered me was some home-made poteen – it was truly powerful stuff! After three liberal portions of poteen, I passed out for several hours and was dead to the world! So if the poteen had been intended as a prelude to seduction, she must have been disappointed!

At the same time Tessa was visiting the major and his family. They had something to eat together and some drinks (but not poteen!) and, with one thing and another, it got rather late. Since it was dark, the major offered to see Tessa safely home to our cottage. What she did not reckon with was that the major's wife was madly jealous of her. Tessa was very attractive and had always had a lot of sex appeal and the major's wife smelt danger. So she picked up her rifle and quietly followed them down the road. Just as the major was saying goodbye, bending down to try and kiss Tessa, with Tessa strenuously resisting the attempt, a shot rang out. Fortunately, she missed. Tessa bolted towards our cottage. Then she could hear a lot of arguing between the major and his wife, but Tessa was safely inside!

Later that evening I woke up, thanked my old lady friend for the drinks and staggered back to our cottage, only to be greeted by a not very happy Tessa, who was blaming me for neglecting her! Ireland being Ireland, we still remained good friends with the major after this episode, in spite of the fact that the rifle had been broken during the evening scuffle. In fact, we called our fourth child Sorrel, after their little girl, who was a very pretty little girl indeed.

During our wanderings over the island we came across a shark-catcher. What was incredible about him was not that he caught sharks but that he had married Brenda, the girl I had used to walk out with when I was a boy at Dartington. They had two little boys, who never seemed to wear any clothes. They rushed around on the rocks outside stark naked, climbing up and down the jetty wall of the little harbour, on the edge of which they lived. Their 'toilet-training' had also been done rather strangely, they used a pot that was always under the table at which they ate. While we were eating with them once, we had to suffer one of the boys defecating under the table at the same time! This reminded me of the 'modern education' that we had both had at Dartington, but I did think that they had added a chapter or two to what they had inherited from Dartington.

Brenda's shark-catcher husband later played the part of a shark-catcher in one of the well-known films made on that topic. Unfortunately, during one of the takes, he was killed by the shark.

There was another major living in another part of the island, who organised parties to which any islander was invited. Most of the time there was country dancing, but there were also drinks galore and I do not recollect having to pay for any of it! There was a knitting factory nearby, where most of the local young girls worked and a lot of these used to come to the major's parties.

One day, during one of these parties, I was standing at the bar next to a good-looking woman and, just to make conversation, I said to her, 'I suppose you are working in the knitting factory?'

'No, as a matter of fact, I am not. I do rather different sort of work,' she replied. 'My name is Eileen Hurley.'

To her utter amazement, the woman realised that I had not heard of her. She was at the time a well-known Shakespearean actress, having played such leading parts as Ophelia and Miranda. When I found out who she really was and what this 'different work' was, I apologised and asked if she would like to join us at Dugort for a meal. She accepted instantly, but insisted on cooking the meal herself. In our kitchen she soon became her real self, the actress throwing her weight about. Within minutes she found herself alone, but nevertheless she toiled on and produced, all by herself, a wonderful meal for all of us, out of whatever she could find in the food cupboards!

During the meal she told us several times that she had come to Achyll to be away from it all and that she was really delighted at not being recognised. Then she disappeared as fast as she had appeared at the bar and we never saw her again.

Nigel, our second child, won an all-Ireland painting competition for children under ten in which all the Protestant schools in the country took part. We were all very pleased about this and became even more friendly with the school teacher who had encouraged him in his artwork.

At one point during the holidays it was decided that I would take Nigel with me on a personal holiday. Our objective was France. We soon found ourselves in Paris and then on the night train to Annemasse, with all its 1932 memories, when Gedeon and I had found ourselves at the end of our budget on our way to the tiny village of Burdignin. From Annemasse we walked or hitched rides from passing cars and soon reached the Lac d'Annecy, where I revisited the farm on which I had played with the dumb boy. We pitched the tent there and stayed quite a while, bathing in the lake and walking in the surrounding mountains.

One day, on our way down from one of the mountains we had climbed, dusk and then total darkness overtook us while we were walking through a forest of huge fir trees. Fortunately, we had a flashlight, so we could see where we were going. At one point there appeared to be two ways to go and I stopped and hesitated, shining the flashlight on to the rather indifferent map I was using.

'Are we lost in this forest, Daddy?' asked Nigel in a somewhat tremulous voice.

'Oh, no,' I lied. 'It's just a matter of knowing which of these two tracks is the quickest way down!'

We reached the valley finally and then our encampment without undue difficulty and Nigel seemed very proud to have a daddy who could find his way in total darkness across a very dark forest!

We lived mostly on fresh food, fruit, vegetables, milk, cheese and suchlike, so we did not often even need to light our little primus stove. Nigel was tickled pink when I told him that he had to have his peaches before he could get on to bread and butter and cheese! Usually, the peaches were the treat you could have only if you had eaten some obligatory dish beforehand! We followed, more or less, the route I had taken years previously, but this time over the Little Saint Bernard Pass and, as before, down to the Aosta Valley.

Nigel really enjoyed seeing the Roman amphitheatre in Aosta. I explained to him, 'This is where the Christians entered the arena and on the opposite side was the gate from which starved lions were let in. Thousands of people would watch while the lions tore the Christians to pieces!

But,' I told him, 'there was a time when the lions came out and, instead of tearing the Christians to pieces and eating them, they slowly walked up to them and licked them in a show of friendship. The Romans watching did not know what to make of that!'

So we played Lions and Christians in the arena, where we took turns being the lions. Perhaps this was one of Nigel's first history lessons! We came to the Aosta Valley with the family later when we already had four children and the other children also enjoyed playing in the arena, looking back two thousand years while they played Lions and Christians.

We walked over the Great Saint Bernard Pass and eventually found a place on the shore of Lac Leman (Lake Geneva) where we camped for several days and enjoyed the warm water of the lake. On our way back, we stopped in Paris. Of course, playing with boats in the little ponds in the Jardin de Luxembourg was a must, as well as walks down by the River Seine, where there was always a busy market, with people selling just about everything, but in particular small and large animals for pets! But the best part of the Paris visit was the shows in the Tuileries, near the Champs-Elysées, where the antics of the French puppeteers kept Nigel engrossed for hours on end!

Of course, the best aspect of this father and son trip was the forming of a strong bond between us. Decades later, Nigel would remind me of that wonderful, idyllic time when he and I were alone together and had tramped across Europe with not a care in the world! This short month we spent together remained with him as a treasure he carried with him all his life.

When we came back to Leicester, Tessa and the other children were still in Ireland. I had some difficulty persuading Tessa to come back, as she had found other interests. But back she came, for the bond between us, established over so many years, starting from childhood, was too strong to be broken. This is what is called 'love'; what we sometimes feel in fleeting ways, in the form of sexual attraction to other persons besides the person we love, is simply a small error in the program and, as every computer programmer knows, such errors can always be debugged, to let the real program work. When Tessa came back with the other children, we simply fell into each other's arms and confessed our undying love for each other. As soon as all the children were in bed, we allowed our passion free rein. Our fourth child resulted from the realisation that, come what may, we were bound to each other and we would stay together 'till death us do part'.

At about this time, my father retired from Birkbeck College and was offered an easy retirement job at Leicester by Professor Goodstein. He came to share our house in Queens Road. Later he bought a little bungalow outside Leicester, where he had a housekeeper to look after him. One of his duties was to run a problem class in mathematics. He did not like doing this very much and I offered to take it over. He thought that I should have a proportion of his salary for this, which I accepted.

This meant that we had some money to have a really good Easter holiday! Our fourth child was well on the way by Easter, so we chose the South of France as a good venue for this unexpected holiday. We advertised for a maid to come with us and a person older than ourselves appeared.

'Can I speak to Mrs Dienes?' asked this lady as she was let into the kitchen, looking at Tessa.

'I am Mrs Dienes,' said Tessa.

'Oh, I am sorry,' said Dorothy. 'I thought you were one of the children!'

Dorothy was engaged to be married to a Sussex farmer and thought she had better have some experience with a family first, so she would know what to do when the children started coming along! So we hired her and took her with us on our holiday.

At that time there were very advantageous fares for what the French called *familles nombreuses*: the head of the household paid full fare, the wife paid half and any children and servants paid one-tenth of the fare! So we booked ourselves on a night train to Menton, where we had rented an oriental-looking building, slightly up the hill, overlooking the sea. Monsieur Bockerey, the owner, showed us round, we paid the rent and established ourselves for five glorious weeks on the Côte d'Azur!

Our Côte d'Azur home was planned in a somewhat strange way. The main living room, for example, had a bath in it! True, there were some curtains you could draw if you wanted to be private. The bath was L-shaped and was made of tiles that looked like ancient mosaics from Byzantium. From the garden we could see the whole sweep of the coastline and it was a wonderful sight on which to feast your eye after the rows and rows of identical houses in an English suburb!

Of course the first thing the children wanted to do was get into the sea. It was April, so the temperature of the water was not all that agreeable, but probably comparable with the kind of water they had been used to in the Hebrides and in Ireland during the summer months. In any case, there was a safe beach and the children were in and out of the water all the time we had them down there.

We soon noticed that all the families kept themselves very much to themselves. Nobody talked to us; no child came round to play. There appeared to be an invisible wall round each family group and this wall was never scaled. This was not how I remembered the seaside at Abbazia! I soon realised that there must be a difference between the French and Italian cultures.

One morning I suggested that we all went to the beach at San Remo, which was only two train stops away on the coastal railway line. We bundled the children into a local train and bundled them out again at San Remo. We made straight for the beach. We had not even had time to put our towels on the sand before several children came running up to us, saying, *'Venite! Venite giocare con noi!'* ('Come! Come and play with us!')

The children did not need any persuading! They made friends at once with a number of vociferous Italian children, with whom they were happy to romp about for most of the day. It did not seem to matter that they did not have a common language. Playing on a sandy shore, racing the waves, playing catch, all have an international body language underneath the sounds children utter as they move briskly through space!

One interesting event which took place during this brief holiday was a journey to Turin on my part. Some people in the mathematics department of Turin University had somehow heard of some of the 'philosophical' work in which I was engaged and invited me to Turin to talk about it. If you have a quick glance at the map of Europe, you will see that the best way to get to Turin from Menton is to take the bus. The alternative train journey would mean going to Savona, changing, then going to Alessandria, changing and then after travelling for most of the day, one arrived in Turin, probably worn out by the journey. Naturally, I opted for the bus.

'You cannot go by bus!' Tessa said to me in a very emotional voice. 'It is very dangerous! These buses crash!'

'Thousands of people travel across mountains in buses and they don't usually crash,' I retorted.

'No! You can't go on the bus! It will crash!' Tessa yelled at me.

'All right then,' I said. 'Since you are expecting a baby and you would worry, I will go by train!'

I did take the train and I was very tired when I arrived. Tessa had made me put a dinner jacket into my luggage, as she said you had to look respectable at a formal occasion. I was met at the station and we went straight to the university where I gave my talk on degrees of rigour. There was animated discussion following the talk, several people at a time running up to the blackboard, drawing figures and writing formulae, talking very loudly and gesticulating in true Italian style.

Finally, somebody asked if I were hungry and, when I said yes, we all trooped out to a nearby restaurant. We settled at a pavement table and went on talking about degrees of rigour while the waiter decided to bring our food. My large case was still with me, as there had not been time to go to my hotel. The conversation turned to how people lived in England.

'The English are very formal, are they not? They go everywhere in dinner jackets, I have heard,' said one of the professors, Ludovico Geymonat.

They all burst out laughing at the very absurdity of the idea.

Then I said to them, 'You will never guess what is in this suitcase!'

Of course nobody did. I placed the suitcase on a nearby empty table, opened it and my dinner jacket fell out on to the floor! The merriment that followed was incredible.

'You see! I told you! In England you have to be formal!' said Professor Geymonat.

I explained that my wife had thought I would need to be dressed formally and so I had agreed to pack my dinner jacket. They all commiserated with me, saying, '*Mamma mia! Le donne! Le donne!*' ('Dear oh dear! That's women!)

The next day, before taking the train back to Menton, I got a morning paper. On the front page was the story of a bus disaster. The bus that I would have taken had fallen into a ravine and all the people in it had been killed. I came back to Menton a chastised man.

The professors in Turin were so intrigued by my talk that they asked me to write it up in a paper for their journal. Within a month or two, I wrote this paper, which was duly published in the *Rendiconti del Seminario matematico di Torino* in 1951.

Before this publication, I had already had work published, the first one being the one I wrote during my Ph.D work. The second one was *Sur les procédés dénonbrables*, presented in Paris by Borel in 1946, after which a more detailed dissertation, entitled *Sur la comparabilité dénombrable des ensembles Boréliens*, was published in the Paris *Journal de Mathématiques* in 1947, followed by an incursion into multi-valued logic, 'On an Implication Function in Multi-Valued Logics' in the *Journal of Symbolic Logic* in 1948. All the above treated problems in well-established branches of mathematics. My talk at Turin was my first attempt at examining possibilities that had not been discussed at any length, such as the degree to which a mathematical proof could be considered rigorous, with ways to test the degree of rigor of any particular proof.

Sorrel Anne, the fruit of the renewed love of Tessa and me, was born the following July. She was born in a few hours at about four in the morning. We were now a family with four children and we could use legitimately the word 'some' in the sentence, 'Will you send some of the children to help with this!' She was a beautiful baby, defying all ideas that newly born babies are ugly. And the other children were thrilled to bits to have a new little sister!

At one point after we came back from Menton, I had the radio on for no particular reason, and there was broadcast: a talk by some minister from Carlisle. I was not particularly in favour of listening to religious talks, but I knew that I was still battling the elements as far as the problems in my relationship with Tessa were concerned. Of course, we knew theoretically that we were made for each other, but how did this work out practically? Maybe what the minister had to say could be of some help. So I listened. It was a short simple talk and it was all about forgiveness. What he was saying touched me to the quick and I burst into copious tears. Of course! I had never forgiven her or anybody else and was possibly nursing grievances that were poisoning our relationship. It was good that I was quite alone at this point, so things could take their course.

I thought that I should rethink the 'opium of the people' theory. I thought back to my childhood, when I dutifully would go to Mass with my mother, went to confession to confess my many 'sins', took Holy Communion and said my prayers. Then I remembered all the mythology that went with it.

'Oh no,' I thought. 'I could never be told what to believe!'

I was in a cleft stick. I knew in my heart of hearts that I must take a radical step. I knew there was a loving God somewhere who wanted me to come to Him with my problems; with the forgiveness message ringing in my ears, for the first time in decades I prayed. There must be a way out, a way through which I could be in constant touch with God but without selling my integrity by pretending to believe in things I did not. Then the following thoughts came to me: 'The Quakers! They believe in what they call the 'inner light', or 'that of God in every person' and they believe that a person can have a direct line to God. A mystical idea! And there was no creed to get in the way.'

I said nothing to Tessa, but we had a peaceful night, our hearts again filled with love for each other, thinking of the baby who was coming and slipping into unconsciousness with my hand in hers.

The next morning I said to her, 'Would you like to come to a Quaker Meeting with me?'

'Yes, I would,' she replied in a voice that showed a little bit of surprise but not as much as I expected.

The following Sunday we went to a meeting. Dorothy looked after the children. Tessa had been to Quaker meetings as a little girl, so it was not as new to her as it was to me. There was some vocal ministry, but I could not now say what it was about. It was silent worship that impressed me. Several dozen people were there and Tessa and I were with them, silently worshipping God, the Maker of the Universe, the Universal Love Principle! I could not imagine anything more beautiful. I felt as if I had come home after a long period of wandering in a dark desert! After some months of attending the meetings, I applied to become a member. Tessa did not, for she felt she was not ready to take such a step, in some sense not feeling worthy to be a Quaker!

Two Quakers were appointed to come and see me. They would report back to the meeting and then the meeting would decide whether I could become a full member of the Religious Society of Friends, more briefly known as Quakers. They asked me a number of questions. One of them was whether I believed that Jesus Christ was the Son of God. I told them that I had not been able to understand what that meant and, if they could explain it to me, I would tell them whether I believed it or not. They could not explain it. So we parted and they reported back to the meeting. A month later I was a member of the Leicester Meeting.

The first three milestones I mentioned earlier had been reached and passed. Alongside my spiritual transformation, I had started to rethink my role as a mathematician. I looked through my published papers. There were not many of these and I wondered how many people had read my papers. Out of those few, how many had enjoyed them? Or made use of them? In all these years I had received only one letter from a reader and that was somewhat critical. He wanted to know how I thought anyone could count up to infinity! Indeed, that was one of the feats that had to be performed in order to 'compare' one Borel set with another! Naturally, nobody could actually do it! It was all highly theoretical!

Then I began to think along other lines. I knew that most people had difficulty with mathematics, for the most part, it was an unpopular subject at school. And how come I did not encounter such great difficulties? Difficulties there were, to be sure, but they formed part of the challenge of doing mathematics. People enjoyed other challenges, so why not mathematical challenges?

Then my thoughts turned to the problems I had tried to tackle in my doctoral thesis. In that thesis I had tried to dig into why different mathematicians had different ideas on how the foundations of mathematics should be laid. Could it be that the difficulties in the philosophical foundations of mathematics had something to do with the difficulties people experienced when they tried to learn mathematics? This was a very new and possibly revolutionary idea.

As the years rolled by, I kept wondering about such possibilities. I had an idea that some people perhaps had a closed idea of mathematics while others preferred to leave things more open. The formalist view, based on precise axiom systems, was closed. The Borelian view was more constructive and so more open. Perhaps the intuitionist viewpoint was even more open? Or was it as dogmatically closed, using different dogmas, as the formalist viewpoint?

If there was an 'open to closed' continuum in people's thinking, this would find echoes in their political views and in their attitudes to social problems as well. I was so fascinated by the problem that I wrote a paper on it and sent it to *Mind*. It came back, with very mixed reactions to my writing. I was somewhat disappointed, but then I thought of the Italians, who were perhaps more adventurous thinkers. I wrote the paper in Italian and sent it to the *Rivista di Filosofia*. They published it. My ideas on 'open to closed', 'implicit to explicit', 'constructive to analytical' were in print, if only in Italian! So the sixth milestone had been reached!

My philosophical paper was all very well, but there was no substance to it. If indeed there were such 'dimensions' in our personalities, they must be measurable. It was up to me to devise some experiments through which it could be shown that these 'dimensions' were really there!

Such things involved a knowledge of psychology. I knew very little psychology. It was obvious what I had to do about that – I had to take a degree in psychology. I registered as an external student of London University and decided to do my 'practical' at the psychology lab at Nottingham. Eventually, my work on personality problems became part of my practical work, which I submitted at the same time as sitting the theoretical and practical examinations in London. So I gained my degree in psychology and I had found out something about the relationship between some personality traits and children's ability to form mathematical concepts, by doing a series of experiments on one hundred ten year old children in Leicester.

I gave a talk on this subject at the Psychology Department of the University of Florence, which Professor Marzi had invited me to do. I had a very warm reception and they published my results in their *Psychology Bulletin* and at about the same time, or a little later, I published the same results in English in the form of a monograph, under the aegis of the University of Leicester. This meant that milestones seven and eight had been reached.

My father had bought a small bungalow outside Leicester in the meantime and had hired a Hungarian housekeeper to look after him. This housekeeper also cooked and cooked in very delicious Hungarian ways, but of course this was not suitable for a person getting on in years. He had a heart attack, from which he recovered, but afterwards he decided to sell his bungalow and go and live with Michael Tippett, the up-and-coming composer, who had a lovely big mansion in Tidebrooke, Sussex. He spent his time writing poetry while Michael went on composing his operas; at that time it was *Child of Our Time*. My father eventually published his poems, on which he had worked for several years and which consisted of a strange amalgam of Christian and Eastern philosophies. He thought that he would make a triumphant entry on to the English literary stage, but his poems were far ahead of his time: very few people understood them, let alone enjoyed them, for they were so full of allusions to ideas and writings of which most people had never heard.

We had begun to feel that Sorrel ought to have a companion but also that there were already too many people in the world, so would it not be sensible to foster a baby? We located a baby in a home. Her name was Barbara. She seemed very ill, coughing all the time, and was thin and pale, not likely to last long unless somebody looked after her. We were told that her mother was not likely to want her back so we agreed to foster Barbara. We tried to treat her as one of our own, giving her the love and attention that she obviously lacked. It meant getting up at night to feed her, change her or just comfort her. We really wanted to treat Barbara as if she were our own child.

Unfortunately, just about this time my father had another heart attack. We simply had to go to Tidebrooke. We called Michael Tippett on the phone and, although he was somewhat horrified at our coming with four children, he said it would be all right. We had tried to explain to him that, even though Barbara were not ours, if we were really to treat her as a member of the family, we would have to bring her too! So we came down, took the children to Tidebrooke and went to see my father in hospital in Tunbridge Wells.

He was in a bad way. He could not talk very well, or move his hands too easily. He wrote us a cheque so we could rent some rooms in Tunbridge Wells and be near him, though he could hardly sign his name! We talked about many things, in particular about politics but especially about the communists. He had changed his whole life because he had given support to the communist movement and he was quite sad to hear from me that I was quite disillusioned by the party.

He said, somewhat sadly, 'You have got as far as that?'

We also discussed religious issues and I told him that I had joined the Quakers.

'They are good people,' was all he could manage to say.

'We have to take the children and Barbara back, then we shall come back and stay around here,' I said. 'So we shall see you on Tuesday!'

'On Tuesday?' he said slowly. 'It's such a long way away!'

And that is the last we saw of him.

A few days later I woke up to a frightened scream. It was Nigel. A minute later the phone rang and we were told that my father had died.

He was buried in the churchyard at Tidebrooke. Gedeon sent a message, which I read at the graveside. Tessa was in a very bad way and kept shouting and sobbing, 'You can't bury Paul! No, you can't.'

She had to be restrained while the coffin was lowered into the ground.

We went back to Leicester, only to hear the news that Barbara's mother would probably want her back. She did not like it that Barbara cried when she picked her up and smiled and cooed when she was with us. But it was a healthy happy Barbara whom we eventually restored to her mother, who had decided to get married, so we all agreed that the place for Barbara was with her mother. We had become very fond of Barbara and it was quite a wrench for us to give her up, but we both thought it was the right thing to do.

In the meantime, without a Barbara to act as counterweight, we thought that Sorrel was getting really spoilt by the other children and should have a little brother or sister! So Bruce was born into this world three years after Sorrel. Now we had a family of five children, halfway to the ten children mark set as a goal before we were married! But Bruce's birth was not a very easy one from the point of view of danger to the mother, as Tessa had high blood pressure and the birth had to be induced. She was advised not to have any more children, since her doctor thought five children already made quite a family! Tessa had done many prenatal exercises in order to have a 'painless childbirth'; in fact, the nurses in the hospital had not believed that the baby was coming since she was not screaming away as a woman in labour should! So there was nothing for it but to scream! She screamed and they wheeled her into the delivery room, just in time to have the baby. She had to do a bit more artificial screaming, when she noticed that the umbilical cord was round Bruce's neck and the nurses were not doing anything about it!

Anyway, Bruce was duly born and brought home to our Queens Road house. He was born on Tessa's birthday, on 18th November, as a special birthday present!

Coming back to my own work, I had some indications that there was more constructive thinking than analytical thinking in children, particularly in girls. It seemed that a ten year old's ability to solve progressively more complex mathematical problems was much higher than generally believed at that time, so I thought I had better do some fieldwork in schools, to see for myself how things stood there. I carried out some close observations in the county schools and realised that there were certainly two areas in which children appeared to be stuck. Could they not be unstuck by providing more constructive situations from which to learn the appropriate mathematics? One such area of difficulty was concerned with the learning and the application of the place value convention; another was to do with the use of brackets – in other words, with the use of the distributive property of multiplication with respect to addition. So I devised some materials which I thought would allow a constructive rather than analytical way of getting over these hurdles. The place value convention, I soon realised, was based on the use of powers, be it only the power of ten. I hypothesised that in order to learn a mathematical concept involving variables and constants, it stood to reason that one would have to vary the variables and keep the constants constant. The mathematical concept of 'power', involved in the use of the place value convention, had two variables: the base number and the exponent.

I argued that they should both be varied in order for the concept of power to be fully grasped. So I invented the material that came to be known as the Multi-Base Blocks, which consisted of pieces of wood, which were the physical realisations of the powers of two, of three, of four, of five and so on; also of ten and of twelve!

For the brackets problem, I invented a very simple type of educational material, which included some strips and squares, peg boards and balance beams, giving different 'embodiments' of the algebraic identities, such as:

$$a(c+b) = ac+ab$$
$$\text{and } (a+b)(a+b) = a^2 +2ab+b^2$$
$$(x+2)(x-3) = x^2 -x-6$$

which came to be called the Algebra Experience Material.

I went to see the county director of education, a certain Mr Mason, explained my problem and asked his permission to use some of his schools to experiment with these ideas.

'Sure, I don't mind what you do, as long as the head agrees with you,' said Mr Mason.

I consulted a headmaster and had a similar reply: 'It sounds fine to me, as long as the teacher agrees with it!'

When I asked one of the teachers, she replied, 'It's fine with me, as long as the children like it!'

So this was democracy!

I started in one class with the material, where I tried to teach them place value and algebra, using another class as a control. The children in the control class got wind of what was going on in the experimental class, in which the boys and girls were having a good time and they wanted to do 'the stuff with the blocks' too!

I could hardly keep up with manufacturing the material as the work spread to several classes and then to several schools. I soon gave up the idea of a control group and reported to Mr Mason that there was a great deal of new and enthusiastic learning of mathematics going on and more and more teachers were interested in doing the same. He asked me to organise some teacher education courses, which were attended by dozens of curious, elementary school and secondary school teachers.

One secondary school teacher came up to me after a session on roots, powers and logarithms and said, 'You know, I have been teaching logarithms for years, but tonight was the first time I really understood what a logarithm is!'

The National Foundation for Educational Research (NFER) heard about my work and invited me to come and discuss possibilities in London. The result was that I signed a contract with them to manufacture the multi-base blocks and the algebra experience materials and they asked me to write a paper to give teachers a theoretical as well as practical foundation for the use of these materials, later to be known as 'manipulatives'.

Concurrently with all these activities, I ran an interdisciplinary seminar, in which the problems of learning were discussed by educators, psychologists, philosophers, mathematicians and even the electronic wizard, Gordon Pask. These were very fruitful discussions, bringing together very varied points of view on what learning really was. One of the members of the seminar was Malcolm Jeeves, a psychologist from Leeds University, who eventually offered me a job in the Leeds University Psychology Department. But later he accepted the chair of psychology at Adelaide University in Australia and asked me to accept a post as associate professor (reader) in the department he was to run. I was a little hesitant about this but in the end I accepted.

The NFER paper did not seem to me to give enough guidance to teachers about the new principles of mathematics learning I was developing, so I thought that I would write a book. In this book I tried to describe the abysmal state of affairs which existed in the field of mathematics learning and enunciated some new principles which I thought would help turn things around. Then I wrote some chapters on how these principles could be realised in practice, in particular in the teaching of arithmetic and algebra. I sent the manuscript to Herbert Read, author of the wonderful book *Education through Art*, whom I thought would be a kindred spirit and would regard favourably my attempts to turn mathematics into an art form! He wrote an enthusiastic preface to the book, which ensured its publication.

The news of my activities eventually reached Harvard. Jerome Bruner, author of *A Study of Thinking*, phoned and asked me if I would like to come to his Center for Cognitive Studies for a year, to investigate these strange new phenomena I had been writing about in mathematics learning. He had read my *Building Up Mathematics* and thought there was a lot of scope for a psychological examination of what went on when children learned mathematics in this 'new' way.

At the same time I was asked to collaborate with Dr Margaret Lowenfeld, of Poleidoblocks fame, in a project she was planning, financed by the British Association for the Advancement of Science. Her methods of clinical psychology involving the resolution of learning blocks in children and the principles that I had enunciated in my recent book, appeared to have something in common and she was keen to investigate.

So I had a problem. Should I continue to work at Leicester? My salary was satisfactory and I was also paid a retaining fee by the County of Leicester for the work I was doing in their schools; my mathematical lecturing load had been greatly reduced by Professor Goodstein, to allow me to have more time for my expanding educational and psychological work. This was very far-sighted of him since I do not believe many heads of departments would have acted likewise. I made discreet enquiries about the possibilities of a readership (associate professorship). It came back to me through the grapevine that many in the senate thought I was a 'dark horse' and it was not possible to predict how I would end up! This gave some negative points to the Leicester option.

Then I had to dovetail the work with the British Association, the work at Harvard and the work in Adelaide if I were not to stay in Leicester.

At this point I was invited to give a talk at Halmstadt in Sweden. I thought this could be combined with an interesting European tour. At that time we had a dormobile, which I filled with four children and their headmaster, my daughter Corin and one of my collaborators from the NFER, John Biggs. We drove through Holland, Germany and Denmark and took the ferry into Sweden and drove to Halmstadt.

Fortunately, all educated Swedes can speak English, so the talk could be given in English. The talk began with a demonstration by the four children, all about ten years old, during which they not only carried out complicated calculations in different bases using my materials but were able to factorise some quadratic and even some cubic polynomials. The Swedish teachers thought I had drilled them in particular examples so I asked them to give the children any such problems themselves. They did; And the children acquitted themselves marvellously, solving all the problems by manipulating the materials. There was great consternation. The Swedes did not want to believe what they saw. In fact, I was never again invited to Sweden! I must have been presenting very dangerous stuff, which would undermine their whole system!

My passengers took the boat back to England and I made for Krakow in Poland, where an international meeting on mathematics learning was being organised. Having landed in East Germany from a ferry from Denmark, the dormobile showed signs of breaking down, so I pulled up at a garage and asked if a mechanic could have a look at the car.

The mechanic, who was also the boss, asked me if I would like to wait in their house while he did the work. So I met his wife and children and we chatted.

The mechanic's wife made some tea, which she served in a very beautiful china set, with slick black figures over a white background. When I admired the teaset, she said, *'Das können Sie haben!'* ('You can have it!')

When I refused, she tried to persuade me by saying, *'Für die Mitwirkung von Osten und Westen, können Sie diese Sachen haben!'* ('For East-West collaboration, you can have these things!')

I could not refuse this and she packed up the china in tissue paper and handed it to me. The car was ready so I paid the mechanic and we said a somewhat tearful goodbye. I hugged the children, then got in my dormobile and drove away into the night. We now have only one of these pieces of china left intact, which we guard with great care!

During the stretch of my journey through Poland, the dormobile was never empty. I transported whole families from A to B and at one point we even shared our food, which I cooked on the dormobile stove! A student came with me nearly all the way to Krakow. He had hailed me with a beautifully carved 'hitch-hiking stick', with a little flag attached to it with the sign STOP sewn on very carefully! When he was leaving, he asked me if he could sign my hitch-hikers' book. When he realised that I did not know what this was, he explained that, as people gave lifts to others, the people receiving the lifts entered the kilometrage into the book and signed it. At the end of the year, the one who had given the longest total of kilometres got a new car as a present from the government! Since he could not sign my non-existent book, the student insisted that I took his stick. I tried to refuse but he insisted. I still treasure this stick and, now that I am much older, I use it not for hitch- hiking but to help me walk up steep hills!

The meeting at Krakow was very interesting. All the active workers in the field of mathematics learning were there. There was my old rival from Chelmsford, Caleb Gattegno, Krygowska, Servais, Papy, Pescarini and many others. At one point we had a discussion about how much language had to do with mathematics learning. I was insisting that one could learn mathematics without the use of language. My colleagues did not agree.

I said, 'Let's get hold of some children and I'll teach them, using my material. The only word I know in Polish is 'tak', meaning 'yes'. That should be enough!'

My colleagues were incredulous, but we went out into the street and collected some children, Krygowska explaining to them that we would play some games with them. They came up and I gave them the multi-base blocks to play with.

At first the children had a ball building castles, but later I tried to set them some problems. They had to exchange pieces of wood for other pieces of wood, so that the total volume would remain the same. This would be used for 'carrying' in bases three and four. We ended up with the children carrying out adding and subtracting exercises using piles of wood even writing down what they were doing. The only word I ever used was 'tak'! I did not know the word for 'No'.

We socialised a great deal during the meetings and got to know each other well. The language of the meeting was French, as that was the language we had in common. I became particularly friendly with Pescarini and we decided that 'one day' we would do wonderful things in Italy in mathematics education!

I visited a Polish countess, who was now an ex-countess, for her castle had been taken away from her. She had been given just one room to live in, overlooking the castle that had once been hers! She eked out an existence, making little figures out of wood and selling them. We had some good meetings. She spoke excellent French, and, as I was taking my final leave, she offered me some of her beautifully carved birds. I was most touched! These birds still adorn the window sill of our kitchen in our Devon house, so when we cook we can think of my generous countess friend, who gave me some of her most beautiful work.

I was due to drive to Vienna after the meetings in Krakow. At the Czechoslovakian border, the road appeared to be closed and there seemed to be nobody about. I knocked on the door of a nearby house and found the border guards were in there, drinking what appeared to me to be some red wine. They were happy to have a customer, stamped my passport meticulously and offered me some wine. It was hard to get away, as the border guards kept pouring more wine into my glass. In the end I excused myself, suggesting that I had a long way to go.

Approaching the Austrian border, I thought that I had better cook myself a meal, so I stopped just outside the border village of Mikulov, put up the dormobile roof and started cooking my simple meal. But I did not reckon with the local children! A number of these came up and wanted to see what I was doing. I asked them in and, when they saw I was cooking, they went to get some food! So we pooled our food and cooked a big feast for I do not know how many children! They all gave me their names and addresses and wanted me to write to them. After we had eaten our meal, we said goodbye to each other and I drove over the border, bound for Vienna and eventually for England.

I told the headmaster of the four children who had come to Halmstadt with me about the Mikulov children and he seemed very enthusiastic about children in his school writing to Mikulov.

In fact, he got in touch with the Ministry of Education in Prague and some exchanges were arranged between the children of Mikulov and those of Husbands Bosworth, where the English school was. Perhaps East-West détente was slowly beginning!

I gave notice to Leicester University. It was the end of an era. I was to spend an academic year at Harvard, the following summer and autumn was to be devoted to work with Margaret Lowenfeld and for the Christmas after that we should be in Adelaide, to start my work in a psychology department for the first time instead of in a mathematics department.

Ten milestones had been achieved and Leicester was to become history.

The account of our Leicester years would not be complete without telling about some of the things that happened to us during our summer vacations. I will tell about these adventures in the next chapter.

Chapter Fourteen
Other Happenings During the Leicester Years

Professor Geymonat, then of Turin University, usually spent his summer vacations in a mountain village called Valtournanche, about halfway between Aosta and the Matterhorn. He would rent an apartment in one of the apartment blocks in the village and move into it for the summer with all his children. For the summer of 1951 he suggested that we did the same so the children could become friends and the adults could have interesting things to talk about! We accepted the idea and reserved the apartment next to theirs. However, there was a problem: without a car, it would be very lonely not being able to go anywhere except the village square! So I had to learn to drive, but, before learning to drive, I had to have a car to drive! New cars were almost impossible to buy and there were waiting lists of three to five years for them! So we had to look for and buy a second-hand car. Even this presented some difficulty, as, in the absence of new cars, even second-hand cars tended to be sold at rather high prices.

In the end we settled on a fish van. We cut windows in the van and covered them with transparent plastic. We painted it all red and called it the Red Devil.

I took a number of driving lessons from a driving school, crashing the gears many times and nearly knocking into other cars much too frequently for my instructor's liking. Before I could take the test, I had to have a great deal of practice. But somebody who already had a licence would have to sit with me while I was learning the ins and outs of weaving about in the traffic in Leicester.

At this point I was giving language lessons to all and any who wanted to learn French, Italian or German and eventually even Spanish (though I had strictly no knowledge of Spanish, I kept just a couple of lessons ahead of my student in the Berlitz book). My dentist wanted to learn French. So, often we had our French lesson concurrently with a driving lesson. All the happenings during a drive gave plenty of scope for French conversation, so we killed two birds with one stone on these occasions. During one lesson, he noticed that I had toothache. He told me to drive to his office. He took me upstairs, sat me in his chair and soon fixed up my tooth. I gave him an extra hour of French in return! Once I even crashed through a gate into a field, as I had mistaken the accelerator for the brake!

It is not surprising that I failed my first test. The children were shattered. Their wonderful father who 'once knew everything and could do everything' had actually failed an exam! They could not get over the shock. Fortunately, a month later I took the test again and passed, just in the nick of time, as it was only a week or so before we had to leave for Valtournanche! There were no autoroutes at this time so it took us quite a long time to drive through France. We entered Switzerland through the Jura mountains and, skirting Lac Léman on the north side, we started to penetrate the Alps. This was a wonderful experience for the children. Nigel, of course, remembered the mountains from his memorable trip with me when he was only seven years old. The others soaked it up with childish wonderment.

At that time there was no road tunnel under the Great Saint Bernard Pass, so there was nothing for it but to go over the top. Tessa got more and more nervous at every hairpin bend and wondered vociferously why I had brought her to such a godforsaken place! However, we managed to make it to the top. It was already getting dark; it was cold and there was thick snow lying everywhere, the road lined by huge drifts. The children were tired so we thought we had better stop for the night.

We were not flush with money, so we thought we would ask the Great Saint Bernard brothers at the Monastery to give us shelter for the night. Amazingly, they were very ready to do just that. We were ushered into a room, full of palliasses set out in a neat row and we were told we could sleep there. Having given the children a little snack, we all dossed down on the palliasses and slept like angels till the morning.

The problem in the morning was that the car would not start. But we put the children and our bags in the car and pushed the car to the frontier post, where they checked our passports and car documents and told us to go. It is a twenty or thirty kilometre downhill drive from the pass to Aosta and we freewheeled down all the way, hoping the brakes would hold! They did – or I would not be here writing these lines! At the first garage we drove in and asked what was wrong with the car. The mechanic mumbled something about the timing, moved a few screws and the car was fine.

We finally arrived at Valtournanche. Although it was a fine day, it was cold. The Geymonats were already there and they welcomed us warmly and showed us to our apartment. It was extremely basic. In the bedrooms there were beds and nothing else. In the dining room there was a table and chairs, nothing else. In the kitchen there was an electric stove. We were warned not to put a saucepan on the hot plate without holding the saucepan with a dry rag, otherwise we would get a shock, which we did on several occasions when we occasionally forgot to take the appropriate precaution! The bathroom consisted of a stove, on top of which was a large cylindrical water tank, with a tap leading into the bathtub. The fuel to be used could be obtained by foraging in the nearby forest for dry branches and pine cones, which became the daily duty of the children to collect, or they could not have a bath!

We stayed in the apartment for several weeks, but Tessa became more and more miserable. Conditions were far from ideal and the weather was cold so Tessa would say, 'This is not my idea of a holiday in Italy! I thought it was going to be hot and sunny! That we could sunbathe and sit outside in pavement cafes and sip wine and feel good!'

'Do you want to pack up and go to the sea?' I asked.

'Oh yes, that would be wonderful! Let's leave this dreadful place and have a real holiday!' she replied.

Although the children had made friends with the Geymonat children, who had actually admitted them to their secret club, called *La società dei monelli audaci*, devised for performing daring feats, we told the Geymonats that we would do a little tour with the children and said goodbye.

At first we camped by the Italian lakes, which was definitely one up on Valtournanche, for the water was always warm enough for bathing and there were lots of people about. We could buy all the food we needed and there was lots of wine to be had cheaply! I recall camping near Porlezza, next to a little stream, which was used by the locals for doing the washing. Sorrel was only a year old, so there was plenty of washing to do for her, as well as for the rest of the gang. Tessa bought a washboard in the local 'Ferramenta' store and joined the women at the stream, rubbing her washing against the washboard, just as she had seen the Italian peasants do so effectively!

In the evenings, when we cooked on our primus stove, some of the local children came round and, as they looked hungry, we offered to share our meal. We finished up feeding half the village children for supper! We did not mind since food was not very expensive to buy and these children really seemed to need more food than they got, a lot of them looked very thin! They thought potatoes were a great treat! We thought that such getting together was a good education for our own children, showing that we must share things.

We eventually ended up at Ventimiglia, on the French border, where there was a good camping ground directly on the seashore. We swam in the sea and generally enjoyed the Mediterranean marine landscape of pine trees, olive trees and vineyards, all growing in close proximity to the shore.

We soon found ourselves back in Valtournanche and realised that it would soon be time to start wandering across Europe on the way home to England. We packed up and started down for Aosta. We stopped at the Roman amphi-theatre, where the children had their last game of Lions and Christians. There we met an Australian tourist who wanted a lift back to England. Rightly or wrongly, we agreed to take him and he piled into the back of the Red Devil with the children.

Sorrel was really a hit with the Italians. Tessa dressed her, Italian-style, in gorgeous frilly little dresses and as she gazed out on to the road from our van, the onlookers would invariably exclaim, '*Che piccolina! Che carina!*' ('How tiny! How pretty!') and it was difficult to drive away because they all wanted to talk to her and admire her.

But now we were leaving Italy and we had the whole of France to cross before we were on home ground again.

One evening we were driving along rather late, when it was dark and, being an inexpert driver, I was not used to the flashing headlights at night, especially since our beam, being adjusted to English left-hand driving, shone right into the eyes of oncoming drivers! At one point I veered too far to the right so that two of the wheels were rolling in a field while the other two remained on the road! I managed to get back on to the road but the car soon came to a stop, fortunately just near a farmhouse.

We asked the farmer for help and he came out in an ancient Renault, fit for the London to Brighton vintage car race and towed us into his farmyard. We then asked him if there were anywhere we could stay. He showed us to a barn, which we all thought was perfect for a tired family.

Fortunately for us, the farmer was not only a farmer but quite a mechanic as well. The next day he took our car to bits and told us that one half-shaft ('*demi-arbre*') was broken. But he took the good half-shaft out as well and took it to a factory where one of his friends worked. He turned out an identical half-shaft, which the farmer then put in so we could be on our way! No normal garage would have done that job, I am sure. They would have needed to 'order the part' and we might well have been stuck for weeks!

We spent several days at the farm, waiting for all this to be done. We were warned about the farmer's dog, which was a very fierce one. '*Il saute au cou!*' he warned us ('He gets you by the neck!') so the children left the dog severely alone!

As a send off we were treated to an extraordinary spectacle. From our barn, where we were putting the finishing touches to our luggage before leaving, we could hear some very noisy shouting. We looked across the farmyard and saw the farmer being ejected from the house, for a crime we knew not what, with practically no clothes on! In a few seconds there followed his underpants, then his shirt and finally his trousers, ejected through the door, which was then firmly shut! We said an embarrassed goodbye to the half-dressed farmer, piled into the car and drove off.

After some further vicissitudes, such as flat tyres, eventually we made it back to Leicester, to resume normal life once more. On looking back, the children came to refer to this particular holiday as 'Mad Holiday Number One'.

Another memorable holiday was the one of the summer of 1953. On this trip we took Roy with us and made for the Yugoslav resort town of Dubrovnik. The first port of call was Abbazia, now called Opatija, since it no longer belonged to Italy but to Yugoslavia. We found a modest little hotel and stayed there for several weeks. I nostalgically revisited the places where Gedeon and I had used to play, as well as the little island where I had learned to sing *Santa Lucia* from my eight year old girlfriend. We found a sandy beach where the children liked to play and most days we spent in truly holiday-like lazy fashion on the beach.

But Tessa was not really a 'beach person' – she preferred to be active. So one day, while I sat with the children on the beach, she asked Roy if he would come with her to do a spot of shopping.

Very close to our little group there were two ladies, speaking to each other in Hungarian. This was not so surprising, since in the north of Yugoslavia there was a fair-sized Hungarian minority left there as a result of the Allies redrawing the map of Europe after the First World War. They started making remarks about us, such as, 'English husbands are really exploited by their wives. Look at that poor wretch, being left with four children, while his wife is off gallivanting with that young man!'

'And he is not even that good at looking after the children,' said the other one. 'Look at the children's hair! Look what a mess they are in!'

At this point I took a brush and comb out of our beach bag and proceeded to brush and comb all the children.

'Look at that little girl's face! It's all covered in chocolate ice cream!' said one of them.

At this point I took out my handkerchief and wiped all the chocolate off Corin's face.

This went on for some time, until the two ladies decided to leave. They never caught on that I might have understood what they were saying! Of course, a group of English-speaking children and parents, sitting on an Adriatic beach, could not possibly understand Hungarian. The idea was so absurd that, in spite of all the circumstantial evidence, they never guessed that I knew what they were saying and I never let on!

There was a one-way system in Abbazia and Volosca of which I was not aware. There was a small road down to our beach from the road which was one-way but, as it was such a small road, there was no sign to indicate that one could not take a left turn when re-entering the main road. One day, as I took this left turn, a policeman waved me down and told me that I would need to pay a fine for coming up the one way street the wrong way.

'But there is no sign, coming up that little road,' I said, pointing to the cul-de-sac leading to our beach, 'so it is not fair to fine me.'

'You must pay the fine at once,' insisted the policeman.

By this time there was quite a crowd gathered round the car. I asked one of the bystanders what would happen if I did not pay.

'He will make you go to the police station and make you pay much more,' replied one person.

The argument went on a bit longer. The policeman was demanding one hundred dinars.

Then I thought I would change tactics so I pulled a five hundred dinar note from my wallet and said to the policeman, 'As a token of East-West co-operation, I offer five hundred dinars.'

The policeman was flabbergasted. He fumbled around for the change, but I repeated my assertion and then, amongst a great deal of clapping and 'Bravo-ing', I drove off.

After Opatija, we spent some time in Crikvenica, in a very modest hotel. There was a good beach there and the children enjoyed themselves. The hotel's clientele mostly consisted of members of the Yugoslav intelligentsia. We got to know some painters, doctors and an orchestra conductor. I asked them why they were not able to go to the five star hotel up the road.

'Oh,' they said, 'that is reserved for the workers and peasants!'

I could not resist having a look at this upside-down situation, so I walked into the five star hotel and the first thing I saw was a burly peasant in huge gumboots sprawling across a divan in the foyer, spitting frequently on the carpeted floor! I did not need to see any more! Our artist friends from the little hotel had obviously reported the facts accurately.

We thought that we would leave the car at Crikvenica and do the rest of the journey by boat. We stopped for a few days on a lovely little island called Rab and then finally took the plunge to sail all the way down the Dalmatian coast to Dubrovnik (the old Ragusa). The boat was packed, probably over its permitted capacity; it certainly seemed to sit very low in the water! You could never get to the restaurant to eat, as the groups of workers and peasants were always having meals, this being their official trade union vacation! I managed to get off the boat at Split (the old Spalato) to buy some provisions, otherwise the children would have starved.

Dubrovnik is (or was) a wonderful place. It is a walled city and only pedestrians are allowed within the walls. The wall is so thick that two or three people can walk abreast on it; in fact, you can walk all the way round the walled part of the city. We rented a couple of rooms in a house near a tram stop. This tram would take us to the gates of the city, after which we had to walk like the rest. This city predates the Venetian Republic and its beautiful ancient buildings are a sight for sore eyes! We often walked round the narrow streets of the old city with the children and sat down frequently in pavement cafes for ice creams (*sladoled* in Croat).

A few minutes' walk from our lodgings was a very scenic little cove, surrounded by cliffs, with many little rocky islands dotted about. This became the children's favourite place. Nigel actually learned to swim there, motivated by his desire to reach one of the little rocky islands. We spent most of the time exploring around our little cove, except when we went to see the sights in the walled city.

Unfortunately, Sorrel developed a fever in Dubrovnik. We called a doctor, who diagnosed malaria and prescribed the usual quinine medication.

When he had finished, I asked him in Italian, '*Quanto Le devo?*' ('What do I owe you?')

He looked around our little family of four children and at Tessa, who was obviously pregnant, expecting our fifth child, and said, '*Tutti questi bambini, sono tutti suoi?*' ('All these children, are they all yours?')

When we answered in the affirmative, he said, '*Non c'è niente da pagare.*' ('There's nothing to pay.')

In a few days, Sorrel was over the fever and was soon back enjoying the water. In Yugoslavia, children up to about the age of seven all seemed to bathe nude, so of course Sorrel did likewise, revelling in the warm Mediterranean water running over her body as she splashed happily in the shallow waters of the cove.

On the way back we spent a little time on the west coast of Istria, where we found a very agreeable campsite, in a beautiful pine forest along the seashore. Then we moved on to Venice, where we stayed on the island of Giudecca, where the youth hostel was situated.

To get to the Lido, we had to take the *vaporetto*. This was always very crowded and Tessa was never very good in crowds, getting claustrophobia easily. At one point, while we were trying to get off the *vaporetto*, she felt hemmed in and started to scream blue murder. Somebody said the magic word, *'Incinta!'* ('Pregnant') and instantly the crowd parted and Tessa walked off the *vaporetto*, gliding through the crowd like a queen.

When we got to the beach on the island of Lido, Sorrel proceeded to strip.

'Oh no,' we said. 'You must have these bathing trunks on!'

'But they will get wet in the water!' retorted Sorrel, very logically.

She could not understand that, just because we had crossed a red line on the map, everything had to be suddenly different!

On the way back we camped at Latte, a little village near the French-Italian border, west of Ventimiglia. There was a little cove nearby where we could all bathe, not quite as scenic as our Dubrovnik one, but it served us well for a few days. One night we heard a lot of noise down at the beach, followed by people climbing up and down the cliff path, one person even walking round our tent and tripping over the guy lines! Tessa seemed very frightened. I tried to calm her and offered to go out and look.

'Oh no! You mustn't!' said Tessa. 'These people might be dangerous!'

Since she was expecting a child, I agreed to lie low. The noise subsided for a while, until at dawn we heard the sound of cars stopping at the little store on the road and then driving away. This happened at least a couple of dozen times and then there was quiet. We eventually worked out what had happened.

There was a boat landing some contraband cigarettes in the cove, which were brought up to the store and then bought by customers who had clearly been told to come early to avoid detection. No wonder the smugglers were nervous about our tent! We might have been the police in disguise!

I went in to the store and asked the owner, '*Avete ancora sigarette?*' ('Do you have any cigarettes left?')

'*Tutte via, non ce ne sono più,*' was the curt reply. ('All gone, there aren't any more!')

So we had been correct in our reconstruction of the 'crime'. We thought it was the better part of valour to pack up and leave, which we did. We had no intention of having a more serious brush with the smugglers!

We had many more memorable holidays, some more adventurous than others. On one of our extended Easter holidays we took a bunch of children, all aged about ten or eleven, to Florence. I had already made contact with the Psychology Department of Florence University, so they arranged for the children to attend the Scuola Svizzera and to show the Italian children and teachers how our children were learning mathematics. It was a fine warm April and lessons were held out in the open in the school garden. Our children used the multi-base blocks and the algebra materials and tried to show the Italian children the ins and outs of the use of these materials. They were surprisingly successful at this, in spite of the language 'barrier'. Apparently there is no such thing as a 'mathematics barrier', since concepts are international and the material speaks for itself!

On the way back to England we went to visit the community of Agape (the ancient Greek word for 'Brotherly Love') in a Waldensian area of the Italian Alps. The Waldensian community is hundreds of years old, predating the reformation by several centuries. Needless to say, they suffered persecution by the official Catholic Church. This particular community was set up after the Second World War, when there were many abandoned children, some of them resorting to prostitution, whom these people gathered together and tried to help by giving them training for work as well as moral training for living. We were warmly welcomed by the community and it was explained to the children (in somewhat broken English) how the community operated and tried to help people.

It was still winter in this part of Italy and thick snow was still lying on the ground; we had to leave our van in the village and walk up to the centre as the road had not been cleared. During the night we could hear the roaring sound of avalanches and sometimes we had to go and reassure the children that there was no danger. Actually, on the day we were to leave, the road out of the village was cut off by a large avalanche. So we had to wait. We posted a lookout at the window, to give the signal when a snowplough appeared. As soon as this happened, we would all troop down to the village, get into our van and follow the snowplough back on the road out of the village.

We did not have long to wait. The convoy was arranged, the snowplough in the lead. We did not have far to go before we heard again the menacing thundering sound of an approaching avalanche. The snowplough stopped and so did the rest of the convoy. The driver of the snowplough asked us to wait while he drove on to investigate. The road was cut off again, huge amounts of snow having hurtled down the mountain. With a plough working from both ends, the blockage was cleared in a couple of hours and then we were off. As we descended into the lower part of the valley, we left the winter behind and everyone gave a sigh of relief.

The next stop was Geneva. I had made contact with a Monsieur Roller, who was a member of Jean Piaget's team at the Institut Rousseau, where all the epoch making research by Jean Piaget and his co-workers was being carried out. It was a lovely sunny day, so we decided to hold a demonstration by the children in one of the beautiful parks of that city. The children did very well and the members of Piaget's team who were present were highly impressed. This paved the way for a period of work at the Rousseau Institute for which I would be engaged some years later.

We had one very pleasant summer in Austria by the Ossiachersee, near Villach. Again we did not take just our own children but some of their friends as well. We arranged to stay in the local youth hostel and invited my mother to come and stay as well. We did a lot of hiking with the children, who were trying to pick up some German by playing with the local children.

On one of our hikes, Nigel noticed a beer advertisement pasted over an alpine chalet which said EXPORT HELL, and said to me, quite seriously, These Austrians are really clever – they even export to Hell!'

Of course the German word *'hell'* does not mean the same as the English word 'hell'. The German *'hell'* means 'clear'! But Nigel did not know this.

I spent quite a lot of time over that summer translating Rózsa Péter's book on Infinity from Hungarian into English. I knew I was not a good translator, so when talking to Dr Péter about the book, I suggested that what I could do was read each paragraph and then render the meaning of the whole paragraph as it would seem to an English-speaking person. I could never do a word-for-word or even a sentence-for-sentence translation. This was agreed and thus this beautiful book became available to English speaking readers.

Unfortunately, some of the children contracted chickenpox! This meant that I had to bring the healthy children back to England in the car, while Tessa stayed with the ones with chickenpox! They moved into a little guest house and we all said goodbye. I drove back through Switzerland and France, while Tessa waited for the quarantine to pass! Fortunately, we had taken out insurance for just such an eventuality, so the catastrophe did not break the bank! In fact Tessa and the chickenpox children came back in style on the Orient Express! She told me afterwards that this Orient Express lark was not all it was cracked up to be. The train seemed to be full of lascivious Turks, who had obviously boarded the train at Istanbul and were on the constant lookout for sensual gratification! So she had to keep both herself and the children under a very strict regime!

We once hopped over to Florence with another group of children, staying in Youth Hostels on the way, but boarding them in Florence with some Italian families whose children went to the Scuola Svizzera. It happened that the parents who sent their children to that school encompassed just about the whole spectrum of social classes. So the children had much to learn and tell each other about their experiences. There were some who stayed with a count and his family. They had thought they would have it made. Not so! They were roped in to do the household chores and were told to be extremely polite and considerate to the servants. Some who went to stay in working-class homes had the opposite experience: everything was done for them and they just sat around like princes and princesses, being waited upon by their hosts and their children!

We left Florence for Carrara (where all the marble comes from) and the Italian children were sorry to let us go. There was a pair of eleven year old twin girls who got so attached to us that they begged their parents to be allowed to come with us to Carrara. The parents finally relented and the girls spent the week with us at Carrara beach, with the result that they became avid correspondents of some of the children we had taken with us. Something working towards international understanding? A guard against so-called culture shocks? Maybe. Anyway, a good time was had by all.

At one point I was asked to be examiner for the Italian oral examination in one of the county's secondary schools. I accepted and suggested that some of those learning Italian could profit from a visit to Italy, which I could arrange. This was duly set up and I filled our Bedford van with about ten teenagers, bound for Florence and Rome. We stayed in youth hostels so the cost per person was very small and nearly everyone taking the Italian exam was able to come.

Two dramatic things happened on that trip. The youngest girl, who was only fourteen, fell in love with one of the boys! Since we stayed in youth hostels, where there are boys' dormitories and girls' dormitories, we did not have to fear too much that a 'fate worse than death' might befall the girl. They were two young people in the throes of early romantic awakening and we felt sympathetic towards them and so did all the other boys and girls. The amount of Italian these two practised was probably somewhat reduced by their passion for each other, but then that is life! The other dramatic thing was the climbing of the Gran Sasso d'Italia. After our visit to Rome, we decided to split the group, Tessa would take those who did not want to do the climb to Florence, to stay in the youth hostel and I would take the 'climbers'.

It so happened that there were three boys who volunteered for the climb. We took the cable car to a hotel about halfway up the mountain (what cheats we were!), where Mussolini had been confined during the last war until rescued by the Germans! This fact led to much conversation with the owner of the hotel about those eventful days and the boys learned some Italian and some modern history at the same time.

In the morning we set out to climb our peak. It was quite a long ascent. We followed a narrow ridge for a time, then had to cross a small glacier (I believe the only one in the Appenines!), after which there was quite a bit of scrambling to do before we triumphantly reached the highest point on the Italian peninsula, not counting any Alpine ones. We lingered for a while, taking in the three hundred and sixty degree horizon, revelling in the fact that every bit of it was below us! But a storm was brewing up, so I suggested we start our descent or we might be caught. I recalled my childhood experience of the descent from the Monte Maggiore with my father and I did not feel like having a repeat performance!

The storm approached inexorably. We reached a small hut on the ridge, probably constructed for the local rangers, on which there were several lightning conductors, with which man had tried to tame the forces of nature, harmlessly earthing the enormous voltages. I looked at these metal rods leading into the earth below and noticed that they were all sparkling!

'Don't touch those rods,' I said to one of the boys, who was just about to examine the rods much too closely for my liking. 'Can't you see those sparks? There might be thousands of volts going down through these rods!'

The boy backed away just in time. We heard an enormous clap of thunder and there was a simultaneous blinding flash of lightning and we were thrown to the ground, though unhurt. The rain was now coming down in buckets and there was a continuous roll of thunder, the countryside being lit up by the rapid blue flashes that appeared all about us.

'Let's get out of here!' I said to the boys.

'Yes, let's!' they all replied.

'But what's that wailing noise?' asked one of them.

'Oh, that's only wolves,' I said. 'They are not in favour of thunder storms either!'

Indeed, the howling of the wolves added another mysterious theme to nature's drama being enacted in the sky. We quickened our steps, as, though I tried to reassure the boys that wolves were really quite harmless, they still thought it would be better to get to the hotel as quickly as possible.

We reached the hotel just before dark, all soaking wet! A shower and a good Italian meal with plenty of pasta was what each of us needed to put us right. After having thus been restored to our previous state of well-being, we all slept soundly into the late morning the next day, when we had to take the cable car down to the foot of the mountain, pick up our car and drive to Florence to join the rest of the party.

The children wanted to visit an Italian school, so we arranged to have a look at one. At that time all the elementary classes were segregated into boys' classes and girls' classes. At the start of the scuola media, when the children reached the great age of about eleven, the boys and girls were put into mixed classes. During the little party which they organised for us, we were able to meet the children from the scuola media. There were lots of goodies, music was put on and we were encouraged to dance!

I asked one of the teachers, 'How is it that you keep the boys and girls separate in the lower grades and not in the scuola media?'

'Well, you know, little boys and girls, you never know what they might do together,' answered the teacher.

'But then what about the scuola media children?' I enquired.

'Oh, they are really grown-up – they do fine together,' was the surprising reply.

We spent an interesting summer in Norway. We drove to Newcastle on Tyne, where we took ship for Bergen. It was a very rough crossing and a long one. All the children were sick and I recall that I was nearly always the only customer in the ship's dining room! We disembarked at Bergen and started driving up the Hardanger Fjord. The children were very happy to be on *terra firma* at last!

I recall saying to Tessa, 'Now we are on equal footing! I don't know a word of Norwegian and nor do you, so on this trip I am not the universal interpreter!'

Unfortunately for my ambitions to be a non-interpreter, I was and still am somewhat quick at picking up languages and it was not long before I could begin to understand some of the things people were saying and begin to be able to form simple sentences through which to express myself. However, Tessa soon discovered a good trick. What she had to do was find a child of about thirteen or fourteen; such children would have had enough instruction in English at school to be able to speak quite passably in this language. Older children or those who had left school were not such a good bet, as they would have forgotten what they had learned! So if she wanted some instructions, she looked for a young teenager!

We had to cross some mountains to get to the cottage we had rented for the summer. This took quite a long time, as it was not possible to make any speed and the speed limit all over Norway at that time was sixty kilometres an hour! We stopped at a lovely hotel overlooking one of the fjords. Fortunately for the children, there was a free supply of unlimited milk and on all the landings there was a bowl full of apples provided for the guests. Apart from drinking lots of milk and devouring a large number of apples, the children ate like horses, since they had not eaten at all during the crossing.

In the morning we had the Norwegian smorgasbord, the real McCoy! Every one could help themselves to whatever they wanted and to however much they wanted! When I went to pay the bill just before leaving, I was given what I considered a very small bill.

'Your children have been so good, we are not charging for them,' said the person at the desk in fluent English. 'We are only charging for the adults.'

I was not about to argue the point, so I paid the bill and we all piled into the car and drove off.

We eventually reached Lillehammer and about twenty kilometres east of that town, now of Olympic fame, we found the little village of Sjusjoen, where a spotlessly clean, sturdily built wooden cottage was waiting for us to move into. The cottage overlooked quite a large lake, surrounded by what seemed to us to be an impenetrable forest. It was a beautiful day and we drank in the fresh mountain air and duly admired the surrounding countryside.

The owner of our cottage had another cottage next door to ours. She had a little son called Hans, who was five years old. Our daughter Sorrel was about that age at the time and we thought how nice it was going to be for Sorrel to have a playmate. We did not have long to wait. Hans invited Sorrel to play in the garden of his house and soon they were fast friends.

After a week or so, we began to wonder how the two children communicated with each other.

I asked Sorrel, 'How do you speak to Hans?'

'In Norwegian,' replied Sorrel.

We knew that children learned languages very quickly but this was really rather incredible.

A little later we were sipping cups of coffee with Hans's mother in their garden, when Tessa said, 'Isn't it good that Hans and Sorrel can communicate after such a short time!'

'Yes, it is,' replied Hans's mother. 'I should not have thought that Hans would learn to speak English so quickly!'

'English?' said Tessa in surprise. 'I thought they spoke in Norwegian!'

'Oh no,' replied Hans's mother 'I am sure they are speaking in English'

Sorrel and Hans were down the garden at the time, playing happily with each other.

'Let's go and listen,' I suggested, to end the argument with the aid of some factual evidence.

We walked down to the children, who were talking to each other in absolute gibberish. Each child thought they were speaking the other's language by making unintelligible sounds, which was what each one heard from the other! We walked back up the garden to our coffee and had a really good laugh. Had they been left to it, the two would probably have developed their own language and had philologists from all over the world come and take tape recordings of their invention. But we were only in Norway for two months and there was not enough time for such a miracle to happen!

Before we left Norway, the local children went back at school. Since I was interested in education, I thought that I would go and have a look at how a Norwegian school functioned. So one morning we drove down to Lillehammer and asked to see the headmaster of one of the Lillehammer elementary schools. He was very pleased to see us and showed us around. I noticed a line of skis along one of the corridors.

The headmaster saw I was looking at the skis, 'Yes,' he said. 'Every child has his or her skis and in the winter we take them out regularly. It is important for everyone here to learn to ski, in case of emergencies.'

He showed us some of the work going in the classes. In some classes the boys and the girls were together, but in some they were separated.

I noticed that the girls were doing carpentry while the boys were doing sewing and knitting.

The headmaster explained, 'You see, the boys will learn carpentry anyway, as in nearly every house there is a workshop and the adult males in each family always do that kind of work. But they would never learn to knit and to sew, which the girls would learn anyway, but the girls would never learn carpentry at home! So the school fills in the gaps!'

For the mid-fifties this was, I thought, a very advanced point of view to adopt!

Then I tried to draw him on the question of discipline. I had noticed that there seemed to be no undue noise coming out of any of the classes and in some of them I saw children working away quietly without any teacher being present.

'What kinds of punishments do you have for misbehaviour?' I asked the headmaster.

'Punishments?' replied the headmaster in a surprised voice. 'These are children, not criminals to be punished!'

'But surely sometimes a child behaves in a way that warrants censure?' I ventured to say.

'Oh yes, of course,' replied the headmaster. 'In such cases we draw the child aside, but never in front of the other children and try to explain to him or her why certain behaviour was not right.'

'But supposing it happens again?' I insisted.

'Then we explain it again,' replied the headmaster logically.

I was still not satisfied. I thought I needed to push this educational idealist, if indeed such he was, further into a corner. I finally said, 'Surely you must have had cases when repeated explanations have not worked. How have you dealt with the situation in such cases?'

The headmaster thought for a minute, 'Oh, yes, I remember a certain boy some years ago and I am ashamed to say that I did administer a very severe punishment.'

'And what was that, pray?' I enquired.

'He was not allowed to come to school for three days!'

At this point I realised that the discussion was at an end. The children obviously enjoyed coming to school so much that not being allowed to come was the ultimate deterrent for misbehaviour!

I am not at all sure if this school was typical of Norwegian schools or if any of them are still like that. But I certainly recall my conversation with that headmaster as one of the highlights of my educational experiences!

During this vacation, I was preparing myself for a psychology degree examination. Tessa was very nervous that I was about to change track and had visions of ending up in the poor house! Having a permanent post as a mathematics lecturer at a university was something she was used to as the normal thing from childhood, since her father had been on the faculty at Birkbeck College. So I really had to hide my psychology textbook, from which I was trying to get relevant information. I often put it under my pillow at night and then, early in the morning, before Tessa woke up, would read some of the chapters on memory or on intelligence. As she began stirring, I would place the book back under the pillow.

We also had some interesting times in the Western Isles with the children. We had one holiday in the Isle of Mull, another one in the Isle of Lewis and yet another on the Isle of Skye. By the time the children had grown into teenagers, they knew a lot about these faraway islands which their father was so keen to explore when he himself had been a child.

For the Lewis holiday we went to Uig and the children went to the local school. We swam in the lochans, which were much warmer than the sea, although the children loved playing on the wide and deserted sandy beaches of Uig Bay.

To get anything that was not on the van, which came weekly from Stornoway, one had to take the morning bus to Stornoway and come back on the evening one. One day Tessa had to go in because she had toothache.

When she finally got as far as the dentist's chair, the dentist said to her, 'Which one do yer want oot?'

'I don't know,' replied Tessa. 'That is what I hoped you would tell me!'

The dentist soon realised that Tessa was not a crofter's wife and started talking to her intelligently! As a matter of fact, one of her teeth did have to come 'oot'!

Another time Tessa had to go in with severe stomach ache. She got on the bus, which was already rather full, though Uig was near the beginning of its run. There was a sheep tied to one of the seats and a coffin occupied most of the middle section of the bus.

'Who is the coffin for?' enquired Tessa.

'It is for Chrissie Mary, the poorrr lass,' replied the bus driver. 'She died yesterday at the hospital.'

We knew that Chrissie Mary had gone in to Stornoway a few days earlier with what they all thought was appendicitis.

They put Tessa in a ward with a very dour Scottish lady. When Tessa started knitting, waiting for the doctor, the dour lady said to her, 'Would you be knitting on the holy Sabbath day?'

Tessa put down her knitting so as not to distress the other patient and picked up a book. The dour lady was horrified and said to Tessa, 'Do you read anything else but the Bible on the Sabbath?'

Tessa put the book down and, as she did not have a Bible handy, she lay there expectantly.

'You know you are lying in poor Chrissie Mary's bed, don't you?' said the dour lady. 'She died in that very bed only yesterday!'

This was not very cheerful news, especially when the next day, which was no longer the Sabbath, Tessa was told by the surgeon that her appendix would have to come 'oot'!

It turned out that Chrissie Mary had not died of appendicitis but of cancer and in due course Tessa had her appendectomy and in a few days' time was sent back to Uig on the bus, this time without any sheep or coffins to obstruct leg-room in the vehicle. We had to be careful with her for a good while, not only because she needed to convalesce, but because Jancis had whooping cough and we had to keep those two separated!

We had taken two young girls with us on this holiday, one thirteen and one fifteen. They ate like horses but they always rose to emergencies and helped us to look after the children. The thirteen year old had never seen the sea before and when we arrived and she saw that huge expanse of sand with the vast Atlantic beyond the bay, she rushed down to the shore and ran into the sea with all her clothes on! Is it not wonderful to be a child and have such delight in a simple thing like the seashore? Or perhaps the seashore is not such a simple thing. It is the work of millions if not billions of years and perhaps we should take a leaf out of that girl's book and stand in awe on the shore, the dramatic dividing line between two worlds, one above and one below water!

Our holiday on the Isle of Skye was a very interesting one. We rented a cottage in Uig, not Uig Lewis, but Uig Skye! Uig is on the west coast of Skye, overlooking Harris, which on a fine day you can see clearly across the channel known as The Minch. The Quirang range towers up behind it towards the east, forming a kind of rocky spine for the northern part of the island. There were many inland lochs in which it was pleasant to bathe, for the water was always warm.

We climbed in the Quirang with the children. We took them out to the inland lochs for the whole day sometimes, letting the children play in the safe waters of the rivers and lochans. We went trawling for mackerel sometimes, which was a very easy way to catch fish. All you do is row while another person holds the net at the stern of the boat and hauls in the fish! So we had lots of fresh fish during our stay, sometimes sharing the haul of some of the other inhabitants of the village.

Groceries beyond the normal run of foodstuffs had to be ordered from Glasgow and a cargo ship would bring and deliver them to the quayside about every two weeks. We got friendly with the crews of these little ships. One five hundred tonner was called *The Hebrides*, there was another called *Dunara Castle*, of about the same size and then there was a small two hundred ton boat called the *Challenger*. Once, while the crew of the *Challenger* was unloading cargo, the captain asked us if we would like a lift to Harris. Since we were having the most beautiful and sunny summer within living memory and could see the reflections of the seabirds as they flew over the water, I decided to accept and take Corin and Nigel over to Harris.

The captain provided the children with food on the journey, which they ate happily, sitting on oil barrels tied on to the deck. He dropped us on the island of Scalpay, which is very close to Tarbert, in Harris. We camped there for a few days and made friends with a number of families.

Some of the children asked us if we would go with them to see the lighthouse on the other side of the island. So we trooped over, in Indian file, following paths through high bracken and heather known only to the local children and had a very instructive time at the lighthouse, where the keeper explained how everything worked. We got a ride in a fishing boat to Tarbert. We then retraced my steps of long ago, when I had walked from Tarbert to Borisdale with Joy. Every time we passed a croft, we were invited in and given a cup of tea, some oatcake and some eggs! After about three of these sessions, I recall walking over a desolate part of the island when Corin spied a croft.

' Oh Dad!' she said. 'Let's go the other way – I couldn't possibly eat another egg!'

We finally made it to the MacDonalds at Borisdale, where we stayed for several days. Mrs MacDonald very tactfully never enquired about what had happened to Joy, but the two strong, healthy children accompanying me on this walking tour, showed her that something must have gone right somewhere!

We eventually got a lift back to Uig Skye in a fishing boat and resumed our lazier lifestyle.

My father was with us on this holiday, so we two had a number of interesting mathematical talks. I was working in multi-valued logics, having already written a paper on three-valued logic, which I was going to submit to a mathematical journal.

'I have bad news for you,' said my father when we arrived back from Harris. 'The Dutch logician Heyting has just published a paper which is more or less the one you were going to send in!'

Although this was disappointing, from another point of view it was not. If I could work on the same lines as the well-known logician Heyting, then I was not such a poor mathematician after all!

During the time we were on this holiday, I took Corin and Nigel on another walking tour, this time on the mainland of Scotland, staying in youth hostels. We did about fifteen miles of walking each day, with a lot of stops to look at the local flora, admiring the landscape and having refreshments along the way. There was one hostel, known as Craig, where we arrived one afternoon and found the hostel empty; not only were there no other hikers, but the warden had departed leaving a note, saying:

> *Gone to the Highland Games. Will be back tomorrow. Put the money in the box for the night's stay, as well as the money for anything you take from the store. The items are priced.*

We made ourselves at home. We cooked our supper on the stove provided, ate it with gusto and, after a little walk round the bay (the hostel was built on the shore of a little bay), I put the children to bed, told them a story and they fell fast asleep.

Soon some other hikers arrived. They also cooked and had their meals, after which we had a little sing-song and exchanged tales.

Then I said, 'Excuse me a minute, I'll just see if the children are all right.'

'Children?' they asked incredulously. 'But we are at least twelve miles from the nearest road! How did you get here? Fly?'

'Well, come and see the children,' I suggested.

We all trooped up to the dormitory and the hikers were amazed to see two angelic-looking little children, fast asleep.

'But how old are they?' asked one.

'This little boy is just five and the little girl is eight,' I replied, pointing to the sleeping children.

Apparently, to walk twelve or fifteen miles during one whole day, in those days, was reckoned to be beyond a five year old child.

My educational work in the County of Leicester started to develop well after I had 'learned some official psychology'. A certain Len Sealey was the mathematics adviser for the county and he became interested in my experimentation. He showed me round some of the schools into which he had tried to introduce a more active form of mathematics learning. I saw electric trains with which the children played measuring their speeds up and down inclines; there were 'shops' where some of the younger children were supposed to learn how to buy things and count the change and thus learn the dreaded 'pounds, shillings and pence'!

At one point I said to Len, 'I am sure the children are having a ball, but have you actually thought about what, if anything, they are really learning through these activities? I see no system or hierarchy of concepts being acquired.'

'Surely they learn the mathematics while they are engaged in these activities,' replied Len, a little uneasily.

I thought we were not getting anywhere, so I tried to explain myself.

'Mathematics consists of a whole hierarchy of concepts, woven into structures at different levels, each level based on the levels below it. In my experimental classes I have tried to *embody* concepts in physical form, so that the rules of the game correspond to the mathematical rules we want the children to learn. When they have learned some concepts, then they can apply them and that is where *your* games would come in.'

We had many discussions on the above lines and finally we agreed to co-operate. We had a lot of multi-base and algebra material made and distributed it in the schools where the teachers were interested in moving forward in the direction indicated. I worked in most of these schools myself, teaching the children while some of the teachers watched, who then took over. We had meetings after school and weekend courses at the Teachers' Centre supporting the practical work in the schools. I introduced some transformation geometry, which is old hat now, but it was practically unheard of then, particularly for elementary school children.

One day an HMI (His Majesty's Inspector) came into the class in which the children were learning about the rotations and symmetries of the square. He noticed much activity: the children moving a lot of cardboard squares about and some of them were moving each other about on the classroom floor.

He said to me enquiringly, 'What are all these children doing?'

'Why don't you ask the children?' I suggested.

He went up to a little girl, who cannot have been more than ten years old.

'What are you doing with that piece of cardboard?' he asked.

'Well, you see, sir, it's like this,' replied the child. 'When you flip the square over holding it by the red line and then you turn it halfway round, you might as well have flipped it about the blue line, you would have got the square to the same position!'

The inspector listened, somewhat nonplussed.

The little girl, quite unabashed, said to the big burly inspector, more than twice her size, 'I am sorry, sir, you don't seem to understand. But look at it this way. Let me turn you round through a quarter of a turn to your right.' So saying, she got hold of the inspector and swiftly made him do a right about-turn. 'And now take a half-turn.' She helped him again to carry out a half-turn. 'Now, if you had done a quarter turn to your left from where you started, you would have finished just where you are now!'

'I see,' said the surprised inspector.

But the little girl was not finished with him yet.

'Actually, sir, what I was showing you with the cardboard square is not quite what you have just done,' explained the little girl. 'I should really have flipped you! But then I would have had to turn you upside down and stand you on your head and I thought you would be too heavy for me to do it and, anyhow, you probably would not have liked it!'

'Thank you very much, my dear,' said the inspector. 'I can see you certainly know what you are doing. Do you like doing geometry?'

'Yes, sir, I do,' said the little girl, 'but I prefer working on the tetrahedron – the square's too easy!'

Such scenes were quite frequent when visitors talked to the children and received quite sophisticated replies. In fact visitors from other counties started to come and see what we were trying to do in the County of Leicester; eventually they came from other countries too.

During all these feverish activities, what was happening to the family?

Bruce, our fifth child, was born in 1953. So, while I was trying to revolutionise the teaching of mathematics in elementary schools, I also had the responsibility of a growing family and the two responsibilities sometimes resulted in conflict. Corin had become quite problematic at home and we wondered whether it would be good for her to go to boarding school. But where was the money to come from?

At that time the City of Leicester was giving grants for parents who needed to send their children away to school, but the parents had to make out a very good case. We applied to the city council, stating that, all the past moves and uncertainty about future moves had made Corin unstable and that a steady education in one place might be what was required. We had made enquiries at Wennington School, which was run by Quakers, although it was not officially a Quaker school. Our application was accepted and we got a grant on a sliding scale, depending on our income and the number of children in the family. In our case, at that time, this meant a full grant. So Corin went to Wennington, near Wetherby in Yorkshire, where she stayed for five years. Even today she tells us that they were the happiest years of her life and that she learned not only academic subjects but how to care for other people, understand their problems and practise 'conflict resolution'. We often drove up to Yorkshire to see her and she and her teachers always gave us glowing reports.

After Corin had taken her first examinations, we brought her home, as she had agreed that it would be easier for her to do advanced subjects in the sixth form of a secondary school. Also, the grant was reduced with every increment in my salary and sending her to Wennington had become more and more difficult financially.

She had some problems settling down, for she was not used to having rules that had to be blindly obeyed, without explanations. It seemed to her that the new school tried to 'mould' rather than 'educate'.

The school would invite outside lecturers to talk to the children and Corin always enjoyed this. After one of these lectures, she was so pleased with the lecturer's way of presenting her subject that she rushed up to her and hugged her, thanking her profusely for the talk.

The headmistress said to her, 'You know you just don't do that kind of thing! Supposing everybody were to do what you did?'

'I know,' replied Corin, 'but they didn't.'

Later we happened to meet this same lecturer, who told us about her talk and said that, although the school seemed to be a very stiff one, there was one redeeming feature of the event.

'And what was that,' I enquired.

'A young girl came rushing up to thank me and hugged me,' he replied.

'Oh,' Tessa said, 'that was our daughter!'

At a certain point in children's school career they have their IQ measured. Corin's and Nigel's were just a little above average. The psychologist warned us not to expect too much of these children, not to 'push' them in any way, as that would be harmful. It turned out that Nigel was dyslexic and in fact he did not learn to read until he was ten or eleven. When he took the eleven plus exam, which was an exam to sort out the sheep from the goats, he was placed with the goats since he did not pass the exam. Somebody on the appropriate board had the insight to suggest that, seeing Nigel was from an academic family, it might be a good idea to interview him before sending him to what was then called a secondary modern school, the destination of the failures. Nigel acquitted himself very well at the interview and so was allowed to go to grammar school. He instantly became very good at everything, remaining top of his class for the next several years and even becoming one of the 'high flyers', who would take the school certificate examination after four years instead of five. It seems that an IQ test at the age of ten does not have a very good predictive value. One shudders to think how many children are condemned at the age of ten to a mediocre life by such an inefficient selection method!

Subsequent governments later abolished the eleven plus and introduced comprehensive schools, more on the American model. This was one step in the right direction, but there was still 'streaming' into ability groups and it was relatively rare for a child to move up to a higher stream.

Sorrel and Bruce learned to read very young. When Sorrel went to school at five, she could read fluently. She would sit in a corner of the classroom at 'quiet times' and read. The teacher thought she was just looking at the pictures. When she found out that she could really read, she started to use Sorrel as an assistant teacher by asking her to read to groups of children while she was busy with the rest of the class. Sorrel was then sorry that she had ever let on about being able to read!

Bruce was already a good reader at the age of three. While we walked around, he would ask me what the letters were on the street signs and what this or that word was and I always told him. Then I made him some little stories, with simple pictures, which he loved. I could hardly keep up with the demand for such books! Finally he agreed to read 'real books'.

Once Professor Jeeves saw him read and noticed that he was holding the book upside down.

'Oh, can you really read?' he asked Bruce.

'Oh yes, I am reading this one now,' replied Bruce.

'But you have the book upside down!' retorted Professor Jeeves.

'Why? Does it really matter how you hold the book?' asked Bruce innocently. 'I can read any of these books and hold them in any way I like!' he went on.

He demonstrated his skills in this direction by holding the book upside down, the right way up and even sideways and reading the story just as easily. He certainly enjoyed the admiring audience, though he did not see what was so special about it. Bruce did not have any trouble at school. He soaked up everything very quickly; in fact, he tended to be bored for the pace was clearly too slow for him.

Jancis was the middle child. I believe the middle child usually gets the worst of both worlds. There is no spoiling, as you are not the smallest and there are fewer privileges, as you are not the oldest. In fact, there were six years between Jancis and Sorrel, so Jancis was asked to do a lot of 'looking after'. At first, Corin, Nigel and Jancis regarded Sorrel as a kind of live doll. They washed her, fed her, dressed her up prettily and generally spoiled her. I remember once raising my voice to tell Sorrel off about something, when they cried out in chorus, 'Dad! You can't talk like that to Sorrel!' So Sorrel became the 'holy child' and continued to be spoiled!

Let me end with a few more words about our lodgers. The three university lecturers who were our first lodgers did not stay with us for the whole time, although we remained friendly with them. We had a great deal to do with a batch of Hungarian 'freedom fighters' who came over in 1956. During that uprising, the prisons were thrown open and of course not only the political prisoners but the criminals came out. They simply walked over into Austria, said they were freedom fighters and were taken by plane to all sorts of different countries, Great Britain being one of them. We looked after one of these phoney 'freedom fighters' in our house. We had another Hungarian lodger at the time, an ex-hussar officer, known by the name of Zoli. One day our 'freedom fighter' stole ten shillings from Zoli's purse to buy Tessa a present, in order to thank her for being so kind! Soon, he left for what he thought might be brighter horizons and was lodged with a vicar nearby. Some weeks after he arrived at the vicar's, he stole the money from the children's Mothers' Day box and disappeared, never to be seen again. The police even dragged the nearby lake, but he was never found, alive or dead.

But Zoli himself was a strange customer. He was very keen on what he called 'honour'! There was a local journalist who was keen on Tessa, so Zoli kept an eagle eye on him. Once he offered to take Tessa to the theatre, as he often had complimentary tickets to be able to review the plays.

Zoli was furious and said to him in a very loud voice, 'It is disgusting of you to try to take somebody's wife to the theatre. Since Dr Dienes does not seem to think it is a matter of honour, I do! I challenge you to a duel! I will expect you on the corner of Belvoir Street at dawn tomorrow morning!'

'All right, all right,' said the journalist, edging out of the house before anything violent happened.

During the night there was a big break-in in Belvoir Street and the police were there in full force, some of the robbers having been caught red-handed and shut into Black Marias. Zoli arrived on the scene at about four o'clock in the morning and noticed the heavy police presence.

Later in the morning he came in looking furious.

'That journalist friend of yours told the police about the duel! What a coward! He cannot even defend his own honour!'

Then Tessa showed him the front page of the local paper, with pictures of the break-in, one showing a robber being bundled into a Black Maria by two policemen.

'There was a burglary,' Tessa tried to tell Zoli. 'My journalist friend, as you call him, is probably sound asleep and I am sure he had no intention of fighting any duel! We don't do that sort of thing in England any more!'

Zoli slunk out of the room, mumbling something that sounded like 'Cowards!'

We also did quite a bit of 'social work' during those years. There was a voluntary organisation known as the Family Service Unit, which we supported both financially and by working with them. Members of the unit would sit in at court cases in which children were about to be separated from their parents on account of their home life being unsatisfactory.

The judge would ask the unit member, 'Would you be able to take this family on and, if so, do you think there is a chance of some improvement?' If the answer were in the affirmative, then the members of the family would have to promise to co-operate with the Family Service Unit and the case would be reviewed in six months' time. In such cases unit members, which at times included Tessa and myself, would go and visit the families. The policy was 'No moralising!' but to offer help. The first look at such a household would usually give one an impression of utter chaos. We would do the dishes, sort out the food, throw away the rubbish and generally clean up, naturally trying to get family members to join in. Usually the children would enjoy such a clean-up! Then there would be discussions about how to keep house, how to organise the family budget, some talk about food values and so on. It would be uphill work, but, as often as not, we would be able to report back in six months' time in a very positive way and we would often become very friendly with such families.

In fact, so many of these friendships developed that I offered to take some of the children on a holiday in Lewis. I took about ten boys with me and we rented the same croft the family had rented some years previously, when Gedeon came to see us on his honeymoon. We went fishing in the lochs, we climbed the mountains and, what is more important, the boys learned to co-operate with each other to run the household.

Another aspect of our 'social work' was trying to look after and befriend people who were in bad situations. We once let our front room to a pregnant girl called Malvena. She told us that she was a nurse and that the doctor for whom she had worked was responsible for the child. She told us so many other unlikely stories that we thought we would check up on her. She was well known to the local social services! We went to see the doctor, if only to warn him about what Malvena had said to us, but, having talked to him, we were convinced that Malvena's stories were all fabrications. We bought a pram for the expected baby, as well as baby clothes and really tried to do the best we could, but I am afraid that in this particular case we failed. Tessa once found a whole lot of faeces, all wrapped up in paper, in one of her drawers! We both thought this was really the limit and we asked her to leave. We heard afterwards that her baby died at birth but after that we lost touch with her. We engaged in more varied pieces of 'social work', partly because we felt that we should share what we had with those less fortunate and partly because we wanted our children to grow up with the idea that we were all to some extent accountable for what happens to our fellow human beings.

Our religious life developed with the help of the Quaker meetings, to which we took the children, who would stay in the meeting for the first quarter of an hour and then have one of the Quakers take them out to another room for what we called 'First Day school'. We took turns running the First Day School. I really enjoyed this work with the children, during which I tried to give to them all the guidance that I myself had lacked as a child. I also represented Leicester Meeting at the Friends' Service Council for three years, during which I was instrumental in raising money for the Friends' school for rural girls in Northern Greece, as, according to new earthquake regulations, it would have been closed down had new buildings not been built which were more earthquake-proof.

So the Leicester years were very busy and formative, for Tessa and me as well as for the children. It was a big step to take to move away from where two of our children had been born and where we had found so much that was meaningful, both professionally and personally.

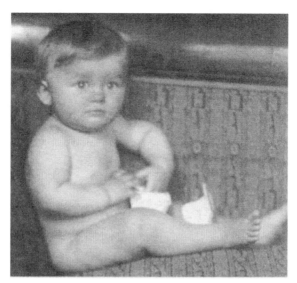

Zoltan Dienes, one year old in Budapest

Zoltan aged twelve

Tessa

'Young married bliss' – Dienes, Tessa and Corin (later Jasmine) in 1940

'Wartime source of milk' – Dienes and Corin (with Tessa's shadow) in 1940

Dienes the elder (Zoltan's father), on the left, visiting 'The Palace' at Eastbury, with Tessa at the front and Corin (later Jasmine) on the right, in 1941

Dienes with his mother Valerie in Canna, in 1946

Dienes travels to the isles (Hebrides), in 1947

The Dienes family at their holiday home in Norway

Dienes enjoying a few days rest on a world tour in Dubrovnik in 1969, with his mother Valerie and brother Gedeon (on the right)

Children in Leicestershire learning addition in Base 5 with the Dienes Multibase Arithmetic Blocks (mid 1950s)

Children in Papua New Guinea learning logic and classification by the Dienes method (1960s)

Send-off at the airport Porto Alegre (Brazil) after several weeks of mathematical work with local teachers and children

Dienes teaching Canadian children about functions with his apparatus

'Isn't it great? We are still together and in love with each other.'
Dienes and Tessa (late 1980s)

Dienes with his honorary degree from Exeter University in 1995
(left to right: Sir Geoffrey Holland, Vice-Chancellor; Dienes; Sir
Rex Richards, Chancellor; Dr Paul Ernest, public orator)

Chapter Fifteen

The Interregnum

After giving in my notice at Leicester, I had to live for a while with a certain feeling of uncertainty as to what would happen next. One permanent job was over and the next permanent job would start about eighteen months later. I had to organise these eighteen crucial months as best I could, looking at the needs of the family as well as my career needs and trying to reconcile the two. Rightly or wrongly, I shall never know, the following schedule was arranged.

Firstly, I would work at the Center for Cognitive Studies at Harvard with Bruner from October 1960 till June 1961. Then I would work as co-director with Margaret Lowenfeld on a British Association for the Advancement of Science project, in which we would relate our respective approaches to each other between June and December 1961. Then we would all move to Adelaide in December 1961, to take up my readership (associate professorship) in the Psychology Department of Adelaide University.

This entailed letting our house in Leicester for the period spent in the USA, because we would need it again for the latter part of 1961 and then selling it at the end of 1961. We found some tenants who took on the house for the period of our absence from England and soon we found ourselves on a ship bound for New York, armed with an

H1 visa, given only to aliens who could not in any way be replaced by US citizens!

The voyage was very pleasant. We put the clock back about every other day by one hour, as these were times before jet lag had been invented. There was lots of amusing leisure things for the children to do and certainly lots to eat and drink.

I spent some of the time writing an article. I had been thinking that there was a certain amount of confusion in people's minds about the processes of abstraction and of generalisation. Most people tended to think the two were really the same. I thought otherwise. I thought that 'abstracting' meant pulling a lot of different situations together into one class, having discovered that they had something in common. So for me, 'to abstract' was 'to construct a class'. 'To generalise', on the other hand, I thought meant that we already had some knowledge of a class of events or objects and, by generalising, we were attempting to enlarge that class. So 'abstraction' was a process leading from elements to a class, whereas 'generalisation' led from class to class, the second class being an extension of the first. If indeed this were so, there were important philosophical as well as consequent educational implications. I tried to write all this down in an article, which I presented to my new colleagues on arrival at the centre at Harvard. It was soon published in the *Harvard Educational Review*.

Going through immigration was not very difficult; in fact, it was practically a hero's welcome with my H1 visa. Apparently they see very few aliens entering the USA with such a visa!

We had our dormobile with us, which was disembarked on the quayside, and then we all piled in with our luggage and started to drive towards Boston. The freeways, rotaries, strange traffic signs, oceans of traffic seemingly rolling along in all directions were all quite a trauma to get used to. We stopped at what we imagined was a wayside cafe but there was nobody there. One of the children quickly discovered some machines into which one had to insert objects called quarters and then press a lot of buttons to 'tell the machine', whether the coffee was to be with or without cream, or milk or sugar or caffeine. We finally worked out how to get what we wanted and mechanically drank our mechanised coffee. The children were quite intrigued by the machines but Tessa was horrified. She thought the whole of America was going to consist of a lot of machines with detailed instructions, to be carried out to the letter! I tried to reassure her that this was unlikely to be the case but she was not convinced!

We had an uncle who was a microbiologist at Harvard Medical School, so we put a call through to him to get some instructions about how to get to Boston. We were to come and see him and his family, before going to a furnished house that had been found for us by the secretary at the centre.

My uncle's daughter-in-law answered the phone, speaking in what I thought was a very American accent (I did not discover till later that all Americans had American accents!), and said, 'When you get near Boston, you will get to a cumplicaded roadery. Take 128 North and then take the third exit.' She went on like that until I was totally confused and said that I would enquire when we got to the 'cumplicaded roadery'. Needless to say, we eventually wove our way through all the traffic, managed to leave the 128 freeway at the right exit and made it into Cambridge, Mass.

The furnished house we were to rent was in Newtonville, part of the town of Newton. It was an old building, seeming as though it might fall apart any minute, but it looked quite picturesque. The furniture was also old-fashioned, but quite pleasant to look at. We had Jancis,

Sorrel and Bruce with us, as Corin had already started teaching school and Nigel was recovering from infectious hepatitis, contracted at Manchester, where he had started his studies in chemistry. He thought it would be great to convalesce in Hungary, so this is what he did. He stayed with Gedeon and his wife Maya, who looked after him. He was supposed to join us later in the autumn (or should I call it fall?) and then fly out to Adelaide to start in the medical school there, as he had decided to train to be a medical doctor. We would then join him a year later; in the meantime Professor Jeeves had agreed to act *in loco parentis*. Jancis was registered at the local high school, while Sorrel and Bruce were taken across the road to an elementary school.

I soon got to know my colleagues and, American-style, we moved on to first name terms almost immediately. Professor Bruner suddenly became Jerry, I was Zed, William Hull became Bill and so on for all the rest of the crew. We soon organised a group whose members would meet regularly and discuss the problems of mathematics learning as they arose. This was the time of the Americans' mad rush to beat the Russians, who had put Sputnik into space before the Americans had managed to do the same, so the 'new mathematics' became the rage. By this time it had become understood that it was no use just looking at the last two years of schooling if improvements in mathematical and scientific achievement were the aim, because the mistakes had already happened much earlier on. We had to concentrate on much younger children!

We had some mathematics material made – multi-base blocks, algebra materials, as well as attribute blocks – which we thought would encourage the development of logical thinking. This material was partly to be used in a private school known as Shady Hill School, where the progress of the children would be monitored, partly under much more controlled conditions at the centre. We selected four children, two boys and two girls, about eight or nine years old and placed each child in one corner of a room at a table, each with a trained psychologist who would observe everything the child did and take copious notes. These notes would supply the raw material for our eventual report on the work. There were no hypotheses, no experimental or control groups. We were simply to observe a certain phenomenon which we hoped we could call into being in this way and we hoped that the psychologists would be able to put down on paper enough of what happened, so we could make sense of everything later on! Jerry would walk around and take photographs and I would provide the tasks and interact with our four subjects, trying to 'create the phenomenon' to be observed.

As you can imagine, this was a very labour-intensive way of looking at the process of mathematics learning, but, since the personnel and the money were available, we decided to try.

During our discussion periods anything and everything was discussed which could be discussed. No holds were barred and sometimes the noise level reached more decibels than I would care to quote! Jerry was very keen on the importance of the role of symbols and language in the learning of mathematics and grimaced when I talked about children having 'symbol shock'! He used to say, 'In the beginning was the word'. I often replied by saying, 'Perhaps in the beginning was chaos'. Jerry thought that it was words that really did the work, the chaos was in our heads; I used to say that you had to have some physical experiences when you were learning something before you could call it anything.

We had a Japanese exchange student working with us so I suggested at one of these chaos-word discussions that I could teach our Japanese friend about complex numbers, of which he was totally unaware, speaking only in Hungarian.

Fortunately, the floor had some squares on it, so I stood the student in one of these squares. I then moved him forwards and backwards through a certain number of squares, writing positive numbers for forwards and negative ones for backwards. He had no difficulty with this. Then I would move him to his right and to his left. I would write on the board positive numbers accompanied by the letter i for steps to the right and negative numbers accompanied by the letter i for steps to the left. We then combined forwards, backwards, right and left steps, writing the corresponding symbols on the board, placing the plus sign between the symbols for forwards-backwards and those for left-right. All this time I would explain everything in fluent Hungarian and he would nod, indicating that he had understood. Soon he was adding and subtracting complex numbers. I had made my point and we all had a good laugh.

I did not know it then but Jerry had found the money for our fares by promising the School Mathematics Project, initiated by Professor Begle at Princeton University, that I would give a studied critique of what they had prepared. So I found myself saddled with the task of reading and then writing something about the initial efforts of these would-be mathematics educators, bent on beating the Russians.

Having read what they had written, it did not take me long to write the critique. Of course, mathematically it was all impeccably correct, seeing it had been written by mathematicians. From the point of view of the probability of children really understanding what was contained in the suggestions, it was very doubtful. I finally suggested that what they offered would probably be very good for the upper five or maybe ten per cent of the intelligence distribution, but the rest of them would simply learn different things by heart without knowing what they were saying, as opposed to all the old stuff which they also did not understand. I suggested that the project should include a mathematics laboratory, with a varied set of materials, mathematical games and stories, so as to take care of those who were less endowed with the gift of intelligence.

It was reported back to me afterwards that Begle really agreed with me, but thought that my suggestions were impractical on account of the implied cost of concrete materials and the cost of training teachers how to use them. Things have changed a little in the past thirty years, but there is still quite a solid body of educators who believe that 'manipulatives' are an expensive luxury. They forget that to try and teach and reteach place value and fractions every September for five years running is not only an expensive luxury but sheer stupidity. But it took several hundred years to persuade people to pass from using Roman numerals to the present Arabic system, so perhaps there is hope yet for mathematics learning in the far distant future!

While all this was being debated by our group, Tessa was at home caring for the children and the children were trying to get used to their new school situations. In England you are supposed to be quiet in school; in the US of A you are supposed to learn how to be assertive, which nearly always implies a higher noise level. This was quite hard for the children to adjust to.

After the first good snowstorm, we saw our children coming back from school in the middle of the road. We ran to tell them that it was dangerous, especially with cars skidding all over the place.

'But they told us not to walk on the pavement!' Sorrel objected.

We then had to explain that by 'pavement' Americans meant the road, which was paved. What was called 'pavement' in England was called 'sidewalk' here. It was not long before they would have to learn that the sidewalk was to become the 'footpath', as that is what they called it in Australia. It must have been somewhat confusing for the children!

Sorrel's new friends were intrigued by her being English.

'Do the English really have a cup of tea at four o'clock every afternoon, which they balance in one hand while they eat a piece of bread and butter in the other?' they would ask her.

'Yes, they do,' replied Sorrel, but they did not believe her.

So one day after school Sorrel came in with a few friends to investigate. They could not believe their eyes when they saw Tessa with a cup of tea in her one hand and a small piece of bread and butter in the other! And it was just before four o'clock in the afternoon! They burst out laughing and we only found out what the joke was when Sorrel told us later about her friends doubting the English tea story.

We were often invited by my Uncle Lajos to his house in Cambridge. He spoke with a definite Hungarian accent, but very softly and with what seemed a very kind voice. You had the impression that he could not hurt a fly. We had a number of discussions about fundamental things. One I remember very clearly is about our perception of the world.

'You see,' he said, 'we shall never know what the world is really like.'

'Why not?' I asked incredulously.

'What you see is not what is *really* out there. It is what your brain turns it into. And there is no way we can contact the outside world without the intermediary of our brain. So the world must remain for ever unknown to us!'

Of course I realised that he was right; I just had not made that connection before. I also discussed my future plans with him, as I thought Lajos was much wiser than I.

But he was wise enough not to advise me. He maintained these things were so individual that no person could think or feel for another with any degree of certainty. In other words, make your own decisions. Echoes of my mother a time back! You make your bed and you lie on it!

In the meantime Nigel was having a fine time in Budapest. In fact, he fell in love with a Hungarian girl and they wanted to get married. They went to the British consul, but, as Nigel was under age, he would not marry them and we did not want to give our consent to such a hurried decision. So, sadly, Nigel left Budapest and his lady love and came to the USA to stay with us until it was time for him to leave for Adelaide. We did some frantic shopping to equip him as far as possible for the year and saw him off at the airport. There had been a number of snowstorms and until the last minute it was not certain whether the plane could take off. Finally it did, rising into the sky from what looked like a canyon carved out of the masses of snow still covering the runways!

We kept in touch by phone and correspondence, but he was really too young to be so far away from his family for so long and he was not happy during his first year at medical school. The first year students were made to go through their paces; the other students doing incredibly cruel things to them. One student actually died when he was forcibly kept under water as part of these 'initiation ceremonies'! They also played 'human chess', in which the first year students were the pieces who were moved about. I shudder to think what happened to those that were 'taken'!

We soon heard from Corin, who said she wanted to marry John, the NFER research worker who had come with us to Halmstadt. This was rather unexpected. I could not go over but Tessa thought she should be there and booked a flight over to London. Apparently, Corin was having second thoughts and did not really want to marry

John. Tessa thought this was just wedding nerves and encouraged her to go through with it. How unwise can one be! They got married but it did not last very long, although they did have a beautiful pair of twin boys after a while.

Mending shoes was very expensive in the USA, at any rate by English standards, so Tessa took my shoes to be mended in her luggage! What with everything, like packets of English tea and the mended shoes, she had too much luggage on the way back and she was asked to pay excess baggage.

She burst into tears at the check-in and said tearfully, 'But this is not stuff that is worth it! And the tea! I'd die without the tea! You can't get it in Boston!'

'I am sorry, madam,' said the official. 'There will be excess to pay on this luggage!'

Apparently, the supervisor heard this exchange and came to investigate.

'We are told that we have discretion to make exceptions,' he said to the official dealing with Tessa. 'Let her pass through and take all the baggage.'

Tessa dried her tears and hobbled into the departure lounge, happy to have my mended shoes and her tea secured.

I recall being invited to a gathering of eminent scientists, more than half of them Nobel Prize winners. Tessa and I were introduced as visiting dignitaries from Europe and were received as such. But after a few drinks the noise level of the conversation rose to incredible heights and I noticed that practically nobody was listening to anybody else, but all seemed keen to have their say. This was our introduction to 'American assertiveness'. After having attended one or two more such meetings, I tried to excuse myself from any more as I really saw no point in everybody screaming away without anybody listening to any of it.

At one of our get-togethers, Jerry was putting something into a parcel when we came in, trying to tie it up. As we came in, he said quite casually, 'Hi there! Just a minute, I'll just have to go and get some Durex.'

Tessa was quite taken aback. She knew that Americans were very forthright and called a spade a spade, but she did not realise that almost as soon as one comes into a house they would start arranging the details of an instant love affair! Of course, Tessa did not realise that in the USA Durex was Scotch tape, whereas in England it is a well-known brand of male contraceptive device! Jerry merely wanted to finish packing his parcel!

In Europe, Americans tend to have a bad press and are regarded often as brash and loud. But let it not be thought that all Americans are egotists like the Nobel Prize crowd we met! I recall that one day we were invited to supper by a doctor whose son was in one of my experimental groups. Tessa was not feeling very well so we phoned up to say that perhaps it was better for us not to come. The doctor was immediately concerned, drove over to our house and had a look at Tessa.

'I believe you have pneumonia,' said the doctor. 'Let me take you to the hospital and we shall see what we can do.'

So saying, he bundled us into his car and then Tessa had a thorough examination, including x-rays.

When we enquired about the bill for the hospital, he said, 'We Americans are honoured to have a person such as yourself working with us to help us solve our problems. There will be no charge, except for the x-rays.'

We went to dinner with this doctor and his family some time later. It was good to be friendly with such considerate and yet cultured persons. He did not have a Nobel Prize, he did not shout to be heard and he was kind and helpful. His family helped give us a more balanced view of Americans!

We got to know another family during our stay at Harvard. They lived practically opposite us in Newtonville. They had a little boy of about five, who became very fond of Tessa and so spent quite a lot of time in our house while Tessa was alone, I was at the centre and the children were at school. We discovered after a while that this family was having some serious marital problems.

One night Tessa woke up and had a compulsion to go over to the house of this family: she felt it was urgent and that she must go at once. She found the husband holding a large carving knife, with which he was about to kill his wife. When Tessa came in, he put down the knife and Tessa talked to him warmly and constructively. The knife was never brought out again and the incident was a turning point in their relationship.

Several years later, after the family had moved to Florida, they invited us to stay with them. They were still happily married, were raising their children and had established themselves in one of Florida's communities. How inscrutable are the ways of Fate? Why was it that Tessa simply *had* to go over in the middle of the night to assist this family in their hour of greatest need?

There was a certain research worker called William Hull with whom we became very friendly. He was teaching at Shady Hill School and so was one of the observers of what we were trying to do there. We devised a set of attribute blocks, based on the original cognition test by Vygotsky and soon found that such materials were extremely useful in getting children to think logically, in terms of categories, similarities and differences. I spent quite a lot of time at Shady Hill School, experimenting with this material as well as with other materials we had devised for teaching positive and negative integers, linear and quadratic equations and many topics that were considered esoteric as far as young children were concerned. The work we were doing at the school aroused some interest, so we were visited by workers such as Robert Davies, who was one of the more go-ahead educators in the field of the 'new math'. This work, along with the work done with the 'gang of four', eventually found its way into my report on the Harvard work, which was published in London by Hutchinson under the title, *An Experimental Study of Mathematics Learning*. The report included many citations from the group of four children who were observed by the four psychologists, some of which were extremely entertaining. The final report in the book simply stated the kinds of problems about mathematics learning which we were now able to identify. There was much to be done to tackle these problems, but at least we had been able to put our fingers on what the problems were, or at any rate what some of them were!

Since Jerry and I appeared to have different interpretations of what had occurred, we decided to write separate reports. I think Jerry was not used to having a person in his group who was not a yes-man and this probably disturbed his ego. He agreed to write a preface to my book, writing it in such a way that he thought I would never publish it. I published it word for word. I thought the public could exercise their own judgements and decide whether they agreed with Jerry Bruner or with myself.

The time was approaching when we were due to leave and go back to England, to begin the next part of this interregnum with the British Association for the Advancement of Science. Word had reached some Canadian Universities about the work I was doing, so they invited me to give a talk in Montreal and another one in Quebec City. So we arranged to ship the dormobile from Montreal and join the ship to England at Quebec, where it would stop to pick up passengers.

The children said goodbye to their friends; Jancis was particularly sad not to be able to attend her own high school graduation. She had imagined herself walking up to the rostrum in a white dress to receive her diploma! It took a long time for her to get over this sadness.

We drove to Montreal in the dormobile and left it at the quayside to be shipped. We found a reasonable place where we rented a room for the night. I remember enjoying the fact that I could do the shopping in French; I bought only milk and cereal but it was such a different atmosphere. Everything seemed much more alive than we had experienced in the USA.

I gave my lecture at the Université de Montreal and then we took a plane to Quebec City, where we were met by a professor from Laval University, who took us to our hotel and then to the university, where I gave them a talk about what we had been doing at Harvard. I remember an outing to the surrounding forests, which were very scenic, but it was the time of the black flies and that fact certainly overshadowed any enjoyment provided by nature!

We were finally able to get to our ship, which had to be reached by motor launch, as there were no mooring facilities at Quebec for the type of ship we were to travel in. We had a peaceful passage and arrived safely back at our Queens Road house, which had been turned upside down by the tenants and was in a most sordid mess! It was so bad that the tenants agreed to pay to have the place cleaned up!

Since we were about to go to the other end of the world, we thought it would be a good idea to visit Hungary with the children. We drove across Europe in our dormobile and arrived tired, but happy, at Gedeon's apartment in Budapest. It was good to see my mother again, who was living with Gedeon, Maya and their two boys. We were taken to the baths for which Budapest is famous and luxuriated in the warm water which springs out of the depths of the earth at many points in Budapest, one of them known to the Turks from the time of the Turkish occupation of Hungary. We indulged in some excursions into the country around, retracing some of my boyhood hikes.

One night we went to see the ballet in the open-air theatre on St Margaret Island. The setting was really beautifully romantic for a wonderful performance of *Swan Lake*. Unfortunately, the performance was interrupted by the Soviet satellite, Sputnik, flying over: everybody began looking up at the sky, starting up a rhythmic applause, so that the performance had to be stopped until Sputnik disappeared among the luscious trees of the island! We also went to the opera house, where sometimes there were operas, sometimes plays. I was surprised that in one of the plays the actor who played the communist official made himself totally silly, ridiculing some of the common anti-capitalistic phrases which he kept inserting into the most unsuitable contexts! Hungarian communism was perhaps already on the decline.

Gedeon took us on a tour of Hungary. We went to Szeged, in the deep south, where we saw Madách's famous drama, *The Tragedy of Man*, performed outside, with medieval buildings in the background for scenery. Then we went to Debrecen, where we visited the high school my father had attended and the local museum.

Tessa said to me, after having had a look at a number of portraits, 'I thought you said that the name Dienes was quite a rare name. In fact, you said that there was really only one Dienes family! Look at all these Dieness on that wall!'

'Yes, I know,' intervened Gedeon. 'They are all called Dienes, but they are all members of our family! Here is our grandfather, for example. He engineered all the canals and dikes for the River Tisza to stop floods. Since then the river has not flooded once.'

Then we went to Tokaj, my father's birthplace. We eventually found the house in which he had lived as a small child. It had been divided into several apartments since then, with a peasant family living in each one. Each of the families treated us to wine and Hungarian goodies and talked about 'old times', although it was hard to pin them down to details. I suppose they were not sure who we were or who might be listening!

When we arrived back in England we had about four months before we were due to leave for Australia. Sorrel was accepted as a temporary student at the local collegiate school, which was the best grammar school; Bruce went back to his old elementary school; and Jancis joined the sixth form at the collegiate, also as a temporary arrangement, since she had already graduated from Newton High School and was ready for university. The children renewed their old friendships and were quite sad to think that they would soon be leaving, possibly never to return!

The work with Margaret Lowenfeld was fascinating. I learned a lot about 'learning blocks' and ways to 'unblock' them. She often invited us all to stay in her country house at Aylesbury, where she was the perfect hostess, thereby teaching us all the little, subtle details that are so important for visiting friends and so easy to provide! We discussed her Poleidoblocks and my games with the multi-base blocks. She was able to tell me some of the interesting clinical aspects I had unwittingly introduced into some of my games and I was able to help her develop her Poleidoblocks, for example by suggesting that she should include eight pieces which, when put together 'correctly', would result in a regular octahedron.

Some of our sessions took place in our Queens Road house. As a matter of fact, Margaret was there with us on our last day in Leicester. We had sold the house and we were selling most of the furniture. While she and I were working at a table, a truck came to take away the table, as it had been sold! Then they took the chairs we were sitting on and I recall finishing our discussions sitting on the doorstep of the house, the entire house having been evacuated except for our luggage, which we were to take on the ship!

So this was goodbye to Leicester. We all knew that it was also goodbye to an epoch in our lives. New horizons were opening up and we all had varying ideas of how to welcome these!

Tessa was very sad about leaving England. She really did not want to go. Rightly or wrongly, I insisted and it was a tearful Tessa and a somewhat quiet crowd of children who took the train to Tilbury before embarking on our long voyage. The University of Adelaide would pay first-class passage for all of us, but they said that they would give us the difference if we went as immigrants at ten pounds each, which we did. So we were shown to a six berth cabin way down below water level! This was to be our home for six weeks! I must say that I had misgivings and I shall probably never know what Tessa and the children thought about it!

The ship was the *Stratheden* and I believe it was her last voyage before going to the breakers' yard! The voyage would take us to Gibraltar, then to Port Said, then via the Suez Canal to the Red Sea and to the port of Aden, then to Bombay, Singapore, Perth and finally to Adelaide.

Once we entered the Red Sea, the water became warm in the ship's swimming pool and we spent a lot of time in and out of the water, or just basking in the sun. This was before we had problems with holes in the ozone layer and we could allow the children to get brown while they romped about in their swimwear. Aden was a wonderful haven for buying things very cheaply. I remember buying a walkie-talkie for next to nothing, as well as tape recorders and many other such 'desirable' items. The ship was also transporting some Dr Barnardo's children, who were being sent to Australia to start a new life on some farms, where I believe they were exploited as cheap, child labour. We made friends with some of these children and I tried out some of my newly invented mathematical games on them.

I also wrote the book, *The Power of Mathematics* on that trip, which was a sequel to *Building up Mathematics*. I used to sit on the deck with the typewriter on my lap and often some child would come up and ask what I was doing. I would sometimes sit them on my lap and allow them to do some typing, which a number of them thought was a great treat!

We stayed a day in Bombay. We thought this was a depressing city. There were a lot of very poor people, and very dirty streets bespattered with beggars, and whole families living on the pavement. Some of the passengers were interested in seeing the 'caged women'; these were prostitutes who were put out on show behind bars, I suppose so that they could not escape. Some of them looked as though they were only children. I thought it was all too terrible and we soon went back to the ship to try and put those sights out of our minds.

At Singapore, a look at Change Alley was naturally a must, but we did not buy anything. Tessa and Bruce were both feeling unwell, running high temperatures, so we all wished to get to Adelaide as quickly as possible.

Upon our arrival at Adelaide, a TV crew came on board and wanted to know where I had come from and what I was going to do in Australia. Apparently, it was quite an event in those days when someone from England joined the University of Adelaide! Our arrival figured as part of the news that same evening!

Nigel welcomed us warmly. He was certainly pleased to see us! He had arranged, through the university, for us to have some temporary, furnished lodgings in a suburb called Glenelg. This rather dingy-looking place seemed a terrible anticlimax, so we had to summon up some courage to put our things into the various rooms.

Nigel then proudly showed us a bottle of wine he had bought to celebrate our arrival.

'Not an Australian wine, surely,' I said somewhat doubtfully, remembering the so-called Emu wines, 'empire wines' that one could get in England which were absolutely undrinkable.

'Now, Dad, don't be prejudiced! Just try it before you pass judgement on it,' replied Nigel, chiding me.

I have forgotten the kind of wine it was, but it was surprisingly good, standing equal to some of the best European wines. So we drank to our arrival and to our new life in Australia.

We were about to make sure that everybody was settled, or else at least ready for bed, when we realised that Bruce was not well. We called a doctor.

When he arrived on the scene and saw me he said, 'Oh, I am honoured to be coming to a film star!'

He had just seen us on the news on the local TV channel! He wanted to whip Bruce off to hospital with suspected appendicitis, but, as we all objected, he decided it was not serious enough. He left some medication for him and said to call him if the pain persisted. It did not. The next day Bruce was as right as rain, ready to face anything.

Chapter Sixteen

Getting Used to Australia

So we had arrived in the Antipodes! And it was now up to us to make a go of it. It certainly seemed a very different world from that which we had just left behind. There were several obvious but perhaps trivial details of which we all became aware almost immediately.

One was the prevalence of corrugated iron roofs! Most houses had roofs which would only be considered fit for chicken huts in England; as a matter of fact, our 'chicken hut' near Winchester, in which we had lived for a year, had a more solid roof.

The other immediately noticeable and somewhat jarring detail was the accent. It seemed to us as if everyone were a cockney. It even sounded as though they were putting it on in order to sound comical! Apart from everyone speaking with a cockney accent, it seemed to us strange that just about every other word was the word 'bloody', which also sounded somewhat comical to 'bourgeois' ears, if not a somewhat pointless waste of breath, as, being so frequent, the word lost its original shocking effect.

A rather important difference was the dry, dusty heat. I could not understand what they meant when the locals talked about 'centuries'; of course they meant a temperature of over one hundred degrees Fahrenheit!

But we found the Australians extremely friendly. As soon as it was known at the University that we had arrived, there was not a single day when we were not invited to some party, where we were wined and dined until well after midnight!

We soon found a partly furnished house in a suburb known as Panchito Park and moved in with the few belongings we had brought with us. These were mostly antique pieces of furniture which we valued and we thought that they would remind us of our previous life.

We were advised by our colleagues at the university that the Presbyterian Girls' College (known as PGC) would be the right school to which to send our daughter, Sorrel, and they recommended that Bruce should go to the local state elementary school, as it was the Australian custom, in the case of boys, to 'toughen them up' in such places. Nigel was to start his second year at medical school and Jancis was admitted to do a university degree in psychology on the basis of her American High School Diploma, obtained at Newton High School.

Lectures would not start until the end of February, since in the Antipodes everything happens 'upside down' including the academic year, since the seasons are reversed from those in the northern hemisphere. Even the sun travels across the sky from right to left, as opposed to the left to right we were accustomed to! And of course the pole star was no longer visible at night; you had to do some complex calculations, starting with the Southern Cross, to work out which way north was if you did not already know it. When you were looking at houses, you had to realise that it was the north windows that were getting the sun. It was all very confusing at first!

I remember once, during our Leicester years, that Bruce asked me why people did not fall off the other side of the world. I replied that they had a lot of hooks and they went from place to place hanging on to these hooks, else they would fall! This was much easier than giving him a lecture on the force of gravity and on the fact that the Earth was round in spite of all appearances. Of course, I did not realise that he would believe me but he did. When we landed in Adelaide, Bruce remembered this conversation and said to me, 'Dad, where are the hooks?'

Parents and teachers have to face the problem of how to answer children's questions. You either make up a funny story, which is what I did, or you attempt to explain things. I recall such an explanation from the time I was teaching my six year olds at Dartington. On the other side of the building the nursery school teacher, Nora, was responsible for the four and five year olds. Once, they had just been to the nearby farm and were lucky enough to see a calf being born. This was very exciting, but, unfortunately for Nora, there was a logical little girl in her group who, upon returning to the classroom, asked Nora, 'Nora, how did the first cow come?'

Nora, wanting to be 'modern', went into a long story of evolution, about the sea creatures and how they came out of the sea and then other animals started to develop, until much later, millions and millions of years later, animals like cows and creatures like ourselves appeared! The children listened spellbound to this story; you could have heard a pin drop.

When Nora had finished, the same little girl said to Nora, 'Nora! That was a wonderful story! But will you now tell us how the first cow came?'

Another example of the first kind of reply to a child's question was given once by Roy, one of our Leicester paying guests, to Sorrel, when she was about three. The adults in the house often talked about highly intellectual topics, using long words the children could not understand. Roy and I were discussing Bertrand Russell's idea that the whole of mathematics could be based on tautologies. So the word 'tautology' naturally came up several times in our discussions. Sorrel was listening intently to what we were saying, not making much head or tail of it on account of her tender age.

After one more reference to a tautology, she chimed in, 'Daddy! What is a tautology?'

Before I could think up a tall story to satisfy her curiosity, Roy placed one of his hands on the table, making a shape that looked somewhat like a dinosaur, using his fingers as legs, making the creature walk across the table, saying, 'This is a tautology, Sorrel! See if you can make one!'

Sorrel had a good try at making her 'tautology' as like Roy's as she could and made it walk across the table. She still remembers exactly how the 'tautology' has to be constructed and corrects me if I make the 'wrong tautology' with my hands!

So you see, you can take your pick. You can tell a tall story about hooks or animal tautologies, or you can give a course on Darwin.

Our stay in Australia was to last four years. During these years, many interesting things happened and our stay certainly made us into very different people from what we had been when we arrived. It took us some time to get used to Australian ways, some of which were quite foreign to us. One golden rule we soon learned was never to talk about anything that was not Australian, as they simply hated to hear how things were done elsewhere: the Australians nearly always regarded such talk as an attack on their ways.

Another aspect of life we found strange was how men and women had pretty much separate lives. At a party, men would congregate together at one end of the room and talk about football (the local teams were North, West, South and East; not very imaginative we thought!) while the women huddled in another corner and talked about babies and scandals. At our first party, when Tessa saw a whole lot of handsome men at one end of the room, she went straight over to entertain them. This was unheard of! But we did live it down and perhaps gradually, as the years went by, this forced separation began to be relaxed a little.

Another strange arrangement was that all the pubs closed at six o'clock, though most people did not leave work until about five. A logical consequence of these two premises was the phenomenon known as the 'five o'clock swill'. We found this arrangement singularly stupid and often said so. However, the 'five o'clock swill' was a sacred cow and we were branded iconoclasts for calling it into question! But before our four years were up, the pubs were allowed to open late and the sacred cow was quietly laid to rest.

Sorrel and Jancis settled fairly well into their respective academic activities. Bruce had a tougher time. The local boys thought he was trying to be posh judging from the way he spoke and he was branded a 'bloody Pom' and had his shirt torn off his back. It took him some time to find his way into this exaggerated rough and tumble which was supposed to 'toughen him up'. Sorrel made some friends fairly quickly, as, fortunately for her, the girls in her class were mostly daughters of professional people so they had some things in common.

Tessa found the Australian scene extremely trying for years, in fact practically until we finally decided to leave! She longed for England and English ways and found it particularly hard that such ways tended to be denigrated by Australians. She was even asked once how many hours each day she practised in front of the mirror to get her posh English accent absolutely right!

My schedule at the university was not very onerous. I was given the statistics course, being the only member of the department who knew any mathematics, as well as the course on experimental psychology, which consisted of the students carrying out a lot of experiments and writing them up. I nearly always took them into one of the local schools, where each student would pick a child or a group of children, with whom to carry out any experimentation. They observed the children developing strategies, using the various forms of the Nim game, or special card games aimed at examining certain types of thinking, sometimes suggested by me but sometimes invented by the students. We also studied the development of language in two year old children. We sat behind a one-way screen and each student picked a child for observation. Everything the child said had to be recorded. After the recording session, each student had to construct a generative grammar which would account for the constructions used by his or her subject and predict the kinds of things the child would say at the next session. This would go on for three or four sessions and in the end we obtained a fairly accurate generative grammar for each child observed, as well as a substantially complete word list. We found that, as a child approached the age of three, the grammars became so complex that we had to give up! In fact, most three year old children speak very much with the same grammar as the adults in the child's environment.

Malcolm Jeeves and I soon got down to devising experiments for working on the problems thrown up by my year's work at Harvard. Some of these were to be tackled in the field, simply by ascertaining if certain approaches worked or not; others were to be tested laboratory-style, with experimental and control groups. The structures we thought of using in the laboratory were mathematical groups. In fact, in our first round of experiments we used the 2-element group; the cyclic 4-group; and the 2 by 2 group (or Klein group).

We worked on hypotheses such as whether it was more effective to learn a more complex structure first and then learn a simpler one which was a part of it, or the other way round; or whether subjects would tend to predict 'symmetrical situations' in spite of evidence to the contrary. We also studied the relationship, if any, between the strategies used to solve a problem and the ways in which the solved problem was verbally explained to the experimenter by the subject. We also tried to ascertain whether children 'went beyond the information given', suggested by Bartlett as being one of the distinguishing features of higher intelligence.

We used adult subjects, students in the psychology department, as well as children from the local elementary schools. I tried some fieldwork in the local schools after I had obtained permission from the local education authorities to work in a number of Adelaide schools. My introduction to such schools happened in quite an interesting way. Jeeves and I thought that the best way into the local school system would be through the so-called demonstration schools, in which future teachers were given their practical training for classroom management. I asked the headmaster of one of these schools to let me work in a class.

After a few minutes of conversation, he led me into one of the classrooms and said, 'You can do what you like with these children!'

I did not know it then, but he had given me what was known as a class of 'retarded' children.

I started off by bringing in a number of concrete materials, mainly multi-base blocks and attribute blocks and just told the children to do what they liked with the material, while the teacher of the class and I walked round and observed.

I would throw out the occasional remark, such as, 'Wouldn't it be nice if those steps were more even?' or, 'Your little house does not have any doors or windows!

How would you have to alter it so as to make a door and some windows?' or, 'Does your pile of wood or your friend's pile have more wood in it?' In this way I started to get them to think about what they were doing. They soon realised that the ratio of the size between a piece of multibase material and the next piece up was always the same. In base three, each piece was three times as big as the next smaller piece; in base four this ratio was four; in base five it was five and so on. It was not long before they learned to add and subtract quantities of wood, some even wanting to write down in symbols what they were doing. To a casual observer they certainly did not seem 'retarded'. Anyhow, I did not know that they were supposed to be 'retarded', so I treated them as normal human beings!

A week or so later the headmaster came in and was surprised to see all the children sitting on the floor 'playing with blocks'.

'What are they doing?' he asked me somewhat doubtfully.

'They are learning,' I replied simply.

'They are just playing on the floor! How can they be learning like that?' said the headmaster.

'Some of them are learning to add and subtract in different bases,' I ventured to say.

The headmaster had clearly never heard of 'bases', except as something to do with baseball, so I continued, 'Why don't you ask them to do an adding exercise?'

'All right then,' he said, agreeing to the challenge.

He went to the board and wrote up a sum such as this:

Tons	Hundredweights	Quarters	Pounds
2	12	1	13
+1	10	3	16
?	?	?	?

Then he asked a little boy to come to the board. He said to the boy, 'There are two trucks carrying coal. The first one is carrying 2 tons, 12 hundredweights, 1 quarter and 13 pounds and the second one is carrying 1 ton, 10 hundredweights, 3 quarters and 16 pounds. The two truckloads are delivered to a customer. How much coal does the customer get?'

The boy was not deterred but asked, 'Please, sir, I can tell you that if you tell me how many pounds there are in a quarter, how many quarters in a hundredweight and how many hundredweights in a ton.'

The headmaster gladly furnished the required information, upon which the boy soon gave the correct answer to the sum.

Apparently, this same boy had not been able to do any arithmetic before. He simply had not had any practical experience in exchanging quantities for equal quantities but differently put together. The work with the multi-base blocks had compensated for this lack so he was able to work out the problem, in spite of the fact that he had never before solved a problem like it.

This was the turning point in our relationship. The headmaster said to me, 'If you can teach these kids by letting them play on the floor, I am sure you can teach anyone anything! Let's see you do some work in the other classes!'

Of course, 'anything' was somewhat of an exaggeration but the work in the 'retarded class' had given me free entry to any class in that demonstration school. This is how my Adelaide Mathematics Project started, which soon developed, encompassing several other schools.

I was taken over to the infant department of the demonstration school, which was run by a very dour-looking, strict headmistress. We walked into a classroom with her and you could hear a pin drop, so deathly was the silence! I immediately realised that I had a lot of persuading ahead of me. But it did not take me long to get these very young children out of their desks and on to the floor, and they started to make very rapid progress in logical thinking, using the attribute blocks and learning about regular sequencing using fairy stories and nursery rhymes. Believe it or not, within two weeks I had the 'dour headmistress' sitting on the floor with the children playing logical games with them! She said to me after one of these lessons, 'Dr Dienes! I am retiring in three years, but you have given me a new lease of life! I never knew teaching could be such fun!'

She had apparently understood the difference between imposing on children and letting them grow.

The mathematics I was bringing into the schools was really a Trojan horse. It was not just mathematics, it was a way to look at what learning is all about, or, even more fundamentally, what knowledge is all about. To 'know' something surely is to know how to handle it. Handling means action: present action or at least past action, remembered accurately, burnt into our person as internalised action. So if knowledge is internalised action, then learning must be the process of internalising such action. If there is no action, then there is nothing to internalise, so no learning of any permanent nature can happen. It is philosophies such as these that climb out of the Trojan horse once it is smuggled into the educational system under the guise of essential learning, such as the learning of mathematics.

Many people seem to confuse knowledge with ability to answer questions such as, 'How long is the River Danube?'; 'What are the capitals of all the states in the USA?'; 'In what year did the Battle of Hastings occur?'; 'What is five times three?'

Knowing the answer to the first question does not mean that you have any idea of what the Danube is like, what kind of people live by it or travel on it, whether they are all friendly with each other. Knowing the capitals of the states in the USA does not mean you have the slightest idea of what Americans are like, what they do, how they feel and what they think of the rest of the world. The year 1066 has no meaning by itself if it is not accompanied by a 'knowledge' of what a clash of cultures and languages is like. 'Knowing' that:

$$5 \times 3 = 15$$

does not mean you are aware of how many people come to your party if they come in five cars and each car brings three children. So it is easy to see that I had taken on quite a battle!

We looked at the symmetry problem in our fieldwork, in particular as it applied to logic, where:

If A then B was often thought to be
If and only if A then B

The complex-simple problem was also looked at in the classes with which we worked. Several times I successfully taught the children about complex numbers before teaching them about positive and negative integers, the latter coming out as a particular case of complex numbers. Naturally, all the work was done in 'embodied situations', any symbolic language coming only at the end of a long series of experiences through which the children had 'internalised' the actions they had performed with the materials and with each other.

The normal curriculum was not neglected either. The speed with which children picked up seemingly difficult ideas surprised a lot of teachers. I soon ran out of my small budget provided by the university for mathematics materials. Parents began to wonder why some classes had the benefit of this interesting way of learning mathematics while others did not. So I called a parents' and teachers' meeting.

Before anybody arrived, I wrote on the blackboard the exact specifications of all the materials it would be good to have in every classroom. When the parents arrived I gave them a short talk about what we were doing and why and explained that I had run out of money for having any more material made.

'We could make some of the material,' chimed in one of the fathers.

'I thought that might be the case,' I replied, 'and I took the trouble to write on the board the specifications for the materials we would need.'

There was great enthusiasm for making the materials! We organised a whole working week, during which several of the classrooms were turned into workshops. The place echoed to the sound of hammers and electric saws; the children were roped in to paint the attribute blocks; the mothers typed out the children's instruction cards on carbon paper sheets so they could be duplicated. At the end of the week we had enough material and instruction cards for the whole school and we had a number of offers to make more as and when required.

Many teachers from other schools came to see what we were doing and wanted to know how to do these things for themselves. I then wrote to the state education department, asking if we could organise some courses. The reply came back, saying the powers that be had opined that there would not be much demand for such courses so permission was refused. The teachers were furious!

'I know what we can do,' I said to the teachers in one impromptu meeting. 'We shall organise a course at the university. Some of you already know enough to be able to pass on your skills to other teachers. I think it will have to be a labour of love though! But do you agree to make the effort?'

Several teachers offered their services *gratis* and I did likewise. The headmaster was the president of the local teachers' union, so he had no problem in inserting the following notice in the *Teachers' Journal*:

> Course to be held at Adelaide University on active and highly motivating ways of teaching mathematics in elementary schools. There will be no fees and there are no credits.

Amazingly, we had over one hundred replies. So much for the opinion of the powers that be about there being no interest in such a course! We organised three courses, running concurrently, using contiguous lecture rooms at the university. Each course was run by one of the teachers from the demonstration school and I would go from one room to the other, making sure there were no hitches. The 'course' did not consist of lectures but of hands-on work by the teachers, who learned first-hand how to manipulate the material and so realise how children could learn concepts more effectively through action than through talk.

Eventually the state education department gave in and realised the truth of the axiom, 'if you can't beat them, join them'. They even offered me a retaining fee, which I gratefully accepted. The Adult Education Department at the university got interested in the courses and offered to subsidise them. This would mean that those who gave the courses could be paid, but then a nominal fee at least would have to be charged. We put the situation to the teachers attending, who agreed unanimously that they could pay a small fee in order for those organising things to be paid for their labour.

The interest of the Adult Education Department of the university in our work made one of the local TV channels interested in making some programmes, to let the general viewer see what this new mathematics was really like. So Channel 9 made six, half-hour programs, most of which were taken up by showing how children tackled problems using concrete materials. After we had taken the shots of the children, the head of the Adult Education Department volunteered to be devil's advocate and asked all the embarrassing questions he thought most people would like to ask and I tried to answer them, thus leading into the work with the children. After showing work with the children, we wrapped up each programme with some tentative conclusions about how teachers or even parents could help children to understand mathematics through passing from the concrete to the abstract more gradually, rather than through verbal instruction.

These programmes were shown at prime time and certainly did a lot to inform the public about what we were doing, thus creating an atmosphere favourable to the prosecution of our work.

I was eventually asked to do short workshops in other cities such as Melbourne, Sydney, Canberra and Brisbane. It was during one of these lectures at Canberra that I was asked whether I could accelerate the learning of native children. I replied that I did not know but it was very probable that native children were not so different from children of European origin as far as their ability to learn was concerned and I suggested that they came to have a look at what was in progress in Adelaide. The minister for territories, a certain Mr Barnes, sent some observers to Adelaide, who duly took copious notes of everything they saw in our experimental schools and reported back very positively to him. This is how I came to receive a telephone call from Canberra, from Mr Barnes himself, who asked me if I would be willing to try my hand at improving things in Papua New Guinea.

So now we are back to the beginning of this narrative, which I began by telling what my experiences were when I first landed in the Australian mandate of New Guinea, now Papua New Guinea.

I do not recall now how many times I actually visited Papua New Guinea but it most certainly happened quite frequently. Each visit lasted at least two weeks, some longer. I could not stay too long, as I had my work at the university with my students, as well as the work I had undertaken in Adelaide schools. During each visit I tried to build upon the work done during my previous visits, which entailed concentrating my personal work on rather a few areas. Goroka, Lae and Mount Hagen eventually became the areas I visited most frequently, so that the native teachers could get used to new ways of approaching the teaching of mathematics. We had most of the material made by prisoners, who worked hard to make us multi-base blocks and attribute blocks, as well as sets of cards, so that there was never a budgetary crisis as far as materials were concerned. But getting through to the native teachers was uphill work. It was very much ingrained in them that teaching was telling, not letting the learner experience something which he would learn.

I was once about to visit a school on the coast, which was not in my frequently visited areas and, as I stood on the porch, I heard the teacher say very loudly, 'This is a large, thick, red square! Now, everybody say: "This is a large thick red square".'

As he held up his 'large, thick, red square', the children mechanically repeated that same sentence a number of times.

'My goodness! What have I done?' I thought. 'This teacher has absolutely no idea about how to use the material!'

I introduced myself while the children all stood silently to attention. Then I tried to show this teacher, as gently as I could, how the blocks ought to be used, letting the children ask questions, letting them make mistakes, simply being among them instead of in front of them, trying to be their friend rather than their strict overseer. Here was another cultural barrier to be scaled!

During the time I was making these visits, the university was founded in Port Moresby. One of my tasks was to bring together university faculty, state education administrators, head teachers, members of teacher training colleges and members of various missions of very different denominations. I was told, long after I had left, that one of my chief accomplishments was to get all these people to come and sit down around the same table and talk! Since I did not know this was supposed to be impossible, I just got on with the job and managed to put together a working committee to run the process of reform after I was no longer able to be present. This naturally brought me into contact with head teachers, administrators and missionaries. It will be amusing to tell about some of the events that happened as a result of these contacts.

I was working with the children in a class in a village a few miles from Goroka, where the school consisted simply of a thatched roof supported by some posts. The floor was just mud and all they had for equipment were some slates. I brought my usual blocks in and got the children to 'play', but also to think. Some of the parents were watching the proceedings from the edge of the building. At the end of the lesson, the bell went and all the children disappeared, as well as the teacher and all the parents, excepting one, who was carrying a large bow, some arrows and an enormous spear.

He came up to me rather menacingly and said in broken English, 'No good play. When I child, I kill. These no kill, just play with blocks!'

He banged the floor with his spear and then he said, 'Look necklace!' He showed a necklace made of human fingers. 'I killed all these! My son no kill! No good!'

There was nobody else around, so I thought I must think of something to talk myself out of this. He was getting louder and more menacing every second and I was imagining one of my fingers soon adorning his necklace! Fortunately for me, a small plane, a Cessna I think it was, was just flying overhead.

I said to my wild friend, 'See plane? Your child play blocks, learn fly plane.'

'Play blocks then fly plane?' he asked somewhat incredulously.

'Yes,' I replied. 'Blocks make more brain. Need brain to fly plane!'

'Oh,' sighed the parent, feeling as if he had just been done out of a perfectly good kill.

But he picked up his spear and slowly walked away into the bush. I also sighed, but it was one of relief. But then, I thought, this sort of thing went with the job.

In those days, the punishment for murder there was just a few months in jail. Otherwise I suppose there would not have been enough room in jail for all the people who had killed others. I had heard that a certain man had bought his wife at a heavy price, a great number of pigs, but the wife did not like the husband, so she ran away. The parents, however, would not give back the pigs, so they took her to the husband. This happened several times, until the wife became tired of the procedure repeating itself, took an axe and killed her husband. She ran about twenty miles to the nearest police post and announced to the waiting policemen, 'Please put me calaboose. I kill husband.' She had a summary trial and was put in jail for three months. This was just as well, as otherwise the husband's relatives would have made short work of her. But when she was released, she could not return to her village since she would have been killed almost immediately. So she wandered up into the higher mountains.

Once, I was working at a Lutheran mission station on the far side of Mount Hagen, as far as the road had been cut out of the bush.

One day a woman came to the door of the missionary's house, where I was staying and said, 'I kill husband, no go back village. Come mind children, all right?'

The missionary was just about to send her away when his wife intervened, 'We are supposed to be Christian! You cannot send her away! She won't harm the children – she probably didn't like her husband and so killed him.'

The missionary saw his wife's point and the woman was asked to come in. But before she was introduced to the children, the missionary's wife put her in a hot bath, at which she screamed because she did not know what was happening. She had to get years of pig grease off her body!

She became an excellent worker and a very solicitous nursemaid to the three children and they never had any problems with her.

The mission had been carved out of bush in the past five years, so there was nothing there except a few native huts from five years previously. The natives had never seen a white man before. I was amazed at the banana trees and vegetable gardens, which were producing food in abundance after such a short time!

One head teacher from the coast and myself were at one point detailed to meet a team of linguists studying the languages of some of the tribes that lived in the Highlands. We went by truck from Goroka as far as Mount Hagen.

It was too late to proceed so we thought that we would stay the night and look for the researchers the next day. There was a little 'hotel', with about three bedrooms, probably the only one, where we booked a room. It was soon supper time and, as we went in, we were stopped at the dining room door.

'You cannot dine without having a tie on,' said the person who was impeding our entrance.

'But we only have what we have on,' we objected.

'There is a small store a few doors away. You can buy ties there!' replied our impeder.

We assumed that there must be some sort of pact between the owner of the store and the owner of the hotel so we bought our ties and went to have our supper.

The next day we set out, trying to follow our directions for finding the group of bush linguists for whom we were looking. We found them eventually, encamped in a thinly wooded area. It was very interesting to talk to them about their work.

One of them said to us, pointing up a long valley to the north, 'You know, the tribe over there has a language that has no equivalent of either-or!'

It seemed to me that the valley they were pointing towards was just the area in which the Lutheran mission was situated, where I had been working.

I said to one of linguists, 'Now I come to think of it, I remember the children being totally nonplussed when two different solutions to the same problem were reckoned to be correct by myself! Of course, it was inconceivable for them that 'either' this block 'or' that block would fit the bill!'

We told them the story of the tie at Mount Hagen.

'Oh, yes, we know,' they replied. 'We bought some ties there too! I wonder they have not run out of ties yet!'

'Why don't we have a formal dinner, all dressed up?' suggested one of the linguists.

It was getting towards sundown, so we decided to stay at their encampment and have a 'formal dinner'! There was a large and quite level piece of rock, which we appointed as our table. We put a large sheet on it for a tablecloth, securing it with smaller rocks, and set the table for all of us, while my headmaster friend and one of the linguist researchers busied themselves cooking dinner. We all put on our ties and sat down to dinner. We kept having a good laugh at ourselves, as it was certainly a strange sight to see ourselves having dinner as formally as that in the depths of the New Guinea jungle.

There was one school where the children were really very friendly and wanted me to share some of their lives. We thought it would be good policy as well as good ethics to go along with this sort of tendency. One little boy, of about seven or eight, once asked me if I would like to come and visit his grandmother. I agreed and we started out into the bush. We appeared to be climbing higher and higher.

The forest was getting thicker and I kept asking, 'But where does your grandmother live?'

'Not far now,' my young companion kept saying.

After what seemed a very long time to me, we reached a rocky ledge on which a skeleton was sitting. By the skeleton was an empty plate.

'You see, Grandmother has eaten what I brought her last time,' the boy said and proceeded to put some sweet potatoes on the empty plate.

'Grandmother' must have been there some time, as there was very little flesh – she was nearly all bone and I did not even notice any bad smell. I suppose the heat dried the skin, which perhaps kept the bones more or less together, so 'Grandmother' was, at least not yet, just a heap of bones! And the birds no doubt would make short work of any sweet potatoes that might, from time to time, be placed on the empty plate.

The boy talked to 'Grandmother' in his own language, after which we said goodbye and wended our way down to the village.

One trip took me as far away as New Ireland. An education researcher from Denmark wanted to come with me and take moving pictures of what I was doing, so we landed together on the romantic island of New Ireland. It might be worth mentioning that the island seems to be divided into two halves from the point of view of the weather. When it is the dry season in one half, it is the wet season in the other half and *vice versa*. We were aiming at some spots somewhere near the middle, so we knew that we could expect any kind of weather at any time!

At one of the schools, I was trying to introduce some topological ideas into the curriculum. In the case of children of elementary school age, this meant being aware of insides and outsides, connectedness and separateness and suchlike. Paper and pencil work on this did not seem to mean much to the children, so I took them to the shore of a sandy lagoon and asked them to make a harbour. They made several harbours and also made some 'connections', so that 'boats' could go from one stretch of water to another. The whole complex of islands and canals became extremely complicated before we finished the project! I would ask two children to be two crocodiles and stand in the water and then asked them if they could 'swim' to each other without going over dry land. Afterwards they had to be wallabies and they had to decide whether they could get to each other without jumping over some water. After these experiences, the paper and pencil work became much more meaningful and the children really enjoyed making their own mazes for imaginary crocodiles and wallabies.

In another school, which was near a large stream with many waterfalls and deep pools, we worked on the rotations and symmetries of the square by having four children make a square by holding huge banana leaves for the 'sides' of the square, the children themselves acting as the 'vertices'. Rotations were quite easy to carry out, but for the reflections they had to be careful not to let go of the leaves as they twiddled themselves around under and over the banana leaves. Every time a problem was solved, the reward was that they could jump into the pool for a few minutes! Needless to say, they learned very quickly the relationships between the various rotations and reflections of the square!

As another 'embodiment', I asked the children to examine how the leaves grew on a papaya tree (which Australians call pawpaw). They soon realised that there were eight large leaves on each stem, making a beautifully symmetric shape, but that underneath, hidden under the large leaves, was another poor, little, lonely, ninth leaf! They would take this leaf off and carry out the turns and the flips with the papaya leaves which they had done with the banana leaves. This was to encourage the so called 'abstraction process' in the minds of the children! But the ninth leaf also came in as the means of generating the second power of three in a home-made base three set. The unit was the leaf, three leaves growing together separated from the others provided the first power, one whole stem with the nine leaves on yielded the second power and three stems tied together gave us the third power! So we could work on base three arithmetic just by picking a certain number of leaves from the papaya trees, which seemed to abound just about everywhere.

I noticed something else in the coconut tree plantations. The older plantations had been planted in rows and columns, thus making squares whose vertices were the trees. Later plantations were based not on the square but on the equilateral triangle. Of course one could plant more trees that way in a given area! Somebody at some point, from obvious financial motives, must have discovered this simple mathematical fact. So I brought the children to both kinds of plantation and got them to measure a certain land area by pacing out the length of each side of the square-shaped area we needed. We did this in both kinds of plantation, making sure that the two areas were, at least approximately, the same. Then I asked the children to count the trees in each area. They were surprised to learn that there were more trees in the triangulated plantations than in the square-based plantations. We also played triangle games by putting coloured ribbons round the trees, using just three colours, but so that no neighbouring trees had the same colour ribbon.

I got the children to run from tree to tree, noticing each time the colour behind them and the colour in front of them. They were allowed to either proceed straight on when passing a tree, or turn through sixty degrees and make for the next tree. They would start in any position and give themselves the final behind-colour and the final ahead-colour and they had to reach that combination of colours in the smallest number of moves, a 'move' being either a 'go straight on' or 'turn through sixty degrees'. This game was eventually compared with the reflections and rotations of the equilateral triangle – another 'abstraction exercise'.

In one of the schools where I was working, the teacher asked me if I would teach the children about time. Of course, in the tropics there are essentially no seasons since the sun rises and sets about the same time all the year round so what is there to make the passage of time stand out as relevant? I showed them my watch. They were intrigued by the hands going round, but could not link it with what I thought would be 'time' for them. So we went outside and placed a very large pole in the middle of a grassy patch. At the start of school, at nine o'clock, we went out to our pole and placed a rock where the shadow of the pole ended. We did the same at ten o'clock, then at eleven o'clock and we continued until sundown; some children stayed behind to see what would happen to the shadow towards evening. We marked the rocks with the appropriate hour numbers and thus had a primitive sundial.

I also had a stopwatch and we ran races from one tree to another, more distant tree, while a child was detailed to stop the stopwatch when the far tree was reached by a runner. They were surprised to learn that a 'big number' of seconds meant a slower run! This was probably their first experience towards the understanding of the concept of inverse ratio.

Some months later I visited the same school again. The children were pleased to see me but sadly informed me that our 'clock' did not work any more! Of course, how stupid could I have been! In the tropics, the sun 'goes round' from left to right during one half of the year and then it goes round from 'right to left' during the other half. I explained this to the children, who were happy to place more stones on the opposite side of our pole so now they had a 'clock' that would always work!

My work in New Guinea was divided between working in the field to ascertain what was practicable and what was not and working with the curriculum committee I had set up, not to mention giving workshop periods attended by teachers and head teachers. During work with the committee, we tried to put together a scheme for mathematics which would give native children basic mathematical ideas as a result of their own experiences, as well as give them work calculated to improve their reasoning capacity. We would work from morning till night each time the committee sat, with breaks under some shady tree for coffee. After coffee we would often march into work singing the Walt Disney song, 'Hi Ho, Hi Ho! It's Off to Work We Go!'

Occasionally I had a day off, when I would be invited to spend the day with one of the teachers. There was one Hungarian teacher who invited me to sail with him one day. We had a visitor from Sweden (they seemed to turn up from all sorts of places!) so the three of us went out in my Hungarian colleague's sailing boat. What he had not told me was that he had only been out sailing twice before in his newly acquired craft!

I recall trying to land on an island, but somehow the rudder broke. I tried to use an oar as a rudder while my friend managed the sail. However, the wind was coming from the land and we came within a few metres of the shore, but could not make it. There was nothing for it but to make for another island down wind. We did so, although we noticed it was the last island near the coast; to the east lay the vastness of the Pacific Ocean! But I managed to steer the boat into calm waters and we finally landed on the island. I asked my friend for the tools so we could mend the rudder.

'Tools? I don't have any tools!' he said simply.

It was the classical TV 'desert island' situation! So why not do what they do on the TV? We stuck the oar into the sand and placed a shirt on it to act as a flag and waited. The 'desert island' was truly deserted: there was literally nothing on it but a few bushes.

Fortunately for us, another sailing boat came by and noticed our flag. They came into the shallow bay and stopped there. I jumped into the water, swam out to the boat and asked them if they could lend us some tools, in particular a screwdriver. The sailor passed me some tools, which I took and I swam back to the shore, trying not to get the tools wet. We finally mended the rudder and managed to sail back to the bay from which we started.

To celebrate our safe return, my friend and I jumped into the water for a swim. But we did not reckon with the vagaries of the weather. Quite a wind started to blow and our Swedish friend had no idea what to do with the sails! We both started to swim for the boat. Fortunately, my Hungarian friend was a faster swimmer than I and he caught up with the boat and got in. By this time I was a long way away. He made an about-turn and I had to work out which way to swim in order to intercept the course of the boat. It reminded me of the problems we had had to do at school about relative velocity. Lacking paper and pencil in the middle of a lagoon, I had to guess. Fortunately, I guessed correctly and soon all three of us were on board the sailing boat, making for home. We were definitely ready for a hot supper! The tools eventually found their way back to the Port Moresby Sailing Club, where they were received by the rightful owners!

The news of our work in New Guinea travelled far and fast. One day I had a call from Saipan, in the Mariannas, asking if I could come and do a week's work with the local teachers and children. I took a teacher with me from the Adelaide demonstration school and we both had a wonderful week introducing our new ways to the local schools. I am not sure what use they eventually made of what we showed them, as I never heard from them again! Maybe we just shocked the authorities with these novelties!

While my professional life was expanding, the children were growing, making friends and getting used to Australia.

During our four years in Australia, we had four different homes. We never purchased a home in that country, but rented something for the academic year and moved out at the end of it, storing our belongings. The three intervening long vacations were spent in other parts of the world.

The first one was spent largely on the big island of Hawaii, working as a mathematics consultant at the University of Hawaii, mainly engaged in the training of those Peace Corps volunteers who were bound for the Philippines or Borneo. The whole family came to Hawaii with me.

The main reason for the second trip abroad was the fact that I was invited by UNESCO to take part in an international Conference on Mathematics Teaching, held in Budapest. I flew to Budapest via the Far East, but came back via America, so this was the first time I had actually circled the globe. I did this trip alone, leaving the family behind in Australia.

The third long vacation was again a family one, when we circled the globe in the opposite direction, namely from west to east. The chief part of my time was spent at the University of Minnesota in Minneapolis, working with Paul Rosenbloom, but it also included some work with Professor Suppes at Stanford and at the University of Sherbrooke in Canada.

For Tessa, and to some extent also for the children, this kind of life was bound to be somewhat disturbing. It is all very well for a man to follow his career, wherever that takes him, but such frequent changes do cause problems for a family, since getting used to places is a slower process for them than it is for the man whose job takes care of the major part of his interests. Members of the family must get used to changing details in their everyday lives as well as breaking and making relationships as the family moves from place to place.

One thing about Australia that some find hard to get used to is the local fauna. We were told by the locals that we would never see a snake, that they were really very rare. But, sure enough, while Tessa was sunbathing in our garden she saw a large snake rearing up quite close to her. She screamed loudly and our son Nigel came running and instantly killed the snake. He had already killed a 'lazy lizard', because he did not realise that these curious-looking animals were quite harmless. We had the garage converted into a bedroom for him so it was easy for such animals to get in there. 'When this thing came in,' explained Nigel, 'I thought it looked like a baby crocodile, so I killed it!'

Another time, while I was giving a course to teachers, Tessa and the children took a walk amongst the sand dunes surrounding the building where I was working. She saw a tiger snake when Bruce was just about to step on it! She picked up Bruce just in time and fortunately the snake just slunk away quietly.

We got into trouble on account of Bruce another time, when he decided to practise shooting with the bow and arrows I had brought back from New Guinea as payback from the parents of the children I had worked with. He was shooting the arrows from the back garden, over our roof and the arrows were landing in the road in front. These were hunting arrows, meant to kill! When I had explained this to him, he stopped shooting them over the house.

Even I had some trouble getting used to Australians' very frequent use of the word 'bloody'! I was once giving a talk about the state of mathematics education in South Australia, strongly expressing my views on it, filling it with almost as many 'bloodys' as an Australian would have done. When I had finished, there was very enthusiastic applause, one of the teachers saying very loudly, 'You are really one of us now! That was really good!' So, it seems, it takes the embracing of such behavioural details to accomplish the process of adaptation to a different culture.

In the meantime, the two older children were becoming involved in various degrees of 'love lives'. We had a note from Jancis one day, in which she announced her 'entanglement' with the son of the conductor of the local symphony orchestra, while Nigel had become very keen on a very pretty, young girl with gorgeous, long hair. The 'entanglement' lasted quite a long time, but they had their ups and downs. Nigel's girlfriend soon became his wife and they had a little boy very soon; and so, for the first time, Tessa and I became grandparents. Unfortunately, they did not get on at all well and soon separated, the little boy remaining with his mother. To this day we have no idea how he has fared in life, in spite of efforts we have made to trace him. Corin, having married John, came out to Australia because John had accepted a post at Armidale in New South Wales and she eventually became pregnant. Life seems to go on inexorably, does it not?

Tessa and I and some friends were having a little party one day, when Tessa complained of having some bad pains. This went on for some time and then suddenly they stopped. For no reason she could think of, she went to the phone and called the hospital in Armidale where she knew Corin was going to go when the baby arrived.

She got through to the sister in charge who said to her, 'Your daughter has just delivered some beautiful twin boys and she is just having a blood transfusion. She and the boys will be quite all right, so don't worry!'

So Tessa had had the labour pains of our daughter and the time they ceased was precisely the time when she was giving birth to the twins! Wonders will never cease!

It was not long before we went to Armidale to see Corin and the new babies. We had purchased a second-hand car, quite a decent-looking one, and drove to Armidale, stopping just once for the night on the way. It was great to see Corin again and we were very proud grandparents.

We did not know at the time, but her marriage was not to last very much longer. One day Corin and two very bedraggled-looking twins, nearly two years old, suddenly appeared on our doorstep and asked for shelter. So Corin's love life was having its first deep crisis; there were many to follow. As parents, we wondered what we had done or not done for such things to happen to our daughter. I suppose we can guess but we shall never really know!

During the third year of our stay in Australia, we rented a really old house, with a wonderful, wild garden at the back. There was also a big veranda at the back, covered with a grapevine and one could sit and pick grapes while sipping passion fruit juice, which also grew in the garden. We had fig trees, peach trees, orange trees, lemon trees and even a small kumquat tree! The conveniences were not very modern: there was an outhouse in the back garden, in which the toilet was situated. One day Jancis and I decided to smarten up this aspect of the property and painted it in new, gleaming colours. We called it Santa Pidella, a parody of my childhood song *Santa Lucia*!

When we left, the wonderful old house was pulled down, our 'secret garden' was bulldozed and a block of flats constructed where our romantic abode had stood. Before we left, we saved a black and white reproduction, probably a photo, of the Mona Lisa, which adorns the entrance to our bedroom in our Devon house to this day. When we look at it, we think nostalgically of that year we spent in our old house and romantic garden.

After three years in the Psychology Department, I was offered a 'personal chair' in the university Education Department. I could never find out why this could not have been organised in the Psychology Department; maybe there was red tape unknown to me, but, anyhow, there it was. So I accepted my transfer to education, with the understanding that my duties would not change; in other words, I was still to work in the Psychology Department as though I were a member. A personal chair was a considerable honour, as it was created for a person on account of his or her accomplishments and was not to become an established chair in the university. This worked well for a year, after which the professor of education decided to give me a very heavy load of lectures in his department while I had to continue with my work in the Psychology Department as well. When I reminded him of the undertaking that I was to keep to my original schedule, he told me that that arrangement had no legal basis. I tried to contact Professor Jeeves, who was away in Europe, but was not able to. I had an interview with the vice chancellor about my problem and he informed me that 'legally', the professor of education had every right to determine my schedule since I was 'legally' in his department.

At about this time I had been offered the headship of a research institute at the University of Sherbrooke in Canada, whose aim would be to research the problems of mathematics education and would be called the Centre de Reherches en Psychomathématiques. This was extremely tempting especially since I did not relish the idea of continuing to work under such an unreliable head of department, which I considered the head of the education department to be. So we held a family council.

I announced to the rest of the family, 'There is a possibility of our moving to Canada to live. Or we can stay in Australia. If we stay in Australia, I shan't want to hear any more about things being this way or that way in Australia and isn't it awful! If we decide to stay we stay and we take the good with the bad. If we go, we go to something unknown, but it is a new adventure. Which is it to be?'

There followed a lengthy discussion amongst Tessa, Sorrel, Bruce and myself. The other two would certainly want to stay, as they were getting established and they had their love life and their friends all mapped out. In the end the children voted for going to Canada; Tessa was somehow a little uncertain; I voted to go too. It seemed that Tessa did not want to impede another interesting step in the furthering of my career so she did not say no. Often I wish that she had! Had we all stayed in Australia, possibly many unhappy things which happened to us later would not have occurred. We would not have made money and I would not have become famous, but, then, what is more important in life? Unfortunately, one cannot put the clock back. If I had the same choice to make again, I think I would make the Australian choice, hoping that somehow, with good will, I could handle the problems at the university!

So I went to the vice-chancellor and gave in my resignation.

Chapter Seventeen
The World Trips

The first one of these took place after one academic year in Australia. The professor then responsible for organising Peace Corps work in the University of Hawaii, a certain Jack Stalker, invited me to come and do some work in the islands, mainly for the Peace Corps, but also for some of the so-called neighbour island school systems. Tessa, at that time, was still extremely nervous of flying, so we decided to send her and the children, including Nigel and Jancis, by P&O Liner. I would fly out to find suitable accommodation and meet the boat in Honolulu. The Adelaide University vacation lasted from the end of November until the beginning of March, so we had three months to play with. I found myself in Hilo at the end of November, after having put the family on a transpacific liner from Adelaide to Honolulu.

I soon found a furnished house well up the hill, out of Hilo, on the edge of the jungle, as it was only a few years after one-third of Hilo had been destroyed by a tidal wave. There was a very pleasant view from the house, overlooking Hilo and the bay beyond, with the jungle reaching right up to the back of the house. As I walked up the steps leading to the veranda, I had not reached the halfway mark, when I fell right through! The termites had been at their work for some time it seemed. These interesting creatures have a way of consuming the entire inside of a wooden structure, leaving it looking quite unaltered! This is what had happened to our steps. Fortunately, there was another entrance, which had not as yet been attacked by these clever white ants, so I went in and made myself at home.

Jack and I discussed what work I should do and I suggested that the first thing to do was to make mathematics materials. So within hours of my arrival, several carpenters were on the job making enough materials not only for the Peace Corps kids but for use in some of the local schools, in which we intended to set up some demonstration classes which the Peace Corps kids could observe and eventually handle themselves. So I was introduced to the principals of two of the nearby elementary schools and they seemed pleased to let me come in and do some mathematics with the attending children. After taking on some of the classes in mathematics, the teachers of these classes were fired up by the enthusiastic reception of my approach by the children and wanted to know more. So we organised some teachers' workshops and the project grew quickly, somewhat on the lines I have already described in the case of the Adelaide work.

The family duly arrived a week or two later, after a very pleasant voyage had been enjoyed by all. We installed ourselves in our furnished house and Bruce and Sorrel were sent to the nearest elementary school, Bruce going to grade three while Sorrel went to grade six. They soon made friends, so they had no problem adapting to the new situation. There was just one little problem which bothered them: all the children went to school with bare feet, Sorrel and Bruce being the only ones who wore shoes! Children always have to be like their peers, so they were unhappy when Tessa told them that they had to wear shoes to school. They solved the problem with truly childish logic by leaving the house with shoes on, taking them off on the way and hiding them in a safe place so they could arrive at school just like the other children. Since they always came back properly shod, we never discovered this ruse until years later, when the children told us about it. Sorrel was in one of my experimental classes, but I never put two and two together since I was trying to occupy all those barefooted children with interesting mathematical games, Sorrel being one of them!

My schedule was quite a heavy one, as I had to teach the Peace Corps volunteers mathematics methodology, as well as go round the two schools teaching the children, and hold teachers' workshops. There would be buses waiting with their engines running after each workshop session with the volunteers, which took them to the schools where we worked with the children. Then I would go back to the centre and teach more volunteers, until the end of school, when, as often as not, we had workshops for the local teachers. So it was good to have some leisure occasionally when I could relax with Tessa and the children. We had been lent two cars for the time we were there; Nigel drove one and I drove the other so we were able to get around the island on free days.

One Sunday, a student telephoned our house asking if we would like to go to the beach with a group of students. It was bucketing with rain so I said no thanks, in such weather we thought it was wiser to stay at home.

'Oh no!' insisted the student. 'We are planning to go to the other side of the island. It hardly ever rains there!'

I could hardly believe this when I looked out of the window and saw the sheets and sheets of rain pouring down from a very grey sky. But within minutes the students were there, so we got into our two cars and followed them. We crossed the mountain range in the middle of the island and reached the eastern shore. By this time it had stopped raining. A few miles farther on, the sun came out and continued to blaze down for the rest of the day!

Looking round I also noticed a sudden change in the vegetation. Instead of the luscious, jungle-like vegetation of the other side, the place seemed littered with cacti and other such desert-loving vegetation. This was my first introduction to microclimates, which appear to abound in tropical areas of the world.

Later on there were some suggestions made about my joining the faculty of the University of Hawaii and I remember going to an estate agent on Kalakaua Avenue in Honolulu and asking if we could have a look at some property.

'What kind of climate would you like, sir?' was the first somewhat strange question asked by the agent.

'What do you mean?' I asked, somewhat nonplussed.

'Well, you see,' he replied, trying to be very patient with this ignoramus of a customer, 'up the valleys you get much more rain, so your lawn will be greener and you don't have to sprinkle so much. The nearer you get to Wakiki, the drier it gets.'

We never had to face the climate problem in Honolulu because the negotiations about joining the faculty broke down at a certain point, so our stay in Hawaii was destined to be only temporary.

The work with the volunteers also included taking them out into the bush. Once or twice we had to use mules, just to introduce them to this time-honoured method of transport, which might be what they would have to use in Borneo or in the depths of the countryside in the Philippines. The mule train took us along quite dangerous-looking trails, sometimes following narrow ledges with a vertical cliff on one side and an equally vertical drop on the other, with just enough room for one mule! We all held on for dear life and hoped that the mules knew what they were doing. They did!

Another exercise was getting used to local food, where the word 'local' meant Filipino local! Apparently, one dish which is served as a delicacy for honoured guests is lightly cooked, fertilised egg! One was not sure to what extent the egg was still an egg or whether it was already becoming a chicken! But eat it we did! The students had to learn how not to offend the feelings of the locals who would be entertaining them. There were several other interesting dishes, some of them cooked in blood, which also took some getting used to. But it was all part of the job, though it had never been mentioned by Jack in the job description!

The work went on to some extent by improvisation, but we had to do a certain amount of planning. This often took place at the Steak and Lobster, which was a restaurant built on stilts over the lagoon. The 'fog machine' had to come by quite frequently, in the shape of a truck spewing anti-mosquito fog, since lagoons are the favourite haunts of these unpleasant creatures that delight in drinking our blood.

I remember one evening when we sat down to a planning session and Jack said to the waiter, 'Bring two bottles of white Almaden and lots of good cheeses. And when the bottles are empty, just bring some more and when the cheese tray is empty, fill it!'

After about an hour of planning our work, accompanied by frequent refills of both the bottles and the cheese tray, Tessa tracked us down and insisted in joining the feast. By this time we had consumed more than ten bottles of said Almaden between us, so Jack was hardly in any state to say no! I think the total number of Almadens consumed reached the astronomical total of thirteen and I lost count of how much cheese we consumed. But the work for the following week was all planned and we could go back home and sleep on it. We hoped that the American taxpayer would not think that we had wasted any of his tax dollars!

Jack wanted me to show what I could do on some of the other islands, so occasionally we made short trips to Maui and Kauai.

I recall Jack making a phone call to Kauai saying to whoever was at the other end, 'Now you will make sure that Dr Dienes is properly looked after, won't you? You know what I mean, don't you?'

I was not to find out the significance of that phone call until Tessa and I eventually landed on the airstrip of that island and were welcomed by a beautiful and graceful Tahitian woman, dressed in a flowing *mumu*, with equally flowing, long, black hair.

'Aloha! Aloha!' she greeted us as we descended the steps from the Hawaiian Airline jet. 'How nice to see you!'

After we exchanged a few more platitudes, she said, 'I will take you to my shack. It will be better for you there than in the hotel.'

We wondered a little what it would be like to sleep in a shack. We need not have worried. It was a beautifully equipped little house, built on a lagoon quite by itself. In the middle of the main room was a large sunken area, made up as a bed. Our hostess had obviously not expected Tessa to come as well, yet everything seemed to be prepared for two, including the enormous bed in the sunken area. I suddenly realised what Jack had meant when he said on the phone to make sure that I was 'properly looked after'! Indeed, such romantic arrangements went far beyond the call of duty!

The beautiful Tahitian woman glided out of the house and out of sight and left Tessa and myself temporary masters of the 'shack'. We could not fail to have a romantic time in such a setting!

In the morning I thought I would have a dip in the lagoon. I tiptoed out of the house so as not to wake Tessa and walked on to the shore of the lagoon. It was as still as a mirror; I hardly liked to disturb nature's wonderful masterpiece reflected in the still dark water! But in the end I did so, though getting immediately on my back, so as not to disturb the sea bottom. I swam for a few minutes and came ashore as I had entered the lagoon, using my hands as oars while approaching the edge of the water on my back.

I noticed our Tahitian hostess striding down the path amongst the trees. She also noticed me and, when she saw me at the edge of the lagoon, she shouted very loudly, 'You mustn't swim in that lagoon! There are animals on the sea floor that would bite your foot off!'

'I am just coming out – I've already had my swim,' I replied. 'And I still own two perfectly good feet!'

'Thank God for that,' sighed the beautiful woman. 'I had forgotten to tell you about not swimming in the lagoon!'

When I explained how I took my dip, she realised how the 'miracle' of my not being bitten by one of these fierce creatures had occurred.

I gave my talk later that day to an appreciative audience. In the meantime, Tessa had been introduced to the conductor of the Honolulu Symphony Orchestra, who lived on Kauai, so they went shopping together. When it was time to go back to the airstrip to catch our plane back, she was loaded down with some gorgeous *mumus*. I am not sure who paid for the *mumus*; I somehow doubted that Tessa had enough funds with her for such luxurious shopping! Then I supposed that was also part of the perks of the job I was doing.

During our stay in Hawaii I was invited by the University of Chicago to work with the university Education Department for about four days. So I had to leave the family in their tropical paradise and fly to Chicago. This was very interesting work, for I realised soon after I had made contact with the faculty, that they were probably ahead of most American universities in their educational thinking, and they welcomed my ideas. I gave some workshops with children and teachers which were well received and, as a consequence, they asked me to write an article, which I did. The article was published in *The School Review*. I was looking over a reprint of this article the other day while I was sorting out things in my files and I felt that the thinking expressed in the article had stood the test of time.

Before I knew where I was, the Chicago assignment was over and I found myself in Hilo again. I resumed the work I had begun and really tried to lay the foundations for some serious work in the future, since we were having serious discussions about coming to Hilo (or Honolulu) on a permanent basis.

I had missed a very important traditional Hawaiian feast known as a *luau* as a result of my trip to Chicago. All the family had enjoyed the *luau* put on by the Hilo faculty for the members of the team training the Peace Corps volunteers, at which a pig was roasted in the ground in traditional style and there was much hula dancing! I shall never know now whether it would have been better to attend the *luau* or to spend some days interacting with the Chicago University Education Department!

Sometimes we were invited out by friends we had made in the neighbourhood, sometimes we ate in restaurants, but mostly we ate at home. In the local supermarket we could obtain everything Hawaiian, in particular items such as mango juice or papaya juice. One night we went to dine at the 'drive-in volcano', which was built right on the edge of a crater. But this was not the main volcano, with the cinder cone in the centre of the island. The large crater must have been formed many centuries ago as a result of a very large eruption, but within living memory the crater produced only small eruptions, though quite frequently, so one could actually have dinner while watching lava fountains shooting into the air.

One night we awoke to the sound of the tidal wave warning. Tessa became very nervous and wanted to wake up all the children and climb high on to the mountain to avoid catastrophe. It took me some time to explain to her that any tidal wave which could reach as far as where our house was would need to be comparable to Noah's flood. So finally she agreed to sit it out. In the end it turned out to be a false alarm, although there were some very high waves hitting the shores of the Big Island that night.

All these exciting events came to an end when we had to leave, so I could resume my duties at the University of Adelaide. Goodbyes are always sad, especially when a period of time we have enjoyed comes to an end.

The children in the experimental classes had become very fond of us, particularly of Sorrel and Bruce, and there were some tearful goodbyes accompanied by the passing of *leis* around necks. During the bestowing of a *lei* you are allowed to kiss, otherwise it is absolutely taboo, so a lot of children made use of the lifting of the taboo at our farewell ceremonies. We had to promise that we would be back, after which they sang us some Hawaiian songs accompanied by ukuleles, which the schools provided for every child. So we said goodbye and hoped it would not be for ever.

I wrote up the work of the Hawaiian Mathematics Project in several articles for the new *Bulletin of the International Study Group for Mathematics Learning*, which was intended for a group of mathematics educators whom I tried to pull together from different parts of the world in order to coordinate such work internationally. This bulletin later became the *Journal for Structural Learning*, in which we published much of interest to mathematics educators and psychologists working in the field.

So this was the end of the first 'world trip'. I think we all learned some valuable lessons in cultural adaptation and I took another lesson about how to adapt my developing methodology for mathematics teaching to different situations. Tessa and the children were booked back to Australia on an ocean liner, while I wrapped up the work in Hilo and Honolulu and finally took the plane back to Adelaide.

The second world trip arose out of an invitation by UNESCO for me to attend an international meeting on mathematics education in Budapest. UNESCO paid my first class round trip fare from Adelaide to Budapest and back; when I found that I could exchange that ticket for a round-the-world one for an additional twenty dollars, I decided to do just that and plan some more work in Europe and in the USA, including a return to Hawaii. In Budapest I met almost the same gang I had got to know in Krakow a couple of years previously, with one or two additions. The additions were mainly from the Soviet Union, who seemed very keen to learn from the rest of us about stepping up the efficiency of mathematics learning, though at the upper echelons they were already doing better than any of us!

The language of the meeting was French, although there was simultaneous translation into English and Russian. Papy gave some rousing contributions, accentuating his discourse by what I can only describe as a ballet, moving up and down the podium with great rapidity and making full use of his arms to emphasise what he was trying to convey. I reported on the work in Leicestershire, Adelaide and Hawaii, stressing the necessity for controlling the environment of the child rather than lecturing him if we wanted him to learn. This description of the dialectics between a controlled mathematical environment and the learners seemed to appeal to the Marxist-Leninist members of the group and Tamás Varga asked me if I could show some of what I meant in Hungarian classrooms. I visited some classrooms and persuaded some teachers to start working towards an experientially-based mathematics curriculum, which would be guided by Professor Varga on the spot; I would be their intermittent consultant. This meant that Professor Varga joined the International Study Group for Mathematics Learning and, for years to come, he would religiously attend all our meetings. The Hungarian Ministry for Public Instruction supported the movement and soon we had a fair number of schools working from specially made instruction cards prepared by Professor Varga and his team (Munkalapok!), using concrete materials, all provided by the ministry.

During the course of this conference, I asked Pescarini whether he had started to establish any such classes in Italy. He replied in glowing and poetic, flowery Italian about the plans he had for doing so, but nothing had been started as of then.

I stayed on with my mother and my brother Gedeon a few days after the close of the meeting and thought about Pescarini's flowery speeches. I decided to call him on the phone and suggest that we started something right away.

I got him on the phone and said, 'Look here, Angelo, we have had enough of *bei discorsi* (beautiful speeches). I think it is time to start some real work. I am coming to Italy so please meet me in Ravenna tomorrow at the station. I shall fly to Milan and take the train to Ravenna!'

He was a bit taken aback but agreed to meet me.

I duly arrived the next afternoon and Angelo Pescarini was there on the platform waiting for me.

He took my arm and said, 'We are in luck! A very prominent citizen of Ravenna is having his funeral this afternoon. Everybody who counts will be there. Let's go to the funeral and see whether we can get permission to start work.'

I remember walking slowly and solemnly after the coffin, having left my bags at the left luggage area at the station, but, as soon as we were able to get a decent distance away from the remains of the prominent citizen, we made some discreet enquiries from the provveditore of the district, who was walking solemnly and slowly a little farther back in the procession. He made some non-committal remarks about how interesting it would be to be in the stream of international advances in the teaching of mathematics. This was enough for Pescarini!

The next day we went to a small village some distance away from Ravenna, San Pietro in Campiano and met a teacher called Pasini, whose class was to become our first experimental class. Pasini was a born teacher. He put his heart and soul into his work and the children ate out of his hands. He was immediately impressed by the enthusiasm with which the children greeted my suggestions and gladly joined in the 'games' himself. This was the start of a long period of collaboration between Pasini, Pescarini and myself; in fact, to this day, although Pescarini is now retired, we work together for the betterment of mathematics education in Italy. So after a few days' work, I thought I had started something and I was satisfied that Pescarini and Pasini would keep the fires burning until my next visit, of which there were to be a great many.

The next stop was the Rousseau Institut in Geneva, where I was invited to come and work for as long as I had time. This was a very fruitful and interesting time for me, even though it only lasted about a month. The same Monsieur Roller who had looked after us on a previous occasion, when we had visited with a bunch of children and shown them mathematics in the park, looked after me this time. He found me some reasonable lodgings and introduced me to Jean Piaget, Inhelder, Papert, Greco and a number of other very well-known workers who were collaborating with Piaget in untangling the developmental secrets of children's intellectual growth. I went to see some of the experiments in progress and attended the sessions at which the experimenters came to report on the results. Everyone had got used to some typical 'Piagetian responses' that children would give when they were still 'pre-operational' in whatever was being tested. One day a student reported that a child subject, upon responding 'Piagetian style', looked at the experimenter, who seemed perplexed and asked, *'Mais, monsieur, ce n'est pas ça que vous voulez que je dise?'* ('But Sir, isn't that what you want me to say?') We all burst out laughing and Piaget gave us a lecture about having a poker-face during such experiments, so as not to let on anything, even unconsciously, to the subject or the experiment would not be valid!

I also tried to do a little work in Geneva schools. When I was first taken to an elementary school, I was struck by the big notice over the entrance which said, very sternly, DEFENSE D'ENTRÉE AU PUBLIC. I assumed that we were not identified with the general public, so we marched straight in. It is part of the Piagetian gospel that, during the so-called concrete operational stage, children could learn effectively only through experiencing concrete situations. I thought, just for the hell of it, that I would challenge this theory, although on the whole I was in agreement with the Piagetian ideas being developed at the institute.

I invited Piaget, Inhelder and some of the students to a demonstration in one of the 'forbidden' schools. I told the children a story about a dance hall where boys and girls and men and women went to enjoy themselves, but where there were certain rules about the persons with whom you could dance. For example, you could only dance with a member of the opposite sex and only with a member of your own age group. If you could not find a 'legal' dancing partner, you waited in the refreshment room until someone suitable arrived. I even complicated the story by telling the children that if people stayed too long in the refreshment room, then a member of the manager's family would come to the window and knock, which meant that they would have to go, but send back other customers. Whom they would send back depended on whether the person knocking on the window was a boy or a girl or a man or a woman! Then several members of the manager's family would come to the window at the same time and they would each have to be obeyed. All this created pictures in the children's minds, which they would then manipulate to solve problems about who would be in the refreshment room after certain events had taken place. The mathematics involved was considerable, but only 'imaged experiences' were provided. The success of the venture seemed to go against a strict interpretation of the 'concrete operational stage' theory.

'*Ce n'est pas possible!*' said the students, as we filed out of the classroom, leaving behind some children who would have gone on trying to solve the strange problems much longer.

'*Qu'est-ce que vous en pensez, Monsieur Piaget?*' I said. ('What do you think of it, Mr Piaget?')

'*Tiens,*' replied Piaget. '*C'est très intéressant!*'

The students who had been indoctrinated with Piagetian ideas could not believe that anything could happen which did not support the theory. On the other hand, Piaget himself, the originator of the ideas, was ready to accept any new fact as interesting and to possibly modify any theory to take in the newly observed facts rather than try to adjust the facts to the theory, as the students were apparently doing.

Piaget always came to the institute on his bike, usually passing a number of students in the traffic, who were trying to beat traffic during the rush hour in their cars! I can still see Piaget in my mind's eye, his white hair flowing after him, cycling past our car, while Roller was trying to get me to the institute in time for the next research session!

One Sunday Roller took me out for the day in his little *'deux chevaux'*. He asked me where I would like to go. I suggested that we look up Burdignin. So we drove to Annemasse, to Pont de Fillinges, where Gedeon had sat at the roadside waiting for me to rescue him, then we drove up the valley to Burdignin, where we had a drink at the same little bistro where my father and I had had our first 'man-to-man' drink. I thanked my chauffeur very much and we drove back to Geneva, each immersed in his own thoughts.

The next stop on this trip was Heidelberg. There I met my publisher Biemel, head of OCDL of Paris, who wanted to discuss with me the next steps for getting my books published in France. He took me to a Montessori school, where the teachers were anxious to have my opinions about what they were doing with the children. They took me into a classroom in which I was greeted by about fifteen or twenty silent children, all sitting in specially-made chairs with a tray in front them, using 'Montessori materials'. I observed for some minutes while the teacher explained that each material had its particular purpose and that the children would have to 'play' with the material in just the right way. I knew at once that I did not have much in common with these 'educators' and thought that Maria Montessori would turn in her grave if she could see what was being done in her name!

'What do you think of our work?' asked the teacher.

'Do you really want me to tell you?' I replied.

'Yes, of course! We have heard all about your work and we should like to know what you think of this work,' the teacher replied.

'Maybe the best way I could answer your question is by taking over the class, so that I could show you how I would do it,' I said.

This was agreed. The first thing I did was to get all the children out of their 'cages', put some blocks on the floor, with which they had been 'playing suitably' previously and told them to do whatever they liked with them.

'Was wir wollen?' the children asked in amazement.

'Jawohl!' I replied simply.

It did not take long before there was absolute pandemonium! But after a few minutes I managed to direct their play, a little more subtly than the teacher had done, taking my cue from what the children had already started. Within a quarter of an hour all sorts of constructive playing was going on. Some children were making colour sequences with the materials, others built walls with regularly occurring window patterns in them and some started braiding some coloured ropes into interesting-looking patterns.

'But surely, this is just chaos,' suggested the teacher and some of the observers present tended to agree. 'What are the children learning? I don't think they are learning anything!'

'I see that there is not a great meeting of minds between us,' I replied. 'If you were to look carefully, many of these children have already started to solve problems that have arisen from this Montessori environment. Learning, surely, means interacting with the environment and is it not what they are now doing? Before, they were only following instructions!'

But, apart from a polite thank you, there was no meeting of minds. Needless to say, I was not invited to Heidelberg again, until several of my books had been published by Herderverlag and German educators had begun to have second thoughts.

We had dinner together, Biemel and I and he tried to persuade me to spend some time in Sherbrooke, in the province of Quebec in Canada, where he thought I would be well received.

'But look at my schedule,' I said to him. 'I have a lecture in New York, then in Chicago, then in Seattle, then in San Francisco, then in Honolulu and then I have to be back in Adelaide. Where is there a spare day?'

'But it is very important!' Biemel would insist.

'Look,' I said. 'Here is my diary of engagements, with telephone numbers of those who have arranged it and here is my round-the-world ticket. If you can disentangle it and make some time for Sherbrooke, I'll go, but you will have to reschedule all my other appointments!'

Biemel agreed to do this. The next day he came back with my revised schedule and my revised round-the-world ticket.

'It was no problem,' he said to me. 'With a first class round-the-world ticket, it seems anything is possible!'

'How much extra did you have to pay?' I asked.

'Not a cent,' he answered.

So before long I found myself sipping champagne in the first-class cabin of an Air France jet, bound for Montreal. They really do spoil you in long first-class flights: your meat is carved while you watch, as though you were in a good restaurant; you get wonderful vintage wines, not to mention an infinite supply of liqueurs so you can sleep off your big meal!

The new airport had not been built yet so we landed at Dorval. I was met by some very pleasant nuns, who said they were to take me to Sherbrooke. So this was Canada! There was a big snowstorm in progress and you could hardly see the road ahead of you. The autoroute between Montreal and Sherbrooke had not been built, so we had to drive through a number of towns, most of the journey being taken very slowly on account of the bad weather.

We reached Sherbrooke finally, and I was taken to a large, comfortable room in the nunnery to which the nuns who had accompanied me belonged. This was about midnight local time, which was six o'clock the next morning, European time, so I just slumped into bed without too much toilet activity and became unconscious within seconds.

Biemel had arranged for me to do two days' work at the private Roman Catholic school run by the nuns, known as the Sisters of Charity. The mother superior was Soeur Renée. I had spoken with Soeur Renée on the telephone before getting on the plane and had asked her to get me a group of fourth graders with whom I could work, to show the local teachers the kinds of things that were possible using my methodology.

I did not have far to go to give my course as it was to be held in the same building, so the twenty-five below temperatures were not going to be a problem. There were a number of teachers, head teachers and members of the faculty of the newly created University of Sherbrooke in my audience. The place was packed and there was a great sense of expectancy. I had not realised that I had suddenly become famous and I was certainly not used to it, but I thought that I would have to meet the challenge in a positive way. I gave my audience an introduction to my theory of mathematics learning and told them that the best way to explain things was to show how it worked in practice. So I requested that my fourth grade class was brought in.

To my amazement a dozen or so very small girls were brought in.

I asked the teacher who brought them in, '*Est-ce qu'elles sont de la quatrième année?*' ('Are they from the fourth grade?')

'*Oh, non, Monsieur Dienes, elles sont de la deuxième année!*' ('Oh no, Mr Dienes, they are from the second grade!').

I had prepared my 'games' for children two years older, so I had to think quickly. I made a quick decision. I said, '*C'est correct, ça va aller!*' ('It's all right, it'll work!').

I thought that I would play with the two four-element groups, embodying them in very simple ways which second graders could grasp. I knew that there were some university mathematicians in the audience, but I also knew that there were some whose knowledge of mathematics was probably quite minimal. The work had to be mathematically accurate and impeccable, but had to be followed not only by my little actresses but also by the non-mathematical members of my audience.

I put the children into pairs, facing each other and holding each other by the hand. I explained to them that I wanted them to do some clockwise quarter-turns, some counter-clockwise quarter-turns, some half-turns and some whole turns. This meant that, denoting the children of a pair by the letters A and B, they could find themselves in one of the four different positions shown below:

AB	A B	BA	B A

The clockwise quarter-turn was called X, the counter-clockwise quarter turn Y, the half-turn D (for *demi-tour*) and the whole turn C (for *complet*). They soon learned how to combine these moves very well and so knew, for example, that doing X followed by D gave the same result as performing Y.

The children soon had the group table worked out and could tell right away, even without looking at the group table on the board, how they could achieve a change in one move that another pair had achieved in two moves. So far so good.

Now came the second four-element group. This included the whole turn and the half turn, but also what the children called *'sous les bras'* ('under the arms'), called S, which had to be carried out, without letting go of any hands, but changing between facing each other and having their backs to each other. So, apart from the first and the third positions, they also had the following two positions in the 'new game':

as well as:

| :A B: | :B A: |

| A: :B | B: :A |

the dots showing which way they were facing. In the first two positions, the children were still holding hands but facing away from each other.

There was another 'combined move' in this game in which the pair had to do a *sous les bras* and a *demi-tour* one after the other. These two moves combined were called the single move, T. So the moves in the new game were:

| C | D | S | T |

It took them a little longer to practise these moves and to learn how to combine them.

One little girl remarked, rather astutely, *'Regardez, Monsieur Dienes, si on fait deux des trois mouvements D, S et T l'un après l'autre, on peut toujours les remplacer par le troisième!'*

('Look here, Mr Dienes, if we do any two of the three moves D, S and T one after the other, we can always replace them by the third one!').

The other children wanted to test this statement and after a few tries they were all convinced that the little girl who had spoken had indeed the right idea.

Then we built up the group table of this game and wrote it on the board, next to the group table of our first game.

I took a calculated risk then. I went on to the 'dictionary game'. I suggested that we made a 'dictionary' of the moves of the two games, so that if we 'translated' a sentence such as:

D followed by X can be replaced by Y

into a sentence about the moves of the other game, we should always obtain a correct sentence about the other game.

We tried for some time, but, come what may, there always seemed to be some sentences that came out wrong in the translated form.

At one point, the same little girl, who had chimed in before, came up to the board and, stamping her foot on the ground very crossly, said to me, '*Mais, monsieur, vous perdez votre temps! Ce n'est pas possible! Dans le premier jeu il y a deux mouvements différents qu'on peut remplacer par le tour complet, mais dans le deuxième, deux mouvements différents ne peuvent jamais être replacés par le tour complet, comme on voit facilement en regardant le tableau!*' pointing at the blackboard. ('But, sir, you are wasting your time! The thing is impossible! In the first game there are two different moves which we can replace by a whole turn, but in the second one you can never replace two different moves by the whole turn, as you can easily see by looking at the table!')

'*Tu as raison, mon choux,*' I replied. ('You are right, my dear.').

There was a lot of applause and laughter and I thanked the children for working so well and sent them back to their classroom and then told the audience that I rested my case.

There was a lot of discussion following this epic performance. I had to explain the underlying mathematics to those who had no knowledge of mathematical groups or isomorphisms and it came out that such matters were usually studied in university courses, not by second grade children.

Members of the university who followed the work on both days became very excited about the possibilities and, between Soeur Renée and the Mathematics Department, they invited me to give them a two week course the following year, so that ideas and practices could be studied in greater depth. In the meantime, they would contact the publishing house Education Nouvelle in Montreal and try to persuade them to promote my books, which were already available in French and with a view to my writing some more books especially for Quebec.

I accepted this invitation and it was this work which persuaded the university to invite me to head a Psychomathematics Institute there. So the little girl who had told me that I was wasting my time was instrumental in changing the course of my career, which meant switching from working in Australia to putting in twelve years of work in Canada.

I took leave of the teachers, the professors and the children, telling them that we should certainly meet the following year.

I made a number of stops during the rest of the trip, but one that I remember well was the meeting at Seattle. I had given a talk to local teachers and a seminar to the faculty members, after which I had a kind of 'wrap-up conversation' with Professor Allendorfer, who had written a textbook for the 'new math'.

'How do you find the students who come in with a new math background?' I asked him.

'The bright ones of course have benefited,' he replied, 'but I never had trouble with the bright ones!'

'What about the so-called average student?' I enquired.

'Well, it's like this,' he replied somewhat sadly. 'If I give them a cubic function, they can find their local maxima and minima by differentiating the given function and equating the derivative to zero. But if I ask them how I should build my doghouse of a certain kind, using the least amount of wood yet giving the most space to the dog, they can't even get started.'

'I see', I said. 'And about what percentage of pupils do you think have benefited from the new math?'

'I should say about five per cent, or at most ten,' he replied.

I thought back to my predictions on Professor Begle's new math proposals, which coincided with Professor Allendorfer's findings. Obviously, we had much to do still.

I spent a few days in San Francisco and was shown around by Patrick Suppes. He had some computerised booths that looked like 'Skinner boxes', in which six year old children were learning about unions and intersections of sets. The children appeared to enjoy the novelty of the machinery and were all glad to be selected as guinea pigs.

After one of these sessions Suppes asked me, 'Well, Zed, what do you think of all this? Oh no! Don't tell me, I know! Wouldn't it be nice if the kids had some real objects to manipulate before or during the sessions!'

'You took the words out of my mouth,' I replied. 'Hopefully we can combine the machine method with the hands on materials method!'

Suppes was already a member of the International Study Group, which had just signed a contract with UNESCO to produce a document about the new tendencies in mathematics teaching. A person of the name Adrian Sanford, of Palo Alto, had agreed to handle all the paperwork of the group and he was thus appointed general secretary and treasurer to the group. The group was to organise two meetings, one in the States and one in Paris and then report later to a final UNESCO meeting in Hamburg. So Patrick Suppes, Bob Davies, Papy (France), Servais (Belgium), Krygowska (Poland), Pescarini (Italy), Varga (Hungary), Fischbein (Romania), Williams (England) and a number of others, were all roped in to work on this joint International Study Group-UNESCO project.

During my stay in San Francisco I made friends with Adrian Sanford and his family of three children and got to know a number of teachers, one of whom joined me later in Sherbrooke. I really enjoyed making friends around the globe, so that, as time went on, there were not many airports where I could arrive without a friendly welcome!

The next port of call on the way back to Australia was Hawaii. I had some more discussions with Jack Stalker about coming to work in Hawaii, but he said there were too many political problems, people were 'fighting for their political lives' and it was not the time to rock the boat. I spent a few days in Hilo, revisiting the experimental classes in which I had worked a year earlier and I established relations with the Honolulu Quakers, which had the additional worldly advantage of our being permitted to stay at the Quaker House any time we were passing through Hawaii. I finally said goodbye to Hawaii and to any eventual emigration plans to that part of the world and took the plane back to Australia. Thus ended the second world trip!

The third world trip was taken with Tessa, Sorrel and Bruce, but still in leapfrog fashion, by the family sailing across oceans or taking trains across continents, while I flew and so was able to put in sufficient work while they travelled to pay for first-class accommodation on any method of transport they used. So the trip began rather like the first one, with the family, now consisting of only Tessa, Sorrel and Bruce, taking an ocean liner to Honolulu and San Francisco while I flew directly to Hawaii. There was still enough work for me to do there, even though we had given up the idea of transferring permanently to these island paradises. I gave some talks to the Peace Corps Volunteers who were being prepared for their work, and I visited some of the schools in Hilo again, encouraging the teachers there in their efforts to establish more meaningful mathematics instruction.

Tessa's ship was due in Honolulu early one morning and was to stay there for the day, to allow passengers to look around the island of Oahu, on which Honolulu stood. On the ship Tessa had met a Hungarian man, who wanted to know her 'pedigree'. She told him about our ancient Hungarian heritage, which he did not believe. So during a stop at Melbourne he called on the Hungarian Consulate, or a university library, I am not quite sure which and at any rate rejoined the ship convinced of our noble ancestry in Hungary. Then he appointed himself Tessa's guardian, so that no lascivious shipboard acquaintance could approach her without his eagle eyes observing the fact and making sure that no fate worse than death would befall her.

Before going to the port to meet the ship, I had rented a car, so we could drive around the island and have a romantic time after a couple of weeks of separation. When I saw Tessa and the children coming down the gangplank, the first thing she said was, 'I hope you don't mind if a Hungarian gentleman who has looked after me on board ship spends the day with us!'

I said an embarrassed 'no', but clearly I was somewhat disappointed.

Nevertheless we went to the international market together on Kalakaua Avenue, then drove round to the north end of the Island to see the famous marine show. We even managed to get at least to the foot of the well-known Green Matterhorn, but there was no time or we had no energy to climb it! We all had dinner, accompanied by the ubiquitous Hawaiian music, which would have been very romantic without Tessa's strict guardian but then, you cannot win them all, as they say. I took my wife back to the ship and went to my lonely, hotel room at the Reef, fortunately still paid for by the Peace Corps account!

The next day I took a plane to Seattle, where Professor Allendorfer had invited me to hold some seminars. I put in about three days' work at Seattle, after which I flew to San Francisco where I was met by Adrian Sanford and the children. I remember there was a whole big streamer put up right across the exit area bearing the words, WELCOME TO SAN FRANCISCO ZED!

I spent the night with Adrian's family. He had found a house we could rent in Menlo Park, which was close enough to the Stanford campus, where I was supposed to work with Suppes for about a month.

Tessa and the children duly arrived and we installed ourselves in Menlo Park. We arranged for Sorrel and Bruce to go to the local school, hoping that they would make friends or that they might even learn something. The house was no palace, but it was comfortably furnished and the rent was no more than we could easily afford, since I was getting quite a few fees for the work I was doing.

People in Palo Alto, which was next to Menlo Park, were very friendly and Tessa and the children all made friends very quickly. California seemed a very positive and friendly place.

I got to know Adrian's children really well and would sometimes take them out for walks. I recall, on one of these walks, one of the children exclaiming, 'Look, Zed, the road is full of glass!'

'That isn't glass, that's ice!' I said to them.

They had never seen ice on the road before! They took great delight in jumping on the sheets of ice and breaking them up into little pieces.

We attended a cocktail party once on campus. We soon discovered that there were more Hungarians present than non-Hungarians and publicly announced that the official language of the cocktail party was to be Hungarian! We all had a good laugh and we Hungarians did quite a lot of huddling together and talking about the old days. I soon realised after this cocktail party that there were hardly any universities anywhere in which there were not some Hungarians on the faculty!

Tessa would sometimes take the train and go into San Francisco to look around and do some shopping. On one of these occasions, she met a lady who was going into San Francisco to see her psychiatrist. We were not aware of the fact then, that most wealthy Americans indulged themselves by regularly visiting a 'shrink', so Tessa was quite surprised to hear where she was going. They had a very good talk during the one hour journey from Palo Alto to San Francisco and the lady suggested that she would rather spend the day with Tessa than waste time at the psychiatrist. So she cancelled the psychiatrist (no doubt she still had to pay the fee!) and the two of them had a gorgeous spree, where she rented a room for the day in one of the best hotels so that they could have privacy. We all became quite friendly with her and with the husband and we were invited several times to their house, and met the children. We were beginning to learn how things ticked over in California!

Some of my work, as usual, entailed visits to schools. In one sixth grade class I was having less than my usual success with the children. One teacher who was observing said to me, 'Why don't you get them to move?'

I noticed then that more than half the children were black and, indeed, it must have been against their nature to keep still for any length of time. So I started them on some games in which the positions of their arms and their legs became part of the problem. Almost immediately the whole class was electrified and they all started to solve their problems most enthusiastically by moving their arms and legs, following the rules of the game.

During this visit I organised the first of the two ISGML meetings, during which we discussed the problems of mathematics learning which we wished to put before the 1966 UNESCO Conference, scheduled to be held in Hamburg, Germany. I tried to get together the cream of researchers into the theory and practice of mathematics learning who knew about cutting-edge thinking at the time. The next one, for Europeans, was to be held in Paris.

It was getting near Christmas and it was time to say goodbye to California, for the next work stop was to be in the city of Minneapolis. Tessa, Sorrel and Bruce were to take the transcontinental train and I was going to fly and get things ready for when they arrived.

I arrived in Minneapolis safe and sound and contacted Paul Rosenbloom, the mathematics educator who had me to work at the University of Minnesota for three months. We talked about work as well as about life in general, the latter including where Tessa, the children and I were going to live. He soon helped me find a furnished house, which I thought was quite pretty and romantic and I rented a car for the duration of our stay.

The family was due to arrive on Christmas Eve, but there was no sign of them. I had purchased a Christmas tree and fairy lights and had everything ready for a Christmas welcome, but no family arrived. I had rather a sleepless night and then, just as I was going to sleep early in the morning, the telephone went and a tired, weak voice greeted me. I could hardly recognise it as Tessa's voice.

'We are at the station and we've had a dreadful journey!' said the weak voice. 'Could you come and collect us?'

'Certainly,' I replied in what I hoped was a reassuring voice. 'Just wait, I'll be there as soon as I can!'

The thermometer had dipped to nearly thirty below and I had to clear some snow from the drive as well as from the car before I could drive into the road. But within a quarter of an hour I found my miserable-looking little family waiting shivering at the station, not prepared for such arctic conditions! I drove them home, but I am afraid that Tessa was not overly impressed with my romantic little house. I was not overly concerned with that at that moment in time though. I soon made some breakfast for her and the children and we had a tour of the house, each child choosing a suitable room.

Then Tessa told me her tale of woe! They had not gone very far into the Rocky Mountains when an avalanche cut the railway line and they had to be transferred to buses, which took them along some hair-raising hairpin bends, to another section of the track. I don't know how long they had to wait, or whether the same train eventually continued, but they finally reached Chicago, where again they had nearly missed the connection because Sorrel insisted on buying something at one of the station stores. But they finally arrived, very tired and miserable, having taken the train because Tessa thought flying was just too dangerous!

We had a nice family Christmas. We made friends with our new neighbours; we had time to get used to the wintry conditions, which did not let up during our entire stay in that city. We never saw the grass in Minneapolis, but we got to know about the snowplough dumping mountains of snow on to our drive and blocking us in! The children were allocated snow-clearing rotas, but I joined in to help them with the shovelling. We considered it a fun thing to do together! Bruce started in the local elementary school and Sorrel went to the junior high school when the Christmas break was over. They seemed to settle quite well into their respective classes; at any rate we never heard about any problems they might have had!

Paul Rosenbloom was the director of what he called the MinneMast Project, into which he introduced some of his highly amusing and original ideas. In fact, I can probably truthfully say that he is the only American educator from whom I could actually learn something worthwhile. One thing I remember learning from him is how to introduce wave functions to young children without any recourse to trigonometry. He would do this by combining two number cycles, for example:

10 11 12 10 11 12 10 11 12 10 11 12 10 11 12 10 11 12 10
1 5 2 2 2 6 0 3 3 4 1 4 1 5 2 2 2 6 0

in which we start with a cycle of three, namely 10, 11, 12, and make sure that any four successive numbers in the second row add up to the number above it. So we combine a cycle of three with a cycle of four and obtain a cycle of twelve. The thirteenth number is the same as the first one, the fourteenth one is the same as the second one, and so on.

Of course one would start with simpler waves, such as:

10 11 10 11 10 11 10 11 10 11 10 11
1 4 5 2 3 6 1 4 5 2 3 6 1 4

where we start with a cycle of two, and the sum of every three consecutive numbers in the second row will add up to the number above it and from the 2-cycle and the 3-cycle, we obtain a 6-cycle.

The 'waves' then come out like this:

An interesting fact emerges when we choose two cycles which happen to have a common factor. We still get waves but they get bigger and bigger! It is the case of soldiers marching across a bridge having to break step, in case the vibrations of the metal in the bridge and the rhythm of the marching feet have a common factor, in which case we have what is known as resonance, which can break the bridge!

Paul Rosenbloom and I had a very fruitful three months of co-operation, each of us learning something from the other, and it felt very good. In fact, he already knew at that point that he was transferring to Columbia University in New York and he invited me to spend a year there with him, whenever I could find the time.

We made many friends in Minneapolis, mostly through the children, but also through the Quaker meetings, which we regularly attended on Sundays. We became very fond of one family in which there were four children. They lived right out in the country and on many days the children had to ski to the place where the school bus stopped, leaving the skis leaning against the trees, so they could pick them up again after school and ski back home. The house they lived in was hidden in dense woodland. It all seemed very romantic to us. It also impressed us how helpful and kind they were to each other and how the children's mother managed them with such loving care that she never had to get cross with them, for there was too much love floating about to be spoilt by even a cross word! There was a lake near their house, where they sometimes invited us to come and perform ice-fishing at weekends. We had never done ice-fishing before and it seemed a strange thing to do – to build a little hut on the ice in the middle of a lake, make a hole in the floor and drag the fish out! So we were learning the customs that had grown up there on account of the climate and we found it all very fascinating.

The next stop on our round-the-world journey was Sherbrooke. Again I took the plane to Montreal, where I was again collected by the nuns and I soon found a place where we could board for the two weeks the Sherbrooke course would last. Tessa and the children followed by train and duly arrived, noticing that it was even colder there than it had been in Minneapolis! We sent the children to the convent school run by the Sisters of Charity, which was perhaps a mistake on my part, since they did not know any French and were not happy sitting through classes that they could in no way follow.

I worked some more of my 'educational miracles' with the children in the school and had a number of discussions with members of the university, who were interested in the possibility of my coming to join the Sherbrooke faculty. I told them that before I could think about it, they would have to make a very clear and precise offer of what they would pay and how and what they would expect me to do. I told them that I had a permanent post in Australia and that they would have to match or better it. Many members of the university Mathematics Department attended my course and watched children work on the kind of mathematics they had never dreamt would be possible with such young ones. It is these lessons with the children which really persuaded them that they must try and secure my coming to Sherbrooke. In a sense, they wanted to show *les anglais* what they could do; in fact, the whole point of my coming turned out to be part of the power game between the French-speaking Quebeckers and the Anglos, which I was too silly not to realise until it was too late!

The last leg of the journey home to Australia took us through London and Paris, Tessa and the children taking their ship from England, while I had discussions with my publishers in England, France, Germany and Italy. The four publishing houses, Educational Supply Association in England, Office Central des Librairies in France, Herderverlag in Germany and Organizzazioni Speciali in Italy – struck a deal to share the cost of producing the educational materials and the publishing of the books, so before I took the plane for Australia things were really looking up, from the point of view of the distribution and sale of much of the material I had invented and written. This meant that we were not to have any more financial problems in the foreseeable future.

In Paris I got together the European mathematics educators and we discussed everything which had been discussed at the Stanford meeting the previous December. The entire report was worked on, so I had to rewrite most of it. However, before returning to Australia, I was able to send to UNESCO, Hamburg, a version of a report approved by the American and European researchers who were making the most significant advances in the field.

When I arrived back in Australia, for the first time in our lives we had more money in the bank than I knew what to do with, so I rented a very beautiful, furnished house for the following year, standing quite high on a hill with fantastic views extending down to the sea, where we could sit in a pleasant garden and admire the most wonderful sunsets we had ever seen, as well as enjoy those creature comforts to which we thought we were entitled, having, so to speak, arrived!

I also looked around for a new car. I went to a showroom and saw a beautiful white Holden station wagon slowly turning round on a turntable. It had red leather seats, it was beautifully lined and I immediately fell in love with it.

I went into the store and announced to the manager, 'I want to buy the car that's turning around on your turntable. How much is it?'

'One thousand six hundred,' I think was the reply.

I pulled out my chequebook, wrote the cheque and bought the car. Since the manager did not know me from Adam, he had the sense to phone the bank to enquire whether my cheque was good. It was.

So, when the family arrived after a pleasant voyage in comfortable first-class accommodation, I met them in our brand-new Holden and drove them to our newly rented house on the hill, where we all made ourselves at home. It occurred to me that we had moved on considerably since the first time our ship had docked at the Adelaide quayside.

So this was the end of the third world trip. The next one would take us away from Australia, to England, to the USA and eventually to Canada.

Chapter Eighteen
From Australia to Canada

To make the move from Australia to Canada, we decided to carry out what in military parlance is called a giant pincer movement. Tessa and the children were going to head west and I was going to head east and we were to meet on the other side of the globe. Since Tessa and the children were going by sea, it was natural for them to take our new Holden and we packed up what furniture and goods we thought were worth preserving and stored them in Adelaide, to be shipped off to Sherbrooke in time for us to receive them there. The whole thing involved a considerable amount of space-time co-ordination!

I had accepted a period of work at Columbia University in New York, starting in January 1966 and I was to start at Sherbrooke in September 1966. The family was to leave at the end of November 1965, bound for London, where they were to spend Christmas with Rosalind, Tessa's mother and use the Holden while in England. Then in January, they were supposed to take a ship to New York, bringing the Holden with them and we would spend the Columbia period together, while the children went to a New York school. During the summer of 1966 Tessa and the children planned to spend some time in England and at the end of August officially emigrate to Canada, thus obtaining 'landed immigrant status'. Having 'officially entered Canada' at Quebec City, where they would disembark, I would collect them in the Holden, drive them to New York to collect our belongings from the apartment we would rent and we would all drive back to Sherbrooke, at which point I would 'officially enter Canada' and receive my 'landed immigrant status' and then I would take up my position at the University of Sherbrooke. All this took a certain amount of organisation but in the end we did all reach Sherbrooke, where we rented an apartment and started our 'new life'.

I was asked to pay another visit to New Guinea in December 1965, so I thought that I could fit in this visit *en route* to Sherbrooke, although, geographically speaking, it was not exactly on the way. So, having put the family on a ship bound for London at Adelaide, I said goodbye to everyone we knew in Adelaide, including the teachers and the children with whom I had worked, and flew to Port Moresby. I tried to make it so that after I had left, the committee would be able to carry on the good work, so we did a great deal of curriculum planning; we also had a number of discussions on the practicability of various approaches to implementing such a curriculum. The last week of my stay was spent at the Lutheran Mission beyond Mount Hagen, where I had worked on several occasions and got to know the missionary and his three children. I recall doing some really interesting mathematical work with the local children, almost right up to the Christmas festivities. My missionary friend had started wondering whether he should go on working in the mission or whether he should come to Canada and work with me.

On Christmas Day I left the mission, saying a tearful goodbye to the family. Before I left, I was treated to an early Christmas dinner, which we celebrated together as a kind of send-off for me. I managed to get a ride in a Cessna from a missionary who was visiting the mission, so I could catch the plane from Mount Hagen to Port Moresby.

After we were airborne, the missionary who was piloting the plane, suddenly said to me, 'Oh, I nearly forgot! I am supposed to pick up a pregnant woman and take her to the doctor at Mount Hagen.'

Saying this, he turned the plane around more rapidly than I thought was good for our survival chances and made for a low-lying area in the bush. He was practically skimming the trees when I saw the extremely small, short airstrip right ahead of us, where he made an expert landing. Our pregnant passenger was waiting for us so we picked her up, did some taxiing and took off as rapidly as we had landed, making for Mount Hagen. We just made it in time for my plane was about to take off. We were just able to transfer myself and my things to the waiting DC3, its propellers already going.

These DC3s were then the war-horses of the New Guinea air transport system. They were most uncomfortable: the seats were parallel to the fuselage, the passengers facing each other and the baggage was tied down in the middle between them, sometimes consisting of live animals, besides other cargo and the passengers' luggage. And the company had the nerve to charge first-class rates for this luxurious type of travel! Of course, there was no competition, so they could charge what they liked. For most journeys there was no alternative to air travel, since there were no roads as yet across the Stanley Ranges and even the road from Lae to Goroka could be used only in the summer, as the rivers changed their courses during the winter and sometimes washed away part of the road.

I finally made it to Port Moresby, in time to catch the DC6 for Brisbane. So this was goodbye to New Guinea and all the colleagues and friends with whom I had been working for several years. As a matter of fact, I did make another visit a little later, from Sherbrooke, but since then all the news I have had of New Guinea has been through hearsay from travellers or from TV programmes. It was with these thoughts in mind that I settled in for the night flight from Brisbane to Honolulu. I was served a beautiful Christmas dinner, my second one on that memorable Christmas Day!

I had contacted the Quakers in Honolulu with the news that I was coming and they were there when the plane landed in Honolulu to greet me.

'Happy Christmas, Friend! Come and have Christmas dinner with us!'

Of course! I had crossed the date line and it was Christmas Day once again! It was the only time in my life that I had lived through two consecutive Christmas Days!

So I was treated to a wonderful Hawaiian Christmas dinner, my third one that Christmas, which was attended by many of the Friends who made up the Honolulu Friends' Meeting. I stayed for a few days at the Quaker house, trying to regain my strength after the work in New Guinea and the long flight to Honolulu. It is very relaxing to be with people who have a similar *Weltanschauung* to oneself: there are so many things which are understood, that need never even be said. By this time I had largely accepted the major aspects of the Quaker philosophy of life and had honestly tried to live up to it, although I must admit that I did not always succeed in doing so. Anyway, my short stay in Honolulu, which did not involve any work of showing or persuading, was a balm to my soul and I was a much refreshed person when I took my leave of Hawaii and flew to San Francisco.

I was greeted in San Francisco by the Sanford family and it felt good that the children all rushed up to me, jumped around my neck and hugged me, thus showing that they were welcoming an old friend. I was beginning to feel that I had an international family, that the world was shrinking and it seemed as though humanity might one day truly consist of brothers and sisters, with Almighty God as our Father!

Adrian Sanford and I worked on preparing the ground for the upcoming UNESCO-ISGML conference, to be held in Hamburg quite soon, but otherwise this short stay in Palo Alto was really like a holiday, after which I flew to New York to begin my period of work at Columbia with Paul Rosenbloom.

During this time, Tessa, Sorrel and Bruce had gone round the globe the other way and we arrived in New York at just about the same time. Paul Rosenbloom helped us find an apartment that was sublet to us by a Hungarian member of the Columbia faculty who had just left for New Guinea on sabbatical leave. The apartment was on the corner of Broadway and 116th Street, just opposite the girls' hostel. It was on the tenth floor and Tessa was not used to being up so high so it took her quite a long time to get used to the vertical type of travel, not to mention the horizontal methods, when one had to run the gauntlet of the vagaries of the New York subway system.

Columbia was on 120th Street, just four blocks' walk along Broadway. Five blocks further there was the ill-famed 125th Street, crossing Manhattan from east to west and bisecting the heart of Harlem. We made some enquiries about schools for the children, who were now twelve and fifteen years old. We were not sure about letting them loose in the New York public school system, so I had a chat with the Principal of Dalton School, whose fame as an educational innovator had spread worldwide. Who had not heard of the Dalton plan? I struck a deal with the principal that I would do some consulting at the school for no fee and, in return, the children could come to school and pay no fees. There was of course the problem of transport still. The Dalton School was in quite a different part of Manhattan, but, fortunately, there was a bus that passed our door as well as the door of the school, so we were lucky.

Of course the children had to become 'streetwise'. The usual 'Don't talk to strangers' would be totally inadequate in New York. One thing they were told was to put their money inside their socks, under their feet, except for the bus fare, any time they used a bus for transport. On the subway they were told to keep near one of the doors, so that in any emergency they could escape easily. There were two armed police officers constantly walking through every train, they were assured, although this fact made them rather reluctant to use the subway and they preferred the bus!

Tessa spent quite a lot of time during our first week trying to get rid of cockroaches. We had not realised that these creatures inhabited just about every apartment in New York and you might as well try to get rid of the air. But she washed and scrubbed everything, including the parquet floors, which in the end turned out to be a disaster, since the tenants insisted that we paid to have the place repolished before we left!

My work was divided between having discussions with Paul Rosenbloom and members of his department, working at the Dalton School, at a Jewish school nearby and at one public school in Greenwich Village, in which the principal had a lot of go-ahead ideas. The Village is the equivalent of London's Chelsea or Hampstead or the Rive Gauche in Paris.

It happened that I had a cousin in New York who was a very high-class, fashion model, whose children, twin girls of about ten years old then, were in one of my experimental classes of the school in the Village. So we got to know this cousin better who introduced us to some interesting people in the art world, which led to a very full social life. This included free tickets to shows and to concerts, which both

Tessa and I greatly appreciated! My cousin Jay, with whom I had been friendly as a child, was living on Long Island, working as a solid state physicist at Brookhaven Laboratories and we often went to see him and his wife at weekends.

There was one important interruption of the New York work: I had to go to Hamburg for the UNESCO-ISGML meeting, which had been planned for by our two preparatory meetings, the first one at Stanford and the second one in Paris. The old hands here got together again and we discussed the report I had prepared. We decided not to alter the report that had been drawn up as a result of the two preparatory meetings, but to write a special report, describing the work of the Hamburg Conference. John Williams, of the National Foundation for Educational Research for England and Wales, was detailed to take copious notes and then to write up the conference. Copies were to be sent to all who attended, who would be invited to make suggestions for altering the text or adding things. Thus we eventually finished our contractual relationship with UNESCO with two reports. I was looking over one of these reports this morning and sadly realised how little we have progressed in the past twenty-eight years! A great deal of what is written in the reports consists of suggestions we would do well to heed today and I cannot help but wonder whether we were not wasting our time and public money making those gigantic efforts in the Sixties to improve mathematics education, when the situation is not much better today!

There were several Quaker meetings in New York, one of them being within walking distance from our apartment, so we went nearly every Sunday to keep ourselves supplied with what spiritual nourishment we could gather. We found, however, that some who attended that meeting were not the kind of Quakers we had grown to admire, their ethical standards being somewhat lower than ours, and we also thought that a number of them used the meeting as an encounter group to air their psychological problems. In the Quaker advices we are told to be patient and tolerant with such people and try and glean that which is of God in them. But in some cases we found this difficult. There was one man who was very direct in his sexual approaches and Tessa found it quite difficult to deal with him 'prayerfully'! But this was New York and I suppose we were having our baptism of fire, trying to grapple with various unexpected aspects of the Big Apple! Many of my ideas became incorporated into the curriculum material that our group was producing and I felt that at last something was sinking into the American system. I had not realised at the time that Paul Rosenbloom was regarded by other mathematics educators in the USA as somewhat of a maverick and so whatever went into any Rosenbloom productions was not likely to get any further!

Towards the end of my stay, I became interested in establishing a social club for children in Harlem, at which they could have fun playing mathematical and logical games. It seemed that there was no problem getting quite a big grant for doing this and I was 'elected' to head the project. Then one day, one of the people with whom I had had discussions about the project came to me and produced a list of ten persons.

'These are your consultants for the project,' he said in a suave, persuasive voice. 'Just sign here at the bottom of the page.'

'Just a minute,' I said. 'Who are these people? I don't know these people. And what do they know? And why do I need ten consultants?'

'I see you are quite new to New York,' he replied. 'This is how things are done here. You pay one hundred dollars a day for these people out of the project funds and the rest you can spend how you like!'

I realised now why it was so easy to get funds for the project. These people had no interest in Harlem children: they wanted to feather their own nests with as much money as they could squeeze out of the public purse!

'I think I can do this project on my own, on a shoestring budget and I don't need any of your consultants!' I said, handing him his list.

'Just as you like, Dr Dienes, but our organisers won't like this at all! Not at all!' said my visitor as he left me.

We had our premises in 125th Street, so all I had to do was to take some material and send word round to the local schools that we were starting a club where the kids could play some fun games. I stated the time and place in my notice, went to the premises in 125th Street and waited. I waited for half an hour, I waited for an hour; after which one mother turned up with a little boy. I told her that there had been a misunderstanding and sent her away.

That evening I had a telephone call from a caller who would not identify himself, who simply said, 'If you know what's good for you, get out of Harlem!' Then the phone went dead.

If it were not a death threat, it was certainly a threat. Obviously, if you do not co-operate with the Mafia, first you do not get anywhere; second, you are thrown out. I had learned my lesson. I was not about to take risks. I gave up the club project – the Harlem kids would have to do without their mathematical games.

At the end of the academic year, Tessa, Sorrel and Bruce took a ship to England and I was left to handle whatever turned up in the USA. I had several invitations for the summer months. One was to give a talk in Houston, Texas. I had never been to Texas and had heard a lot about Texas and Texans, so I thought it might be fun. However, it was not to be. When I arrived, there was nobody to meet me at the airport, so I booked into the airport hotel. My talk was to be the next day. I took a cab to the university in the morning and presented myself at the appointed time and place. I gave my talk, at the end of which I received polite applause and was given a cheque for my fee and for my expenses.

'What happens now?' I asked one of the professors, who had dealt with my expenses.

'What do you mean? You have given your talk and you have your cheque,' replied the professor.

I was flabbergasted. I went to a public phone booth and called my Swedish collaborator in New York, saying, 'Are you free tonight? Yes, good! I am in Houston, Texas, but I shall not be here for long. Let us meet in the faculty lounge at Columbia. I want to paint the town red!'

'I'll be there,' was the surprised reply.

I took the first flight back to New York, took the helicopter to Manhattan and went to Columbia to see my friend, to whom I spilled the beans about 'Texan hospitality'. We had a very good dinner in the Village, talking about anything and everything, though the town was not painted all that red in the end. But I have never been back to Texas since.

An invitation came to spend a week in Atlanta, Georgia. This was a much pleasanter experience. Atlanta is a beautifully green place – there seem to be trees growing everywhere – and the people are very friendly. The Atlantans welcomed me with open arms and, although I had to do a little work, most of the time was spent socialising and going to shows. Of course Atlanta is in the Deep South. I am not sure if at that time Martin Luther King had made his speech 'I have a dream', but the problems between whites and blacks were very much in the foreground.

A day or two before I was to leave, my hosts took me to a show in which there were a great many very funny scenes.

About halfway through the show, one of the professors sitting next to me said, 'I am so pleased that you can laugh at all the jokes! Most Northerners would not understand our Southern humour after only a few days, the way you seem to!'

So I was a Northerner! And I had had a pat on the back for not being too 'Northern' to laugh at their jokes! Indeed, I was pleased about this, as it showed that my peregrinations all over the globe had taught me something about adapting to a new environment. I flattered myself that now I could probably put a pin down anywhere on the map, go there and live there happily ever after!

So my impressions of Georgia remain just as positive as my impressions of Texas were negative. Of course, I had been in Texas only a few hours when that fiasco happened. How could I learn the local mores in that time? Maybe I should have invited all the professors to dinner and paid the bill. I do not know to this day whether such an offer would have been appreciated or even accepted. So my distaste for Texas remains at the back of my mind, however illogically!

While I was 'playing games' in the USA, Tessa and the children were having their holiday in England. Tessa rented a room in her mother's house in Putney, which had just been arranged for letting and everything was brand new, so Rosalind was very keen that nothing should be spoilt.

This and other conflicts made Tessa's blood pressure rise and, when she had her medical examination for Canadian emigration, the doctor said that with such high blood pressure she could not be admitted. But, he suggested, it could just be that the high blood pressure was caused by emotional factors, which could be checked by her going into hospital for a few days and resting, during which time the rise and fall of her blood pressure could be observed. This was done and it soon transpired that there was no inherent blood pressure problem; anyone's pressure can go up suddenly from worry, which is what had happened. So Tessa was given the green light by the doctor and the Canadians gave her and the children the visa which would permit them to go to Canada and have 'landed immigrant status', which meant the right to permanent residence.

At one point Sorrel wanted to bring a male friend to stay at the house, but Rosalind put her foot down and said no. Tessa suggested to Sorrel that she stayed with our ex-neighbours in Leicester, the Perrys. So this is what the pair did, but then, within a short time, the two of them simply disappeared. This was a big problem, since our immigration was conditional on *all* of us coming to Canada, not to mention the problem of having a sixteen year old daughter loose in Britain with a young boy.

Tessa had to find Sorrel somehow. She had no idea where she might have gone. Escaping lovers usually go to Gretna Green in Scotland, where they can get married by the Gretna Green blacksmith. This was the remnant of some ancient law that had existed for centuries, but it had recently been repealed. So Gretna Green was not the place to look. But perhaps Scotland was a good bet. But where in Scotland? She supposed it was logical to start with Edinburgh, this being the capital of Scotland, but then it was just as likely to have been a simple hunch.

So Tessa went to Edinburgh. She enquired at police stations about the pair, with no effect. Then she thought of contacting the press. One paper agreed to run the story, since it was a good one. A story was run, entitled, 'DESPERATE MOTHER SEARCHES SCOTLAND FOR MISSING DAUGHTER' with a photograph of Sorrel accompanying the article.

Fortunately for all concerned, she was in Edinburgh and the newspaper story did help to find her. Of course, Sorrel was furious to be found and had to be kept very close to Tessa so that she could not escape again.

Tessa contacted a shipping line and told the agent the story, saying it was imperative that they went to Canada immediately, before anything else was attempted by her daughter. But there were no berths available on the *Empress of Britain*, which was the only ship sailing in the next day or two. Then, by chance, they realised that one first-class suite was still not sold for that crossing and Tessa was offered that at a very favourable low price. She took it. They embarked the next day and made themselves at home in their suite, which had its own private terrace and proper beds, not just bunks; in fact, just what the doctor ordered for an exhausted family. The ship called at Liverpool on the way to Canada and Sorrel had to be confined all the time the gangplank was down, just in case she tried to leave the ship!

I met the ship at Quebec City, where Tessa, Bruce and Sorrel went through emigration and obtained their landed immigrant status. There was a lot of luggage, since Tessa had bought some rather nice English china and many other things, which we were allowed to import without duty upon immigrating to Canada. So it was just as well we had the Holden there, into which we loaded all the luggage and our little family and drove down to New York. It was then possible to import a car each into Canada, so I bought a Plymouth in New York and we drove back to Sherbrooke with the two cars! I drove the Plymouth and Tessa drove the Holden.

We eventually arrived in Sherbrooke, where we had rented an apartment in Rue Dominion, where we unloaded everything, left the cars in the driveway and wanted to collapse, but could not, because there was no furniture! Our own furniture was still on the way from Australia. It should soon arrive but it had not arrived yet. In any case, we had to have beds, as no beds had been shipped from Adelaide.

We found a bank, La Banque Provinciale du Canada, where we deposited what money we had left, which was about two thousand dollars. I did not know anything about Canadian banks, but I chose that one because they had plenty of parking space for customers! Then we had to go and buy furniture. We had some beds delivered and a few other basic items. The tradesmen were not bothered about being paid when they learned that I had come to join the university.

So we had finished our giant pincer movement and we were ready to tackle our new life in a totally French environment, working in a French-language university, settling in our new apartment, with a minimum of furniture and a very modest balance in the bank!

Chapter Nineteen

Getting Established at Sherbrooke

There were a number of tasks ahead of us, all of which had to be tackled with a certain amount of urgency, if our stay in Sherbrooke was to be a success. Firstly, I had to find my place at the university; then I had to contact the local school system (commission scolaire), as the schools would play a key part in any future work; thirdly, we had to establish our home in Rue Dominion and make it into a pleasant place to live and establish ourselves as a part of the social milieu; fourthly, we had to place the children in suitable schools.

I was of the opinion that the children should attend French schools, where they would become integral parts of the Sherbrooke scene and receive a good classical education. Sorrel and Bruce objected violently, remembering the two weeks the year before when they had attended French-language schools. Tessa sided with the children; she thought she would not like them to be picked on because they knew so little French. It was three against one and, rightly or wrongly, I did not insist. So we registered them at the only English-language school in Sherbrooke.

I realised much later that this had been a serious mistake. My French-speaking colleagues, including those in authority at the university, must have drawn the conclusion that we had no faith in the French-speaking people of

Quebec, which of course was not at all the case. As a matter of fact, we did change our minds about Sorrel's schooling a year later, but we did not send her to a French-language school, but to a Quaker school in British Columbia, located in a tiny mountain village called Argenta. Her Edinburgh experiences had made her very unhappy and consequently difficult and we hoped that a Quaker environment would have a healing effect on her troubled soul. In fact, she was very happy there for a year, but she could not bring herself to observe even the very few rules they had and the school did not want her back for the following year. We sent her to another boarding school in New Hampshire, just across the border, where she did not make brilliant progress and left school altogether at the end of that school year.

Tessa felt that Sorrel's problems had arisen out of being taken around the world during her early teenage years. She had had to adapt to new schools and leave friends behind that she had just made, so there did not seem to be any solid foundation to her life. She was, as Tessa would say, sacrificed on the altar of the furthering of improvements in mathematics education, or worse, on the altar of my career choices. And it did take her a long time to put herself together, at which point it was already too late to start an academic career. After much soul-searching, Sorrel decided to accept Jesus Christ as her Lord and Saviour, was baptised in water and in the Holy Spirit and became a born-again Christian.

There were no great problems in fitting out our home. Tessa found a very good furniture store, run by the Wilson family, with whom we eventually became very friendly; we still visit them, as we have stayed friends with them ever since. Mr Wilson sold first-class furniture and he delivered everything free of charge. He took back any piece we did not like and replaced it with another piece and he never worried us about paying.

'You pay me when you can, I won't charge you any interest – you must have had a lot of expenses moving here,' he would say.

This reminded me of the record shop where Tessa and I bought our records when we were first married. We must look honest!

My salary at the university was good and we soon paid off what we owed Mr Wilson, so the setting up of a home base in Sherbrooke was not really a difficult process.

I soon contacted Soeur Renée again and she suggested two schools in which I could work. The private school run by the nuns had been taken over by the Sherbrooke Commission Scolaire, but they kept on the nuns as teachers and I was warmly welcomed into the classrooms, where the teachers were prepared to continue implementing the curricular and methodological reforms I had shown them during the two week course the previous year. This school was the École Eymard and the other school that was added was the École Sainte Famille. During the first few weeks of the school term, I spent most of my time in these two schools, trying to get something moving, which could then be 'studied' by the university and used as a venue for teacher education. I had never had so much time to devote to such work before and I felt as if I were really in my element, enjoying myself immensely in getting the children worked up about mathematics.

The mathematics adviser was a certain Monsieur Hamel, who followed these events with interest, often observing the classes and coming to see me to have talks about which way things were heading. He soon became a staunch ally of mine in the Commission Scolaire, where he had to do some persuading of a few doubting Thomases.

After a few weeks I had a talk with the dean of science. He had made some discreet enquiries in the town and he said to me, *'Vous avez une bonne presse dans la ville!'*

He suggested that we should establish a research centre, of which I would be director. The suggestion was soon passed by the Senate that I should direct the Centre de Recherches en Psychomathématiques and that I should apply for federal and provincial funds to run it. I applied to the Conseil des Arts (Canada Council) and to the National Research Council for funds.

It was not long before I received a positive answer from the Canada Council, a grant of some twenty-four thousand dollars having been approved, which was quite a bit of money in those days. It appeared in all the papers, since it was one of the largest ones granted up to that time. But the simple-minded people of Sherbrooke did not understand what a grant was: they thought I had been awarded a prize so several people came to me asking if I would like to contribute some of my prize to certain local charities. I had to explain as tactfully as I knew how that the money had been allocated for research and could only be spent on research!

I soon obtained a much bigger grant from the National Research Council, over a quarter of a million dollars. This was for running the centre; I could appoint research assistants, travel to conferences, purchase materials, pay consultants and so on. Clearly, this was the beginning of a very exciting period in my professional life. When the dean came over to my new office suites and said, '*Félicitations, Monsieur le Directeur!*' I truly savoured the word *directeur*, although as yet I had nothing to direct! It reminded me of the first time I heard myself called Dr Dienes by the military selection board in Oxford, whose members had recommended me for research, thus saving me from becoming cannon fodder.

I asked our lab technician from Adelaide, Mr Parkanyi, to come over and join the project and become my electronic lab technician, since I had big plans for making electronic machines to study the learning process. He was the first person on my staff. Then I had to have some secretaries. Luckily I found a most efficient one in the person of Madame Lessard, who was bilingual in English and French; I appointed another secretary who was bilingual in English and Spanish, as I had plans to make contacts in South America.

By this time I had started some work in a number of European countries, namely England, France, Italy and Hungary, to which Germany was soon to be added, as the German translations of my books were finding their way into the hands of German mathematics educators. This meant that I would have to be absent from Sherbrooke from time to time, in order to keep things going in these countries as well as learn from their successes and failures! The International Study Group for Mathematics Learning (ISGML) was also flourishing and we held annual research meetings attended by those responsible for the research projects which were members of the study group. The study group also organised international workshops for teachers, in which multilingual instruction was given. These meetings, the research ones and the teacher education ones, were held in different countries, such as England, France, Germany, Italy, Hungary and even in Canada. The study group was financed by subscriptions from its constituent projects, as well as by voluntary contributions from members in the form of a percentage of money earned through publications on subjects discussed at the meetings.

Tessa sometimes accompanied me on these trips, leaving the children (who were rapidly growing up!) in the care of some colleague who would stay in our house. Sometimes I would go alone if it were to be a short trip lasting only a week or so. By this time Tessa had largely got over her fear of flying, although there were times when she would refuse to come at the last moment. I remember arranging some work in Paris and booking the both of us on an Air France jet from Montreal. We checked in at Dorval and made ourselves comfortable in the first class cabin. The captain announced that there would be a slight delay on account of something having to be done to the plane. Tessa got terrified all of a sudden and told the air hostess that she could not fly and must get off the plane! The hostess realised that she would have problems if she did not allow Tessa off, so she did. As she was leaving, I just had time to say to Tessa that I would be staying at the airport hotel for a night or two before going into Paris and we would do the planning of the work at this hotel.

We took off, minus Tessa and of course the flight was as smooth as ever. The next day I met Biemel and we planned the Paris work at the airport, had dinner at the airport hotel and finally I retired to my room, thinking everything was in order.

The next morning, I was still half asleep when I heard a faint knock on my door. I shouted, '*Entrez!*' although I did not know who could possibly be wanting to see me so early in the morning.

The door was slowly pushed open and Tessa appeared in the doorway, saying, 'I've come! I couldn't stay in Canada without you, so I came on the next plane!'

I jumped out of bed and hugged and kissed her and told her how much I loved her. It was great to be missed so! So Tessa accompanied me on the workshops and we enjoyed Parisian cuisine together in the evenings, until it was time to move on to the next port of call, which in that case was a town in Germany, where I was to hold more workshops for teachers. We flew back together and I do not think that she ever had such a fright again, at any rate not one accompanied by a sudden departure!

I made a number of trips to Germany, particularly to Heidelberg, as a certain Herr Rombach, president of the

German publishing house Herderverlag, lived there. The publishing house would sponsor workshops, in that way getting teachers interested in my approach and so getting them to buy my books. During the first of these visits I met some mathematics educators, who expressed their desire to come and work at my centre in Sherbrooke. A certain Mr Lunkenbein was the first one I invited to come and he became a member of the centre's personnel, taking part in running some of the learning research we were doing in our laboratory as well as working with children in schools.

Eventually, I had people working at the centre from France, Germany, Hungary, Spain, England, Australia and the USA so we had to expand our secretarial staff to four. At the height of our work at the centre we had about twenty people working together in various capacities, some of them paid directly by the university, others having their salaries taken out of successive federal grants.

I invited Ned Golding, the headmaster of the demonstration school in which I had worked in Adelaide, to come and work on a comparative study between the Australian and the Quebec situation. My other Australian collaborator joined the centre in a very strange manner. This is how it happened.

I was working with Professor Jeeves and John Williams on planning a psychological experiment on the abstraction process. We went to a beautiful lakeside inn in North Hatley, not far from Sherbrooke, to do this work so we would not be disturbed. There was a rather lively dog in the room in which we were working and John Williams insisted that Jeeves got the dog to be quiet by hypnotising it! Be that as it may, we did manage to plan quite a good portion of our proposed experiment.

On that particular day, Tessa went into Sherbrooke in our Holden, which she parked quite prominently in Rue Wellington. Of course, it was an unusual car, partly because it was a right-hand drive, partly because it was really very beautifully made, so many people used to stop and admire it. When Tessa got back to the car, she saw a man, his wife and two small children admiring the Holden.

'Is that your car?' he asked Tessa in an obviously Australian accent.

'Yes, it is. Are you from Australia?' she replied.

They had a little chat and Tessa learned that the family had come to Canada to start a new life, but that they had run out of money and he was looking for a job. He said he was an artist.

'What can you do?' asked Tessa.

'I can paint, or I can teach people to paint,' replied the Australian.

'Why don't you come home with me and meet my husband?' suggested Tessa, 'I believe my husband might be looking for an artist, as I heard him say the other day that he was going to start a project to examine the relationship between art and mathematics.'

She took them to our apartment in Rue Dominion and plied the grown-ups with beer and the children with whatever she could muster, while they all waited for me to come back. I returned eventually, rather tired and I was not overly pleased to find a whole family had taken possession of the place.

When the situation was explained to me, I said to the man, whose name was Cantieni, 'Why don't you come and show me what you can do from the point of view of teaching art to a class of children?'

'But I don't know a word of French!' replied Cantieni.

'Surely it doesn't matter. Art is international, isn't it so?' I suggested. 'If you can do something with the children and learn French in a couple of months, I might be able to give you a job.'

So we arranged for Cantieni to come and do his thing. He came to the École Eymard the next day and I explained to the children that this monsieur did not speak French, but that he would try to help everyone to do some painting. And he did just that. He was very good with the children, able to motivate them, never putting them down but encouraging them. I decided that he would be all right.

'Come to the centre tomorrow and we shall give you a contract. We shall also arrange for you to have a crash course in French – we will pay for the course,' I said to him.

So Cantieni was hired as the artist-in-residence. He became a very useful member of our team. He quickly learned how to shoot films, so with his help we made a large number of teacher education films in which we showed children learning mathematics the new way. He also helped run an experiment in French immersion in an English-speaking 'island' near Sherbrooke, where mathematics, movement and art were taught in French in the morning and the other subjects were taught in English in the afternoon. We paid the French-language teachers, the school board paid the English language teachers. We used the École Eymard as the control school and at the end of the year there was strictly no difference in mathematical achievement in the two schools, but of course the immersion children also learned to speak, read and write French!

Almost every year while I was director of the centre, I hired a Hungarian from the Budapest project of the ISGML, starting with Professor Varga, who was the initiator of the reform in Hungary, begun as a result of the 1962 UNESCO meeting in Budapest. Professor Varga learned a lot of practical details while he was working with us and he put such knowledge to good practical use when he finally returned to Hungary. Another quite fascinating co-worker I had invited from Hungary was Klára Kokás, a

Kodály musicologist, because we wanted to study the music-mathematics connection. This work led to a number of different conjectures about links between mathematics learning and music learning, which eventually came to fruition not in Sherbrooke but in Florence, where I went to work very much later. In this work I tried in practice to put together series of learning activities in which the learning of music, language, movement and mathematics were creatively woven together in fascinating ways. This eventually spread to Ravenna, where there is an excellent teacher who, to this day, practises some of these ideas, which she has taken a great deal further by creating ever more new and more exciting learning situations.

I made one trip to New Guinea from Sherbrooke, for which expenses were paid by a trust we set up in New Guinea, into which funds would be paid, as my mathematics material was constructed by prisoners and others, in lieu of paying me royalties on sales. Anyway, there were enough funds in the trust to pay my first-class fare from Sherbrooke to New Guinea via Vancouver and Hawaii and I could return via Tahiti, Buenos Aires, Porto Alegre and Rio de Janeiro, so I set up some 'workstations' along the route.

On the way to Vancouver, I stopped at Calgary, where my cousin and childhood friend Aga lived with her husband and children. It was great to renew our old friendship and to get to know her children, with whom I instantly made friends. We had some misconceptions to clear up, one of them to do with Aga having married a Hungarian Hitler Youth worker who had worked with the Nazi Youth organisation, which was called Nyilasok. At the end of the war she had written to me to ask me to guarantee her and her husband as 'politically reliable'. Having just fought a war against the Nazis, I could not bring myself to grant them this guarantee, so they moved to Venezuela, where many other ex-Nazis had already gone to settle. They stayed there for several years. All their children were born there, but, as he was working in oil, he was eventually transferred to Alberta, where he continued being an oil man. There was quite a lot we had to tell each other, but the bond we had established in childhood helped us to overcome all these rather sensitive issues.

Aga took me round to Lake Louise, Banff and other beauty spots. We climbed modest mountains with the children, swam in the lakes and generally had a relaxing time before moving on to Vancouver. I spent a couple of days at Simon Fraser University, where several people I had worked with in England were now teaching. Then, without giving any further nostalgic thoughts to Hawaii, I made a beeline for New Guinea.

I just dotted the Is and crossed the Ts in New Guinea, where the mathematics work appeared to be ticking over under its own steam, some of it directed by members of the University, some by the headmaster of the first bush school I visited after crossing that memorable crocodile-infested river! My missionary friend had passed the mission over to another missionary in the Highlands and had moved to Lae. I spent a few days at his house there. I remember that he had managed to fit an air-conditioner in his own study, fixing it into the frame of the window; the rest of the house was provided with a few ineffectual fans and was always very hot. One night I woke up to a big roar. The house was shaking and there was a noise of furniture falling about from the study! It only lasted about a minute and then there was silence.

The next morning all the children came running up to me, saying, 'Did you hear the earthquake last night? Wasn't it great?'

'I was scared!' chimed in the youngest one.

So that was what the noise was! When we went into the study, we noticed that the air-conditioner was no longer in the window but lying, in rather a sorry state, on the floor. My missionary friend had obviously not fixed it up in an earthquake-proof manner!

He was really thinking of giving up the vocation of being a missionary and asked me if I would invite him to the centre. I left him an official letter of invitation, asking him to come to the centre in the capacity of a member of the New Guinea project of the ISGML. This eventually enabled him to get an immigration visa into Canada (he was a US citizen) and to this day he works at a CEGEP in the Province of Quebec. He sent his children to French-language schools, where they all picked up the language very quickly and now they have all made satisfactory careers for themselves.

After leaving New Guinea I made for Sydney, where I was to meet our daughter Jancis, who had married an American and had insisted on moving her whole family to the back of beyond in order to live healthily. They purchased a two hundred acre smallholding, where they grew everything and also kept animals. Jancis, however, was not allowed to have any 'mod cons' – she had to live as they must have lived hundreds of years ago. She had two children by then, a boy and a girl and I was to meet them at Sydney airport on my way to Tahiti. When I arrived and saw them, I could hardly believe my eyes. Jancis looked haggard and bedraggled and the children looked like waifs, crying intermittently. Apparently, the manager of the airport restaurant had already given them some food without requiring payment, because they looked so miserable.

Jancis poured out some of her tales of woe to me while I tried to amuse the children. I bought them more food and a toy each and then it was time to leave, for my flight to Tahiti had been called.

One French member of my centre, a certain Bernard Héraud, had put me in touch with the teachers' trade union in Tahiti, so on the way to South America I could stop there and possibly do some work. As it happened, I had only a few hours available, during which we discussed a possible two week workshop course to be held for Tahitian teachers at some time in the future.

We stopped at Easter Island on the way to Santiago de Chile, where there was really too little time to have a look at the famous statues whose origins are to this day a subject of controversy. I arrived eventually in Buenos Aires, where I had been invited by the British Council to talk to the teachers for a few days about my methods. My Spanish was not exactly fluent, but I did remember some Spanish from the time I had taught Spanish to the lady who sold vegetables in Leicester market place! Anyway, Spanish is so close to Italian that I did not have much trouble making myself understood.

What with my preoccupation with the work I was doing and my problems with the language, I forgot to reconfirm my flight to Porto Alegre. However, somebody did tell me that I had to have a valid smallpox vaccination certificate and I noticed that mine had expired! So I was taken down to the docks, where there was a doctor who did this for all the sailors. I recall climbing up some rickety stairs and being asked to come into a very insalubrious-looking room, where the doctor gave me the appropriate scratches on my arm, filled in with the appropriate poison and asked to be paid a ridiculously small number of pesos! That was the last time I was ever vaccinated, as smallpox soon became extinct after that year.

When we eventually reached the airport, I was told that there were no seats on the plane for they were all sold – and that is when I realised that I had not reconfirmed! I was told to stand by and wait to see if there were enough 'no shows' to permit me to fly. Since this was not at all certain, the British consul, who was with me at the airport, said he was arranging for me to be taken by armed escort across Uruguay to Porto Alegre and that I would be there in time to commence my work. He said the armed escort was necessary because the place was infested with *tupamaros*, or 'freedom fighters' but that they would be unlikely to attack an armed escort. I was just imagining myself being taken hostage by some wild *tupamaros* when I heard my name called on the loudspeakers. I had been given the last available seat. I said a little prayer of thanks to the unknown passenger who had not made his flight!

When I arrived at the airport in Porto Alegre, I was surprised to find TV cameras trained on me and a whole crowd of people there who had come to greet and welcome me. A journalist put a microphone in front of my mouth and I had to explain why I had come. This was Porto Alegre and thus Brazil and the language was Portuguese! I had just managed to get used to some rudimentary Spanish at the last port of call and my knowledge of Portuguese was strictly zero! So I mumbled something in a mixture of Spanish and Italian while the TV cameras were rolling and the still cameras were flashing. The next day I saw this scene on the front pages of several local newspapers, where it was explained how I had tried to speak to them in Spanish and Italian.

The life and soul of the party was a lady of the name Esther Grossi, who had apparently fixed everything. I was taken to my hotel and then to dinner with a number of people with whom I would have to work for about a week. I tried to rapidly work out the transformations necessary to turn Italian or Spanish into Portuguese while I listened intently to what everybody was saying, trying to make head or tail of it.

The next day they produced a group of children who understood French, with whom I could work and Esther

Grossi would help to explain to the observing teachers what was happening. This worked quite well for a couple of days, but, since the children always spoke Portuguese to each other, I had many chances to learn more while I was getting them to play my mathematical games.

By the third day I had learned enough to be able to communicate with the children in Portuguese and I could also explain things to the audience. Esther Grossi was still there, just in case I could not make myself understood. By the end of the week, I thought I had a good working knowledge of Portuguese, simply by learning and internalising the transformations which altered sentences from Italian into Portuguese. It reminded me of my trip to Corsica, when I had quickly learned to express myself in Corsican, simply by using the transformation method.

The course was a great success and they wanted more. Esther Grossi said that her group would join the International Study Group and that I must come and pay regular visits. I was paid for my pains in American dollars, in cash, no doubt obtained on the black market. Esther Grossi also arranged for me to work in São Paulo at the Leonardo da Vinci School, where all the instruction was in Italian, so the classes and the workshops could be held in Italian. This was a relief, because, in spite of everything, trying to speak a language that I could only speak through transformations was somewhat tiring!

So I did a couple of days' work in São Paulo. In the evening, while I was having dinner at my hotel with some of the teachers I had worked with, the waiter came to our table and announced that I was wanted on the phone. I could not imagine who it could be, as my circle of acquaintances in São Paulo could be measured by the number zero! But I went to the phone and listened.

'Are you Zoltan Dienes?' the voice asked in Portuguese. My rudimentary Portuguese could stretch to that!

'Yes, I am,' I replied, trying to sound as Portuguese as I could.

'Are you Hungarian? And do you speak Hungarian?' asked the voice in Portuguese.

'Yes, I can speak Hungarian,' I replied in Portuguese.

'Did you go to school in Budapest in the Piarista Gimnázium?' the voice said in Hungarian.

'Yes, as a matter of fact I did,' I replied.

'Did you have Belyis as your class teacher for Latin and Greek?' asked the voice.

'Yes, I did,' I replied, somewhat nonplussed.

'And did you sit in the front row near the door?' asked the voice.

'Yes, I certainly did,' I replied, even more surprised.

'Well, I sat next to you!' said the voice.

He then explained that he had seen my name in the paper and remembered that we were classmates in Hungary, assuming I was the same Dienes Zoltán. He also informed me that there were three other boys in São Paulo from the same class and that we just had to meet! I agreed and the following evening, instead of dining at the hotel, I had the only school reunion I ever had in my life; of course we drank to each other's health and talked about the 'good old days'!

After these events I had nothing else scheduled on this particular trip, so I flew back to Montreal and arrived in Sherbrooke, where Tessa had prepared a party to welcome me. I suppose this was real jet set behaviour! You work yourself to the bone on a world trip, you come home and perhaps what you feel like doing is having a good sleep, but no, your dear better half has prepared a wonderful party, so you party! In fact, when you come to think of it, *not* having a party might be an anticlimax, so maybe women know more about these things than men!

When I had the work at the centre organised, we ended up concentrating on particular lines of work. First, there was the study of the psychology of learning. This included the study of the learning of logic and of various group structures, using machines that Mr Parkanyi had made; research into the abstraction process by examining the results of multiple versus single embodiment learning, investigating the non-mathematical results, if any, of mathematics learning; and the constructing of tests for verifying such effects as learning to learn, of preference for complexity as opposed to simplicity and so on. Much of this was eventually published in the *Journal for Structural Learning* of the ISGML.

Secondly, there was practical work in classrooms. This included much 'advanced' curriculum work, through which we could show that children of very tender ages were well able to study parts of mathematics that had been regarded difficult or abstruse, such as groups, rings, fields, vector spaces, matrices, finite geometries, transformational geometry, propositional calculus, predicate calculus and so on. There was also a steady effort going on to incorporate the more obviously viable topics into a practical curriculum, based on the use of concrete materials and the abstractive way of learning mathematics.

The publishing house Hurtubise had taken over from Education Nouvelle and had published a large number of individual and group instruction cards, accompanied by some detailed instructions for teachers as to the methodology thought to be the most effective. This came to be known as the program *Mathématique Vivante*.

Particular care was taken with some of the classes at the École Eymard, in which the children were becoming really mathematically literate. In one class the children insisted on doing three hours of mathematics every day! One boy of about eleven once produced a proof for the completeness of an axiom system from which the relationships between the isometries of the equilateral triangle could be deduced. Only after an exhaustive discussion with my colleagues in the mathematics department did we decide that the proof was indeed correct!

Thirdly, there was teacher education. We organised courses for teachers, in conjunction with the university Mathematics Department, which were good for university credits. These were either short weekend workshops or courses that lasted as long as three weeks, with a six hour day; these were offered during the summer vacation.

Workshops were also asked for by and given for the Université de Montreal, where I also gave some courses for teachers of retarded children, each meeting being preceded by a visit to a 'retarded' class, where I tried to show the attending teachers how much can be accomplished with such children, in spite of the disadvantage created by their being labelled 'retarded'.

The ISGML also organised workshops for teachers, one being given jointly by the ISGML and the Université de Montreal. This one lasted three weeks (during the summer) and we had four workshops running simultaneously, ISGML members from Europe providing their services free of charge as instructors; for example Professor Varga was responsible for the probability workshop. Each day began with about an hour of work with children, during which, in the beginning, the teachers attending observed how the instructors handled the children, gradually beginning to take part themselves. The idea had already spread abroad that such mathematics was fun and so, in spite of the fact that this was a period of school holidays, there were too many children wanting to come. So we asked them to pay five dollars each for the privilege. This reduced the number of applicants by a little. Often, at the end of the children's part of the work, when they would be told to go home, they would say, '*On a payé cinq dollars pour venir; nous voulons rester toute la journée!*' ('We've paid five dollars to come, we want to stay all day!').

And indeed some of the children did stay and they often helped the participating teachers solve their problems! I have kept some of the participants' notes from these courses, and sometimes I look at them and feel nostalgic about the 'good old days' when teachers were hungry for more and enjoyed the challenge and the newness of these approaches!

The teacher education aspect of the work of the centre soon became international in extent, incorporating countries as far apart as Australia, Chile, Peru, Argentine, Brazil, the United States, England, France, Germany, Hungary, Italy and Spain, not to mention other provinces of Canada. Much of it was done under the aegis of the ISGML and the costs were always borne by the customers, namely universities, ministries of education, teachers' trade unions, school boards and such, which had requested the services. Ongoing projects were established in a number of the above countries, run by local people and from time to time nurtured by ISGML workshops or seminars, in every case requiring my personal presence. This entailed a great deal of travelling, which meant that enough responsible personnel had to be left behind at the centre, so that work could continue without interruption in my absence.

While we were still living in Rue Dominion, I received the offer of a contract from the Baüerische Rundfunk in Munich, to make thirteen mathematics education programmes for their television viewers. The scenarios had to be carefully prepared beforehand, down to the last detail of physical material required for the shoots. Once I had written the scenarios and these had been approved, I flew to Munich to carry out the work.

The producer insisted that I should have no contact with the children who were to carry out what was in the scenarios but meet them 'cold', when the TV cameras were already rolling. There would be three or four groups of children ready each day and we would start with the first group and carry out the program. Then we would watch it on video and, if we were satisfied mathematically, psychologically and technically, then we would do an introduction and a conclusion, each one in three languages: German, French and English. In this way they could sell the series outside German speaking areas.

On some days we were lucky and the first trial was deemed a success, in which case we finished early. On some days we had three or even four groups of children go through the work before we were all satisfied. It is only after we were quite sure of it that we passed on to the introduction and the conclusion. There was always a bottle of Sekt on a trolley to keep me going and on harder days quite a few bottles were consumed! But when I heard the cameraman announce, *'Deutsche Auffassung, bitte!'*

I knew we were nearing the end of the day's work. After the German version came the announcement, *'Version française, s'il vous plaît!'* and at the very end came, 'English version, please!'

Before doing the English version I would empty the last drop of the bottle of Sekt waiting for me on the trolley.

On the very last day, I mixed up the languages and started to explain things in German for the French version and when I heard the loud shout of 'Cut!', I wondered what I had done. Then they told me again, *'La version française, Monsieur, s'il vous plaît!'* saying it all in French, to drive it home which language it was meant to be!

In the end we had thirteen half-hour programmes, which were aired in most parts of Germany in prime time. They were also aired in Holland and in Austria and several teachers from Hungary later told me how they had enjoyed these programmes which were beamed over into Hungary from Austrian TV.

The contract for the above meant that I was paid a considerable fee for the work, although I received no royalties. This meant that we now had enough money to buy our own house!

It did not take us very long to look – we found a dream house in one of the good suburbs of Sherbrooke, which had just about everything anyone could want. There were four bedrooms and the master bedroom had a vanity room and bathroom and a Romeo and Juliet balcony, overlooking a small copse at the back of the property. Very romantic! There was a double garage, with automatically opening doors, a laundry section, a built-in vacuum system, an intercom from every room to every room in the house and a beautifully curving stairway for reaching the passage between the bedrooms from the lounge. We placed some floodlights on the roof, so when the 'winter wonderland' arrived, the back garden looked like a real fairyland.

The only problem with the house was the name of the street, which was called Rue Grime! Tessa refused to live in a street with such a name! Fortunately, the house stood on the corner of Rue Adam and Rue Grime and we told the vendor that he would have to change the address first, so it became a house in Rue Adam, no longer in Rue Grime! Amazing as it may seem, all this was done, so we soon moved in.

So after two years in Rue Dominion, we could look forwards to some years of the 'good life' in our dream house, accompanied by a rapidly progressing work at the centre, with all its international ramifications and consequently necessary world trips.

Upon looking back, certainly from the professional point of view, these years were the zenith of my working life.

What happened on some of the world trips I will tell in the next section.

Chapter Twenty

Working Internationally from Sherbrooke

There were four different reasons for travelling about the world to sow the seeds of better mathematics learning. First, the necessity, or certainly the desirability, of fostering international co-operation in research into the problems of mathematics learning, in which the centre engaged under the aegis of the ISGML. Second, the necessity of holding as many teachers' workshops as possible, so as to awaken teachers' interest in and provide them with familiarity with the techniques which were being developed by the research groups. Thirdly, there were the invitations by the many publishing houses who had taken an interest in the reform of mathematics teaching to hold workshops for teachers, thereby increasing the market share of the books that arose out of this type of work. Then there were invitations by various authorities such as government ministries, teachers' trade unions, universities and school boards in order to further the cause of improved mathematics education.

A number of members of the ISGML were at home with a fair number of languages, which meant that the geographical area in which workshops could be held was quite large, covering North and South America and most of Europe.

The research meetings were held every year, though sometimes missing a year, until funds began to run out and raising new funds had begun to be difficult. The last ISGML research meetings were held in the mid-Eighties in Italy and it seems unlikely that any more will be held, since such things take a great deal of energy to organise and realise and, now, in my late seventies, with my health no longer perfect, I feel that other people and other organisms have to take over the carrying of the flag of reform in mathematics education, as a great deal is still very amiss in that area of our cultural activities!

The most memorable ISGML meetings were held in Florence, two near Brisighella, one at Visegrád (Hungary), one at Benodet (Bretagne) and one at Barcelona (Spain). The Barcelona meeting was preceded by John Williams and myself stopping over in the Canaries to 'inspect' the Canary Islands project, run by a local teacher by the name of Caparros. We looked over Caparros's work, which seemed to be every bit as well run as other ISGML projects. This was necessary since some members were doubtful whether the Canaries project was up to ISGML standards. At the Visegrád meeting we discussed the problems that related to music and mathematics and whether it was possible, or even desirable, to make the links more conscious in those people's minds who studied one or the other of these disciplines.

I suggested at one point that, since the pentatonic scale had been so important in the development of Hungarian folk music, it was possible that the mathematics of the group of five elements and its automorphisms (i.e. roughly speaking, the ways in which the group can be 'mirrored' back into itself) might be linked with the way in which accompaniments were written to such pentatonic melodies. So we had a look at some of the accompaniments written for such melodies by Zoltán Kodály and, indeed, we found that for the most part they followed the 'rules' of one of the three automorphisms of the group of five elements, by identifying the notes with elements in the following way:

A C D E G   corresponding to   3 4 0 1 2

where the 'automorphisms' were provided by 'multiplications' by 2, by 3 and by 4, taken modulo 5 (meaning that, after any multiplication, one had to subtract as many fives as possible and the number so resulting would be considered the product. In this way you would never get any numbers greater than 4, so you would always get one of the 'notes').

On another occasion, since I was somewhat tickled by such possibilities, I taught a grade five class in a Kecskemét school the Scottish song *Bonnie Banks and Braes*, which is written in a purely pentatonic scale. I wrote on the blackboard the three transformations of the melody, using each of the automorphisms and taught the children to sing each tune. Then I divided them into four sections and got them to sing the melody in four parts. I am no musician and have absolutely no training in musicology, but the resulting sound seemed to be extremely pleasant, even beautiful. Of course, as we all know, Bach was the first 'mathematical musician' of any note and nobody could object to the music being 'ruined' by mathematics in his case! So perhaps we have yet much to learn in this area.

We also discussed the role of language at the Visegrád meeting, some of us suggesting that listening was perhaps a neglected area of the study of language. Could it be that such a lack were responsible for a great deal of the heartache people experience in learning other languages? We had the scandalous idea of setting up a conversation experiment there and then, in which everybody would speak his or her own language. To make the experiment less absurd than it sounded, we thought we would restrict the conversation to Romance languages. We had present speakers of the languages French, Spanish, Italian and Romanian, so we started discussing the very point about listening and communicating, each of us taking a particular language. I remember that I was allotted Italian, Professor Fischbein from Bucharest spoke Romanian, Pons from Barcelona spoke Spanish (although he could equally well have spoken Catalan!) and I think Professor Varga spoke French. We had absolutely no difficulty in understanding each other!

Of course in no way had we proved anything; naturally, we were all reasonably intelligent people and the languages chosen were certainly similar, but, on the other hand, listening was certainly at a premium! The fact that such a thing is even possible, is already of some interest, as an existence theorem would be in the field of mathematics, when sometimes we find it of interest to show that at least such and such a mathematical creature does in fact exist. Perhaps some future researchers who read this paragraph might feel it was worthwhile giving the matter some thought.

Another problem we discussed at Visegrád and at some of the other meetings was the problem of the dynamics of invention. I was told that I was very good at inventing games, but had I ever managed to teach anyone to invent them? I had to confess that, indeed, I had not been singularly successful in handing down the skill I appeared to possess of turning just about every mathematical relationship into some form of amusing game.

'Why not just get together right now and try and invent something?' somebody suggested. 'We can perhaps observe ourselves!'

'Yes, why not!' we all agreed.

We thought that we would take some logic blocks and just go at them 'cold'! I suggested that we should not take a whole set. First of all, it would be hard to observe what was happening; secondly, it was less likely that we could interact with so many pieces coming into play all at once. So we took eight blocks out of the set: the small blue triangle, the big blue triangle, the small red triangle, the big red triangle, the small blue square, the big blue square, the small red square, and the big red square.

I quietly put the small blue triangle in my pocket so that we were left with only seven pieces.

'Where is the small blue triangle?' somebody asked.

'I've hidden it!' I replied. 'Just to make it more fun.'

So we agreed to try and be 'creative' with just the remaining seven pieces!

We were sitting round a round circular table and, before we knew it, the blocks were placed around the circumference of the table.

'There seem to be three red ones following one another, but there are never three of the same shape following one another,' remarked one of the team.

This was soon remedied by exchanging the positions of some of the blocks.

'But what about the big ones and the small ones?' queried someone.

'You mean, as well as keep the three red ones and the three squares in a row?' someone else asked.

'Yes, let's try,' we all said.

It was not long before we arranged the blocks around the table in such a way that the colours, shapes and sizes all followed the same 'rhythm':

red red red blue red blue blue
square square square triangle square triangle triangle
big big big small big small small

The 'threes', of course started in different places!

'That will make a fun game,' we all agreed.

'But what is the mathematics in it?' enquired one member of the team.

'You have just constructed the basis for the multiplicative group of an eight element field!' I replied.

'Oh,' replied the others, not following a single word.

'Oh yes,' I said. 'There is plenty of mathematics, but perhaps it is not out of any conventional curriculum!'

'But how did we do it?' we asked.

'The round table helped,' I suggested. 'And the vivid colours of the blocks were a help too, not to mention our motivation to get something done!'

What does motivate people to do things like inventing games, climbing mountains and many other things, which have no immediate 'payoff'? It was a long time before I was able to go into these problems a little more deeply; in fact, not until I was asked by Siena University to write a thesis on the occasion of that university conferring an honorary doctorate on me in the Eighties. I told them that I would write a thesis on intrinsic motivation, which I defended on the occasion of the conferral of the degree. It was published under the title *Il piacere della matematica* by the Bologna publishing house of Cappelli.

The ISGML research meetings helped those of us who were doing similar work to keep in touch with each other, not only from the point of view of our work, but also from the point of view of cementing the human relationships between us. Some real friendships were made that lasted a lifetime.

John Williams, whom I first contacted at the National Foundation for Educational Research for England and Wales, came to work for about five years at the centre. He came to all the ISGML meetings and was very good at being the devil's advocate, but he also developed into a true friend with whom one could share problems and be helpful to each other.

Ricardo Pons was always a regular attender from Barcelona. He had trained to be a priest and was a Catholic priest for a long time, but at a certain point in his life he felt a great conflict between Vatican dogmatism and the way his spirit wanted to fly freely and explore. He also came to Sherbrooke, if only for brief periods, but we were able to help him through his crisis. He then left the priesthood and married and had two children. We have remained friends ever since. He has often come to meetings in Italy I have organised, as he was also fluent in Italian.

Professor Tamas Varga also became a good friend and staunch colleague and a supporter of our efforts. He contributed greatly towards the study of the learning of the concept of probability, writing a number of books on it, which appeared in a number of European languages and he gladly shared his royalties on these publications to support the finances of the ISGML. We were always on very friendly terms, on *tutoyer* terms in all the languages we had in common. We regret his recent death. Mathematics education, especially in Europe and in Brazil, where he also went to work in the Porto Alegre ISGML Centre, will miss him greatly. The Hungarian mathematics curriculum and accompanying methodology owes much to the collaboration between us and is one of the most advanced in Europe.

Professor Angelo Pescarini also became a lifelong friend, one of the few Italians with whom we are on *tutoyer* terms! He was director of education for the Emilia Romagna region for several years and used the office to further our mutual interests in mathematics education. Much of the work that was done towards rewriting the Italian mathematics curriculum was due to the collaboration between the two of us. The Italian official curriculum and accompanying methodological guide is, along with the

Hungarian curriculum, among the best in Europe, although it would be untrue to say that the majority of Italian schools have attained the objectives contained therein!

Ermanno Pasini, the teacher with whom I started the Italian work in the province of Ravenna, has also been an important figure on the practical side of implementing the reforms. He became the director of eight schools in the Ravenna area and, during his tenure of this office, I visited his schools almost every year and trained the teachers by working with the children, so that now there is a good nucleus of teachers who are well able to handle a more meaningful way of teaching mathematics. He is now retired but I still go to Italy every year and he is never absent from our meetings. We often stay at his house in San Pietro in Campiano, where we talk about old times as well as make plans for future work. He never lets us go without filling every available space in the car with his wonderful home-made wine and other goodies. His wine and his true friendship are both out of this world!

Sandor Klein has been another staunch friend and collaborator. He came to Sherbrooke for about a year and a half from Hungary and helped us carry out the test construction phase of the work of the centre. This work involved doing test runs in Brazil, Hungary, the USA and Spain, in each case involving the respective ISGML centres. The result of the Sherbroooke work was published in a special issue of the *Journal of Structural Learning*. The main work attributable to Sandor Klein is his part in carrying out the testing for the Hungarian Ministry of Public Instruction in evaluating the new methods being introduced through the Varga-Dienes collaboration. He was probably the first to run a testing programme in which such items as anxiety levels, problem-solving ability in general and preference for complexity, 'learning to learn', were included in the test batteries, apart from the usual tests for the attainment of competence in the field of mathematics. During the development of the programme, Varga and Klein often invited me to come and give workshops and seminars, both to university faculty and to teachers. Naturally, with such long standing collaboration, a friendship developed and was cemented between us, which even the political turmoils Hungary went through could not touch. We are still in touch and I go to Hungary nearly every year to work in some school system or give some seminars at universities such as Szeged and Pécs.

Esther Grossi, of Porto Alegre, Brazil, is another collaborator and friend, although, for financial and geographical reasons, I have not been down to South America for several years. Through Esther Grossi's creative good offices, I was able to make about half a dozen visits to Brazil, most of which were concentrated in the south, in the town of Porto Alegre in the Rio Grande do Sul. She eventually obtained her doctorate at the Sorbonne in Paris, based on a thesis for which she had learned the material at Sherbrooke, as well as during my repeated visits to Brazil and Varga's visits, whose work in Porto Alegre complemented the work started by myself. The subject of her thesis was the development of the concept of multiple; the thesis is a thorough examination and analysis of this rather difficult process, which had been studied very little in the practical ways in which Esther Grossi approached it.

Leonidas Kounguetsoff is another mathematician who has become a lifelong friend. He came to the Sherbrooke Mathematics Department because he knew that I was working in Sherbrooke. He also told the rector that, if ever I left, he would leave. After a number of years, when it had become necessary for me to leave, he was as good as his word and left to take up the work of founding a new university in Greece. He invited me to talk to the faculty and students in Adrianoupolis, in 1996 and he is planning to invite me for a longer period to recreate something of the kind we had at the Sherbrooke Psychomathematics Centre. This may or may not ever materialise on account of my advancing age, but the friendship remains for ever!

When Esther Grossi and her group in Porto Alegre joined the ISGML, many other groups that were then being formed under the stimulus of the Porto Alegre work also wanted to be members. In the end I found myself being called to work with groups as far apart as Rio de Janeiro, São Paulo, Brasilia, Minas Gerais and Bahia. Inevitably, I had to learn to communicate in Portuguese, which I did not find too difficult, as it was a Romance language and they all have rather similar structures. Using the transformation trick, it did not take me long to be quite at home in this language.

After my first visit, which I have already described, I was invited every year for several years to work in the various above-mentioned projects. Let me describe some of the events that took place during these work periods.

Tessa accompanied me on all the Brazil trips except on the first one I have already alluded to. I remember one course I gave in Rio, where they put us up in a good hotel in Copacabana (everybody has heard of Copacabana Beach!), but the work had been planned to take place on the island. The bridge had not been built at the time and we had to take the ferry to get across. This would not have been so bad, except for the ride along the eight lane highway from Copacabana to the ferry terminal! If you have not seen Brazilian driving, you can have no idea of what such a trip can be like. The driving is extremely fast, as driving slowly would almost certainly cause an accident and these cars, driven by seemingly demented drivers, at about 120 km per hour, weaving in and out of the various lanes, do not provide exactly the best mental preparation for a day's work.

'Why can we not have a hotel on the island?' I enquired after a few days of hair-raising trips to and from the ferry.

'Oh, it is best for you at Copacabana,' was the non-committal reply.

'But could we not see a hotel on the island?' I insisted a day later.

Very reluctantly, we were taken to what was the best hotel on the island. We had hardly entered the foyer when both Tessa and I immediately realised why Copacabana was best. The smell as we entered was such that we could hardly stop ourselves from being sick. So that was the end of our island hotel hunt!

We got to know some very enthusiastic teachers, one of whom Tessa called Jungle Jim, who still writes to us from Rio and signs himself Jungle Jim. He became, in the end, the leaven that made things move in Rio! Once I went to give a course in Brasilia, the formal capital of Brazil, where everything is decentralised as they have made every block self-sufficient, so people never have to go anywhere so there is less need for using cars. I soon found a taxi at the airport and gave the taxi driver the address of the hotel I was supposed to go to. Being a Brazilian, he immediately started tearing down the highway, taking absolutely no notice of red lights, driving straight through every one of them. I was getting more and more nervous when he suddenly came to a stop at a green light.

'Why are you stopping?' I said to the driver. 'The light is green!'

'I know,' he replied, 'but you never know. Some fool might be driving right across the red lights!' He pointed at the red lights showing on the road at right angles to ours.

Then cautiously, very slowly, he drove across the intersection. I realised that there was no use arguing. The Brazilian culture was clearly based on very different behavioural principles from the ones to which I was accustomed!

Bahia was a typical tropical place, from the point of view of how people behaved. There are, I was told, three hundred and sixty-five churches in Bahia, one for every day of the year. In front of one of them, within sight of all the holy statues, was the square where once the slaves used to be sold. It seemed so incongruous to have a church for every day of the year and yet have a history of most cruel and devastating slavery.

I had a week to persuade the teachers to change their ways. I remember arranging some tables end to end and putting children on each side of this long row of tables, with lots of material and what I hoped were motivating games for the children to play. The children certainly had a ball and, when they went at the end of about an hour, I had to rope in the teachers to do similar things. It was very hot since there was no air-conditioning and it seemed that I was having to make a great deal of effort to achieve very little.

During the interval, one of the participating teachers enlightened me about the Bahia situation. Bahia was in the tropics. There were lots of banana trees everywhere as well as other fruit trees; it was always warm enough to sleep out, so a lot of people just never did any work and lived on bananas or whatever else they could gather from the richly endowed environment. Apparently, many firms had had to get their workforce from the south, where people were used to working. But, as often as not, these workers, who had been transported at considerable expense by the firms in question, decided work was not interesting after six months and joined the ranks of the sleeping-out banana-eaters! All this made me realise that I was probably wasting my time teaching these teachers anything. I also began to understand the reason for the existence of strict slavery in the recent past! In fact, as far as I could ascertain later, no serious project ever arose as a result of my work in that land of bananas.

I remember one day arriving in São Paulo from Canada and noticing that one of our cases was missing. All my clothes as well as my nightwear were in that case! I asked VARIG what they could do and, since I was a first-class passenger, they just told me to get what I wanted and send them the bill. But, fortunately, one of the participants in the course I was to give was the wife of the president of the Bank of Brazil. The president was actually at the airport when I arrived and had seen my plight. He was about my size so he lent me some pajamas! So VARIG were saved at least the cost of those. It turned out that my case had been sent to Glasgow by mistake and it duly turned up after a few days.

One evening, after a day's work, Tessa and I were invited by the president of the bank to have dinner with him and his wife. We were told that a car would come to the hotel at a certain time and take us to his villa on the outskirts of São Paulo. The car duly arrived, driven by a chauffeur, but there was nobody else in the car. We were a little apprehensive but got in the car.

After driving a few blocks, the driver turned into a narrow alleyway leading to a ramp and then to a subterranean locality.

Tessa said to me, 'I think we are being kidnapped!'

'Relax,' I replied. 'Who would want to kidnap us?'

But Tessa was shaking in her boots. I noticed that the driver was filling up the car with gasoline. Probably they were able to get a better type of gasoline there or maybe it was cheaper, I shall never know. The driver got in, started the car and we drove out into the open. Tessa did relax by this time, but we wondered what else was in store.

We arrived at the entrance of a driveway. The place seemed to be teeming with armed guards! Several of them were briskly walking up and down at the entrance of the drive with sub-machine guns at the ready. The driver showed them a card, the guards all saluted smartly and we drove into the grounds.

We were warmly welcomed by the president and his wife. There was a terrific spread of goodies laid out in several rooms. People were all dressed to the hilt, the women dripping with diamonds and most of the men in formal dress! Tessa was a little embarrassed at first, but, when we sat down to dinner and she was placed on the president's right and I on his left, she quickly drank in the honour that this represented and was able to relax and be herself.

I found out afterwards that the Bank of Brazil was one of the organisations that had provided the expenses of the trip, which must have been considerable, with two round-trip first-class fares and hotel expenses for about a month, not to mention my fees. I had to take my hat off to Esther Grossi for being able to organise enough people and money to make our Brazilian trips possible!

The main thrust of my work in Brazil took place in Porto Alegre, where Esther Grossi was living and heading the Porto Alegre ISGML Centre. Porto Alegre was included in all my Brazilian itineraries. Brazil was run at that time by a military junta and, up to a certain point, one had to be on friendly terms with government officials in order not to be hampered in the work of reform. In a dictatorship in modern times the rulers find themselves in some difficulty over the implementation of the reform of mathematics teaching. On the one hand, in order to cope with constantly changing and developing situations arising out of the introduction of new technologies, it is important to teach those in charge how to deal with new and often unexpected situations. This requires some training in a certain amount of free thinking. From the point of view of the rulers, such free thinking is necessary at the upper managerial level, but positively dangerous at the lower levels, particularly in the case of the general workforce, whose members ought not to think but should simply follow instructions.

The traditional way of teaching mathematics mostly consists of teaching children to follow instructions. It also trains them not to ask the whys and wherefores of these instructions. Speed and accuracy are at a premium; thinking is an unnecessary luxury. So how do you make sure that future managers will be taught to face new and unexpected situations and yet keep the future workforce on a kind of obedience training? Of course there is no solution to the problem: you are bound to get the sheep and the goats mixed up.

The Porto Alegre project encompassed thirteen different schools, stretching all the way from shanty town schools to schools where only millionaires and army generals sent their children. I used to visit schools in all areas of the social spectrum and tried to introduce some freedom of thought into the shanty town schools as well as into the upper class ones.

In one of the shanty town schools, the first thing to happen when the children arrived was that they lined up for a large bowl of soup each, served from a huge cauldron. The principal said that if they did not feed the children first, they could not work because they would be too hungry. School materials were practically non-existent. I went into one classroom where one half of it was taken up by a pile of broken tables and chairs. The children were sitting at ancient desks, each child having a diminutive piece of paper on which they had been asked to draw something. The children looked listless and lethargic. What can one do that is constructive in such a situation?

I suddenly thought back to my experiences in Hawaii and in New Guinea where I had had to be creative and use local materials. What were the local materials here? Obviously the broken chairs and tables!

'Why don't we use all this for making mathematics materials?' I asked the equally lethargic and discouraged teacher.

She had apparently never thought of doing so. We somehow got hold of some saws, hammers, screws, screwdrivers and other such tools and got the children to dismantle the broken furniture and put it together into squares, triangles and various types of boxes – we even made a tetrahedron with some chair legs! The whole classroom was soon quite electrified and there was a buzz of activity instead of lethargic stillness. It was not long before the children began enjoying the feeling of learning. It did not really matter what they were learning, I thought, but it was important that they should begin to interact with their environment, a precondition of learning and so begin to learn, in particular learn how to think, naturally unbeknown to the generals in power! On several visits I taught them how to work out problems of geometry by going outside and drawing figures on the ground, using stones and sticks that were around. They learned about measuring and scale drawing, using large sheets of paper I had brought with me. Finally we persuaded some of the teachers to join the other teachers already in the project.

When it was time for us to leave Porto Alegre, we found that the children had put their pennies together and bought us a box of chocolates! We were told that we must accept it as a token of their thanks and that we must take it home and not share it with the children, for that would be an insult! It is hard to get used to such different customs, particularly in a school where half the children do not come to school on rainy days because they have no coats and a number of them trudge through mud on the way to school without any shoes, so they have to wash their feet before coming into the classroom! So we said our goodbyes and told the children we would be back the following year.

One young woman who was teaching in one of the city schools had just been let out of prison, where she had been taken because somebody intimated to the authorities that she was a member of a 'dangerous' communist cell! Of course it was totally untrue, but they kept her in prison for several months, torturing her to try and get the names and whereabouts of the other cell members out of her. Eventually, they must have realised that she really did not know anything, for most people broke down under torture and gave away other people when they could not stand it any more. The teachers in the school where she worked were happy to have her back and asked her to act the part of the seagull in *Jonathan Seagull*. She put her heart and soul into the performance, as flying high and being free had a very special meaning to her!

One of the children in her class, in which I often went to work to start up mathematical games, invited me to his birthday party. There were many children in the family and they cannot have been at all well off, but they gave us a bottle of whiskey as a present. Esther told us that we absolutely must accept; obviously, they thought that by giving us something so rare and expensive, as whiskey was in Brazil, they were doing something to show their appreciation for what I was doing for them. So we had a great party, but we were not allowed to share the whiskey!

When we visited the school where the rich children went, the first thing I noticed was that there was litter everywhere. Clearly, these children had never been trained to pick things up, as they had servants at home who did it for them. I also noticed more fighting and quarrelling than we saw in the other schools. But, we thought, these children cannot help being rich, so perhaps we ought to help them too.

I recall going into a grade seven class, in which I managed to get the children quite excited about the Bolzano-Weierstrass theorem on infinite bounded sets of numbers having at least one point of accumulation. One part of the proof, which really intrigued them, was to show that in such a set there must be either a constantly increasing infinite sequence or a constantly decreasing infinite sequence. This is roughly how the argument ran: If you suffer from a maths blockage, just skip the proof.

The set in question either does or does not have a largest member. If it does not, we can always pick a larger one after any one we have already picked and so we can always pick out a sequence, each of whose members is larger than the one before. If, on the other hand the set does contain a largest member, then let us consider our original set but without its largest member. This 'truncated set' will either not have or will have a largest member. If it does not, we can pick out an increasing sequence as before. If it does have a largest member, truncate this set now by taking away this second largest member.

The newly re-truncated set will either not have a largest member or it will have a largest member. If it does not, then we can again pick out an increasing sequence as before. If it does have a largest member, take this largest member away and consider the set so truncated.

We can go on like this for a long time! If, after a time, we come to a set which has no largest member, then we can pick out an increasing sequence of numbers.

But if we *never* get to this point, then the successive numbers we take away will form a steadily decreasing sequence of numbers!

So whatever happens, there must always be either an increasing or a decreasing sequence which we can pick out of an infinite bounded set!

It did not take long to prove that, in the case of an increasing or decreasing sequence (as long it was bounded), there was always just one number which is such that we can find members of the sequence as close as we like to it. This was done by successively subdividing the 'distance', between the lower and upper bounds, between which all the members of the sequence were situated, into halves, then quarters, then eighths and so on, thus 'narrowing down' the number towards which the sequence 'tended'. These two 'facts', put together, gave us the Bolzano-Weierstrass theorem. End of proof.

I did this piece of work with these children just to prove another existence theorem in the psychology of learning mathematics, namely that children of around thirteen could appreciate and, indeed, enjoy some of the rigorous thinking required on which the so-called infinitesimal calculus (calculus for short) has to be based. In the case of the rich children, the study had the additional advantage of showing these children what disciplined thinking was like! A properly scientific study of the feasibility of this kind of instruction during the early teens still remains to be done.

We worked a great deal on getting children to understand the mechanical algorithms they had learned in arithmetic. For this we used the multi-base blocks, which give a child a realistic overview of what happens when paper and pencil calculations are done. We used the idea of ratio more and more, even considering multiplication and division as particular cases of the concept of ratio. We found that the study of these operations, of fractions, percentages and decimals, could be pulled together under the one heading or 'super concept' of ratio, thereby making learning a lot easier.

Already, in my Australian work on the problems of learning mathematics, I had found that often it was better to teach the general first and then go into the particular afterwards, rather than teach all the particular cases separately and painstakingly and then burden the child with the much more difficult process of generalising. We found that children who understood the concept of ratio were able to particularise this understanding and thus had no problems with its particular cases. These observations eventually led Esther Grossi to write a detailed study on this subject, which became her thesis for her doctorate at the Sorbonne, as mentioned before.

Our Porto Alegre accommodation was always a very nice suite in a first-class hotel, where we could entertain as well as have a restful time when I was not working. Often we had our meals brought to the suite and had them, romantically, by ourselves. Other times we were invited out by colleagues and collaborators, many of whom were keen to get to know us socially as well as professionally. We got to know Esther Grossi's husband and children, whom we saw grow and develop during their teenage years and we were often privileged by their sharing their problems with us. On Sundays they would often take us out into the country, to show us some of the ethnic parts of Rio Grande. There was an Italian part, where we had a long talk (in Italian!) with the priest who looked after a church and told us about its history. We were also taken to the German part, where one could have afternoon tea. We were once taken to a football match. Brazilians are so crazy about football that we could not make them believe we were not as crazy, so I ended up going to the match, where they had a special area reserved so we would not be bothered by the usual football hooligans! Some other programme was arranged for Tessa, for she told them she could not watch football, or anything, for so long continuously!

There were many beggars living on the pavement under a bridge, almost right outside our hotel. There was one particular woman Tessa liked to care for by taking her the food that we could not eat. Sometimes Tessa would simply divide the food on her plate in half and state that the other half was meant for the beggar lady and she would add a little wine, which she poured into a small bottle; generally, she tried to give her beggar what she called a balanced diet. When it was time for us to leave Porto Alegre, Tessa asked one of Esther's children, who could speak French, to explain to the beggar lady that we were leaving so we would not be able to look after her, but that we were coming back next year. This was done and we took leave of her. The next year, when we came back to Porto Alegre, there were other beggars at her pitch: she was no longer there. We were told that a truck came along early every morning and the dead beggars were simply swept up, put in the truck and taken to a dump. Tessa shed a few tears for her deceased beggar lady, but she had done what she could.

During the same visit, we were invited to supper with a very rich family, whose children were in one of the project schools. Their villa stood on top of a huge cliff, overlooking a shanty town below. There was a large swimming pool and a special building by the pool where we had cocktails. This building was as big as most people's houses!

'What are these people lining up for?' I asked our host, pointing to a number of rather bedraggled-looking people holding large containers.

'Oh, they come for the water,' he answered. 'We are the only people around here who allow them to take drinking water away, so they come all the way up from the shanty town to carry water back to their shacks.'

During supper, Tessa couldn't help thinking of her beggar under the bridge while everyone was consuming or wasting a great deal of very expensive food.

'Why aren't you eating?' asked our host, looking at Tessa enquiringly.

Then she explained about the beggar lady she was looking after.

'Don't worry!' our host said. 'I will tell the cook to prepare a parcel for the beggar!'

When we left, we were given a large parcel. As we were being taken back to our hotel in a car, Tessa opened the parcel to see what was in it, quite excited about getting some nice things. When she saw what was in it, she burst into bitter tears.

'Look what they have given her!' she said. 'It is all gristle and fat, stale bread. These things are not fit to give to a pig!'

We threw the stuff into the first rubbish bin we could find and got some better stuff from the hotel kitchen, which we delivered immediately, for Tessa would not have slept otherwise!

Although work in Brazil took up a considerable portion of my time and energy during the Sherbrooke years, even more was devoted to work in Europe. I can remember a great number of workshops and series of lectures, delivered in countries such as Italy, Hungary, Germany and France, with an occasional stop even in Switzerland. Some of these were short, others lasted for several months, during which times I kept in touch with Sherbrooke by phone and correspondence.

Apart from working in Brazil, I was also invited by the Ministry of Education of the Allende Government in Chile to give some courses to teachers. These visits were short, lasting about a week or so each time, so Tessa did not accompany me on them. I recall my flight south on the occasion of my first visit to Santiago de Chile. I was not very sure of my ability to run a course in Spanish, so I amused myself in the plane from New York to Lima, Peru by reading some of my own books which had been translated into Spanish. In this way I thought that I could get familiar with the vocabulary and acquit myself reasonably. I spent the night in Lima and then took the three hour flight from Lima to Santiago. As I sat down in my seat with the usual champagne and orange juice which accompanied first-class breakfast, I was also passed *La Prensa*, which is a newspaper published in Buenos Aires, of course in Spanish.

'Monsieur Dienes!' I said to myself. 'You are going to read *La Prensa* from cover to cover and not once look out of the window to survey the Andes! Then you will learn Spanish!'

Having finished my drink, I concentrated on reading my homework. I just about read all the important news items and articles when I noticed that we had landed and were taxiing towards the terminal building. The official from the ministry who met me was surprised that I could speak Spanish. I asked him how he had thought I would give a course if I could not. He kept quiet after this and took me to the hotel, which faced the main square in front of the presidential palace. There I met a group of teachers, who welcomed me.

I was taken to a school on the outskirts of Santiago de Chile, where I was given a group of children with whom to work, while teachers sat around watching. I soon got the children interested in some mathematical games and from then on it was plain sailing. The children kept wanting to learn *otros juegos* ('more games') and the teachers joined in the fun very enthusiastically.

I was invited to tea by the Canadian consul, who wanted me to meet a number of local dignitaries. By this time I had become quite fluent in Spanish, having got past my transformation phase of learning how to speak.

The consul's wife drew me apart at one point and said to me, 'I am most impressed with the way you use the subjunctive! I have been here for two years and I still haven't figured out how to use it!'

Of course, I was totally unaware that I was using subjunctives or any other abstruse grammatical construction. I said to her, 'I just try to speak in the way they do! Why don't you try that, instead of thinking of the grammar!'

But I never discovered whether she ever learned how to use the subjunctives.

On the Sunday before I was to return to Sherbrooke, one of the teachers took me in his car to Valparaiso. The sea looked beautifully blue so I asked him if we could have a dip.

'Oh no!' he replied. 'The sea is very cold here. There is a cold current that comes up from the Antarctic.'

'Of course, the Antarctic is not all that far from here,' I said.

'Well, it's not exactly near. We have a team of scientists going to Antarctic from the university next week. If you would like to stay, I am sure they could arrange to take you with them,' he said.

'That would be most interesting,' I replied, 'but I must get back to my work in Canada.'

I have often wondered since what it would have been like to visit that last, largely unexplored area of this globe.

At this point we came to a cart, out of which an ice cream vendor was selling his wares.

'Would you like an ice cream?' my colleague asked.

'Yes, thank you, it is rather hot!'

I must warn potential travellers to South America that eating ice creams from local ice cream vendors is quite a dangerous thing to do! No sooner had I arrived back in Canada than I was sick for several days with dysentery. So I might as well have taken the trip to Antarctica!

On another trip to Santiago, which turned out to be the last one, for political reasons, I was again put in the hotel facing the presidential palace. From the window I could see marching workers furiously waving red flags and singing rousing revolutionary songs, which were sometimes countered by battalions of policemen trying to disperse them. After one of these incidents, a university professor who was working with us in the team came to the hotel and said that it might be wiser to move me to another hotel, more on the outskirts on the city, where disturbances were less likely. So I moved. I was able to complete the work we had planned for the week but the situation seemed to be getting worse.

I stopped at Lima on the way back, as the teachers' union there wanted me to work with their teachers for a couple of days. They said they had no money to pay me but they would put me up in one of the good hotels. So I worked with the Lima teachers for two days, who, at the end, put their pennies together and bought me two beautiful plates with some Peruvian gods painted on them. I still treasure these plates in memory of my enthusiastic Lima teachers, who had so little and yet wanted to give so much as thanks for the help they had received.

After one of my Brazilian trips, I remember trying to decide whether to return via Santiago and Lima, or whether to come via Rio. I opted for Rio. It was just as well. It happened to be the 11th of September, which is my birthday, but it turned out to be the day when a coup was engineered by Pinochet against Allende. The presidential palace was ransacked and Allende killed. The same day thousands of people were rounded up and never seen again, some taken from Santiago Airport. So I might have become one of the *disparecidos* of Chile, but for a decision at the last minute not to take the Santiago route.

I was once asked to give a talk at the University in Zurich about my work, which I managed to squeeze in between various other European engagements. On our way out to the airport, the person who was driving me said that we had several hours before the flight, so what would I like to do?

I suppose he thought that I would suggest going to some expensive restaurant, charging everything to the account of Zurich University, but instead, I said, 'Why don't we have some fun at some local school?'

'Yes, there is a little country school just in a village near here,' he replied, somewhat surprised. 'I have worked there myself with one of the teachers before. Shall we try?'

'Why not?' I said and we made a move in the direction of the aforementioned village.

We found the village school and, after the usual polite introductions, I was ushered into a class of what seemed to me about grade two or three, judging by the size of the children. I was introduced to the teacher and the children.

The teacher said to the children, '*Kinder, nur Hochdeutsch sprechen! Der Herr Professor versteht nur Hochdeutsch!*' ('Children! Speak only High German! The professor only understands High German').

This warning was just as well, since I would not have been able to follow the Swiss version of German, which no doubt these children spoke amongst themselves; they were obviously taught High German, which was the German generally understood in German-speaking areas.

I said to the children, 'I am a space traveller. I have just landed from Mars and I am rather tired. On Mars, people never count beyond three, so I can only count: one, two, three. But I want to count how many children there are in this classroom. How will I do it?'

'We can teach you to count beyond three,' the children offered very generously.

'Not now! It gives me a headache to go beyond three, it is so difficult!' I replied.

So I proceeded to count them, pointing at each child in turn and saying, 'Ein, zwei, drei; ein, zwei, drei; ein, zwei, drei,' until I reached the last child.

'Do I know how many children there are in the room?' I asked them.

'Of course not!' replied the children in chorus.

'Why not?' I asked.

One little boy chirped in at this moment and said, 'Because you don't know how many times you said *ein, zwei, drei*!'

'All right,' I replied. 'Then let us count them!'

I used the fingers of one hand for the *ein, zwei, drei* count and the fingers of the other hand for counting how many times I have counted an '*ein, zwei, drei*'. When I got to the third one of these, I stopped. I told them that I would have to start again with the count because I could not go beyond three, but I went on counting the children. When I finished, I asked them if I now knew how many children there were in the room.

'No you don't, because you don't know how many times you have said *ein, zwei, drei; ein, zwei, drei; ein, zwei, drei*,' volunteered the same little boy who had chirped in before.

'Well!' I replied. 'Let us count those!'

In the end we counted two sets of *ein, zwei, drei, ein, zwei, drei, ein, zwei, drei*, then another two *ein, zwei, drei* and there was one child left over.

The children were now satisfied that, at least in Martian counting, I did know how many children there were in the classroom. Then I allowed them to 'teach' me the word *vier*, which is German, High or otherwise, for 'four', because I told them that on Mercury they could count up to four, so we counted the children again using Mercurian counting.

The person who was accompanying me to the airport kept significantly looking at his watch and I realised that we needed to go, in order to catch my flight. We took leave of the children, who were sorry they had no time to teach me how to count beyond four but they wished me a pleasant space flight and we then drove to the airport.

German is not my best language; in fact, it is the only language I know which I picked up through learning the grammar. Knowledge of grammar seems to be inimical to easy speaking, as one has to secretly think of accusatives and genitives instead of thinking of whatever one is wanting to say. Thus it usually took me a day or two in a German-speaking environment before I could communicate at all fluently with my fellows. I recall one occasion when I was invited to give a workshop in Hanover. I had arrived in Dusseldorf on a Lufthansa jet and been duly met by one of the organisers, who was to drive me to Hanover, which is not exactly next door to Dusseldorf. To make things worse, it had been snowing heavily for the past several days and the going was not what you might call easy on the crowded German autobahn. My driver was trying to get me involved in conversation, but it seemed that no German would issue from my brain. He had never met me before and he had visions of my struggling with the language and of the workshop being a total failure. He told me afterwards that, after he dropped me at my hotel, he said to his colleagues, *'Das wird eine grosse Katastrophe sein! Herr Dienes kann gar nicht Deutsch sprechen!'* ('It will be an absolute catastrophe! Dr Dienes can't speak German at all!').

He need not have worried. After a good night's sleep and some very pleasant demonstration work with some local German children, all the German I knew suddenly came back from nowhere into my head and we had a very fruitful week of workshops, using the children and the teachers alternately, to make the mathematical as well as the pedagogical points that I wanted to make.

The publishing house Herderverlag organised many such workshops. It must have paid them from the point of view of selling their books, as they kept asking me to do more. I worked in Heidelberg, Cologne, Munich, Frankfurt, Hamburg and even Berlin during successive years. My books became required reading in German teachers' colleges and it became impossible to talk about mathematics education in Germany without mentioning my work.

But the work seemed to extend beyond the scope of universities, school boards and publishing houses. The well-known periodical, Spiegel, invited me to an international discussion on the new maths, where there were some for and some against. We were even hooked up by transatlantic telephone to some American educators, from whom they wanted to hear. Tessa and I were given a most luxurious suite at the Munich Hilton, with a bathroom each, champagne cooling in the fridge, flowers and bowls of exotic fruit strewn about all over the suite. We had never been entertained at that level of luxury before and possibly never since!

At one point during the proceedings, I was working with some four year old children with logic blocks. There were some very clear rules which emerged from the way the children and I were placing the blocks, but I would occasionally put a block in the wrong place, to see if they would notice. They always did.

After several such mistakes, one little four year old girl exclaimed, *'Du kannst nie Mathematiker werden!'* ('You can never become a mathematician!') The onlookers laughed heartily and all the while the TV cameras were rolling. The scene was shown on the local news that very evening!

Unfortunately, while I was giving my talk, I had not noticed that behind me was a curtain and behind the curtain was a considerable drop. As I usually tend to walk about when I give my talks, at one moment I walked just a bit too near the curtain and fell down behind it! There was a hush in the audience, but I was pulled out and managed to continue the talk and give a good account of myself in the debate that followed. But when I got back to Canada, I had all sorts of strange symptoms of numbness creeping up my legs and I was taken to the Centre Hospitalier Universitaire at Sherbrooke, where they examined me for a week but could find nothing wrong. They simply put me on aspirins, which I still take every morning. This is supposed to make a heart attack or a stroke less likely. Since I am still around, the theory is probably right, although I am not a conclusive proof thereof!

A number of workshops were organised in France by the Office Central des Librairies (OCDL), the publishing house that had published my books in French and had also distributed my educational materials, now known as 'manipulatives'.

At one of my demonstration classes in Paris, again, I did the 'making of mistakes' trick, to encourage the children to be alert and correct me every time I made a mistake. When I noticed that it was nearly time for the bell to ring, I suddenly started getting everything right.

Then I heard a little boy cry out, *'Enfin, il commence à comprendre!'* ('At last he is beginning to understand!').

Another time I was working with younger children in Paris and tried to show my audience how much could be established without uttering a single word. A little girl of about five and I were playing at putting blocks down in a pattern. I was trying to communicate the pattern to the little girl by shaking my head when she put a block in a position that did not follow the pattern and giving her a pat on the back when it did. It did not take very long before she got the idea and she set out the entire contents of a set of logic blocks in what seemed a beautiful symmetric pattern, looking somewhat like a flower but without the stem.

When she had finished the little girl clapped her hands and exclaimed, *'Comme c'est beau!'* ('How beautiful!').

When we discussed the work later, one of inspectors told me in all seriousness, *'Monsieur! Pour que l'apprentissage soit valable, il faut qu'il soit pénible! Pénible, Monsieur!'* ('Sir! For learning to be valid, it must be painful! Painful, sir!').

I replied to the inspector that under such conditions there was no communication possible between us.

Italy was another venue for meetings to improve mathematics education. The Italian publishing house Organizzazioni Speciali (OS for short) was headed by a certain Dr Abbele, who had sole rights in Italy to distribute my writings and concrete materials. So in the same way as Herderverlag in Germany and OCDL in France, he used to organise workshops in Italy which he would invite me to lead. These were mostly held in Florence, sometimes under the aegis of the Florence University Psychology Department, of which his wife was a member. He always asked me to come to a certain school, called Scuola Bechi, in which the teachers were keen to improve their competence in mathematics teaching. This school, many years later, became of central importance as a proving ground for a mathematics programme for Italian elementary schools, for which the publishing house Giunti-Marzocco became responsible, as after Dr Abbele's death, they took over my rights.

Angelo Pescarini was also responsible for many invitations I received to work in Italy, particularly during the time he was a member of the Emilia-Romagna regional government, being Assessore per la Pubblica Istruzione, best translated as director of education. Many of these meetings were attended by literally thousands of people, the proceedings being transmitted through closed circuit TV. I recall one very crowded meeting in Modena, during which I had a dozen children perform mathematical feats on the stage of a very large auditorium. One of them treated the rotations of a regular tetrahedron, but it was embodied literally by choreographing the children to do a kind of mathematical dance. The audience was enchanted by the possibilities of learning mathematics in such a way and I had a very warm reception. Of course, only the mathematicians in the audience were able to see the link between what the children were doing on stage and the underlying mathematics and this came out during the ensuing debate.

One university professor of mathematics informed the startled audience, 'I would like to point out to you all that what you have seen today is the concrete embodiment, in the form of movement, of mathematical topics which are learned in the second or third years of a university course on pure mathematics! Yet none of the children you have seen is more than ten years old!'

I once gave a two week course in a little town called Follonica. This was during the summer vacation so we had to look around for children with whom to run demonstration classes among the offspring of the participants and their friends. We had no trouble getting about twenty children together. I had brought over some of the new materials we had constructed in Sherbrooke, mostly made and painted by Cantieni. This had the double purpose of finding out how these new materials worked in practice as well as teaching the participating teachers some mathematics and methodology.

The children were enchanted with the work and never wanted to leave, to let the teachers get on with solving similar problems! The favourite toy was the set of twenty-one discs, on which coconut trees and starfishes were painted. There were various ways of ordering discs, either by 'adding' the pictures, or by finding properties common to the discs you thought belonged together. The final problem was to choose sixteen of the twenty-one discs and put them in a four by four square, so that all four rows 'belonged together' as well as the four columns. Then you had to find what patterns other sets of four discs made within that square which also 'belonged together' but for different reasons.

There was a group of four boys who worked on the discs for three consecutive days. In the morning, their first question was, 'Where are the discs?' They were absolutely determined to solve the four by four problem. I suggested to them several times that there were plenty of other interesting games they could try in case they were tired of the discs or if they thought the problems were too difficult. The more I suggested they could stop, the more they wanted to do it! In the end they did succeed in solving the four by four problem and you should have seen their faces when, in the end, everything turned out as it should!

At the ceremonial closing of the workshop, the *maire* made a flowery speech and thanked the children for so bravely coming to learn mathematics in the holidays! Such a difficult subject! He asked them what they would like as a reward. The children unanimously decided that they would like another day of mathematics! The game that so motivated the four boys is described in my *Il piacere della matematica*, in the section on intrinsic motivation in which I have tried to list some of the conditions that must be fulfilled in order for intrinsic motivation to become possible. Would it not be good if mathematics could be learned because people thought it was fun and not for the sake of external rewards or fear of punishment in case of neglect?

During this workshop we got to know the Marchesa Incisa, who let us have some very pleasant accommodation in one of her villas for part of the duration of the Follonica work. The Marchesa's estate stretches between Bolgheri and the sea, just south of Cecina and north of Follonica. It lies next to Count Antinori's estate, who is her cousin. Her entire estate is a nature reserve, complete with bird sanctuary, where migrating birds can rest and feed on their long trips. One is forbidden to pick any wild flowers and shooting any animals is certainly out of the question.

Count Antinori has somewhat different views on how to treat nature and he allows hunting and other activities harmful to nature on his estate. On the beach it is easy to see where the marchesa's beach ends and the count's begins. The marchesa's is always kept spotless, while one might find almost anything strewn around on the count's section of the beach, including topless girls sunning themselves, much to the annoyance of the Marchesa!

Once we became friendly with the Marchesa, she invited us several years running to stay on her estate, when she provided us with accommodation either in a villa near the beach or in one of her houses at San Guido, next to the local school. She also provided anything else we needed, in the way of food, wine, even domestic help; all I had to do in return was to work a little at the school in San Guido. In the recent past she had been responsible for the entire running of the school, paying the teachers' salaries and buying the books and materials, but recently the ministry had taken over. However, she still provided everything for all the children for their first communion or anything extra that the ministry would not provide.

In the same way, we were provided with everything, in return for working in the school. This medieval arrangement suited me well, as I could use the classrooms in San Guido for trying out all sorts of things we had thought up in Sherbrooke and, by keeping in touch with the centre, I could compare notes with those who had been left in Canada. The San Guido children were thrilled with all the fun games I introduced into their lives. They were all peasants' children and their lives were mostly filled with work, since they had to help in the fields when they got home from school.

There was also much work going on in England, encouraged by the NFER as well as by the Educational Supply Association (ESA) which distributed my concrete materials in the UK. In fact ESA founded an association which they called the Zed Association, after the way many people addressed me: to my friends I was, as often as not, simply Zed. The Zed Association meetings were held every year and I was present at most of them; they usually lasted about a week and were attended by teachers from many parts of the country. Perhaps the most memorable ones were the ones held in Canterbury and South London, where many lasting friendships were made, not to mention the mathematics that was discussed and learned by teachers. I met Gordon Jeffery at one of the London meetings, who has remained a friend to this day. He later obtained a job at Dalhousie University in Halifax and from there came to visit the Sherbrooke Centre. He invited me to Halifax to hold workshops under the aegis of that university. He later became inspector of schools in Nova Scotia, then he worked at the Truro Teachers' College and now, although semi-retired, he still keeps his hand in education by being principal of a small village school in the Wentworth Valley. Since we now live a part of the time in Wolfville, Nova Scotia, we have been able to keep up our friendship as well as our professional contacts.

Budapest was another venue for much work in mathematics education. During the many workshops I held in Budapest and elsewhere in Hungary, I discovered one of those rare phenomena: a born teacher. Her name was Maria Winkler. She took to my methodology as a duck to water, for she realised immediately the Trojan Horse quality of my suggestions, grasping that by adopting my methods, the whole dynamics of the classroom, the relationships between the children and between herself and the children would change for the better. She appreciated that in a discussion the point was not persuasion but the search for the truth and this generated respect for your fellow creatures and therefore facilitated the formation of true friendships.

I used to call her *szent asszony* ('the saintly woman'), as it was a treat to watch her handle the children in her class. It was all done with such love and tenderness, that the children simply ate out of her hands! She espoused the connection I had tried to underline between mathematics and music and there was always music in her class, which was never 'spoiled' by being mathematised. She knew instinctively when to let the music grow and let the children drink in the beauty of the melodies and when to start asking questions about how it was all put together. I still visit her class every time I go to Budapest, which I still manage to do just about every year. She was eventually given her own school to run. Even during the time the communists were in charge in Hungary, she obtained the unique privilege from the government of running her school without any interference, using curricula and methods of her own choosing. It is great to know that some of our educational saints do sometimes get the opportunity to do their work and act as shining examples to us all.

During the Sherbrooke years, we were in contact for some time with the Fleming School in New York, which was a bilingual school (French and English) run by the very strong personality of Madame Correa, who wanted to make her school into something different from all other schools in the USA. I used to go and work for two or three days at a time in her school, working with the children as well as with the teachers and a great deal of mathematics was brought into the school that would not even have been heard of by most teachers in other New York schools. The

Fleming School also took part in Sandor Klein's evaluation of the Dienes method, using some of the tests we had developed at the Sherbrooke Centre and some developed by Sandor Klein in Hungary. This evaluation work was eventually written up in a book, where all the experimental data from the various centres can clearly be looked at by the doubting Thomases of mathematics education. This book was instrumental in Sandor Klein obtaining his second doctorate or super doctorate in the psychology of learning.

Tessa and I became very friendly with Madame Correa and she often invited us up to her apartment, which was very near the school, although she always put us in a very nice nearby hotel for the nights.

On one of our visits, Tessa moved to open one of the windows, for it was very hot, but Madame Correa shouted in a stentorian voice, 'Don't touch that window. You'll be shot!'

She explained that she had a private guard, armed with a rifle, who was supposed to shoot and ask questions afterwards if he saw any movement around her apartment. So Tessa stopped in her tracks, nearly falling over a huge dog Madame Correa kept in her apartment.

'How does your dog get its exercise?' I asked.

To answer the question, she opened a drawer, pulled out a sheath knife, unsheathed it, pointed it towards the sky and said, 'I take this dog, usually at night, anywhere I like, pointing my dagger menacingly, so any potential mugger would think twice before trying anything on!'

We knew that New York was not exactly a safe place for nocturnal walks but we had never thought of handling the problem by taking a huge dog and a dagger!

The international work radiating from the Sherbrooke Centre encompassed some of the South Pacific area, including Australia. I made one trip west from Sherbrooke; the next one was taken from Winnipeg. They were both intended to include Australia. The first one was done *en famille*, except that the outward trip was made by Tessa and Bruce by ship and by Sorrel and myself by plane. I could not afford the time away from Sherbrooke which a voyage by ship would have involved, but Tessa and Bruce could certainly enjoy the voyage. So at a certain point Sorrel and I took a flight to Denver, where I had been invited to do two days' work. I remember the occasion well from the fact that I was paid two thousand dollars for my pains, which was more than I had received before for only two days' work. So, armed with this cache, Sorrel and I flew to Hawaii, where we were welcomed by the Quakers and stayed for a few days in the beautifully situated Quaker House. We loaded up with *mumus* to take to Australia, some for Sorrel and some for Tessa and flew on to Adelaide, South Australia. Since we had all that money, we flew first-class and were treated to all the goodies, not to mention the frequent drinks of champagne with which the solicitous hostesses plied us from time to time. I do not know about Sorrel's state of health and mind, but mine was definitely influenced by the overconsumption of alcoholic liquor on the flights.

I was met at the airport by the press and I was asked all sorts of questions, including the challenging one, 'What do you think of Australian education?'

I had temporarily forgotten that I was soon to have an interview in Melbourne for a chair at Monash University, so I answered blithely, 'Australian education is a hundred years behind the times!' and other such thoughts. I suppose I should not have been surprised to see myself on the front page the next day, with my tactlessly outspoken utterances as headlines. Telling the truth is possibly a virtue, but at times silence is golden! It will not surprise the reader that, in spite of a very warm reception and a great interest in my work by Monash University a week or so later, I was not offered the chair! I might add, though, that nobody else was offered it, so there must have been some conflict within the selection board.

Apart from some work in Adelaide, I also organised a week's workshop in Melbourne, to which our daughter Corin came. She was no longer married to John Biggs, her first husband. Her absence from Adelaide, where she lived now with her twin boys, resulted in a further change of husbands. I was finding it a little difficult to get used to this idea of the next generation which seemed to involve a change of 'partners' (as they are now called) in the way we used to change our clothes. But *plus ça change, plus c'est la même chose* as the French would say, but then, of course, the French do not trouble about new husbands and wives – they just have a regular succession of lovers. Probably less unsettling for the children, *non*?

We had an apartment in a block for the week of the Melbourne course and we rented another one, next to ours, so that Jancis and the children could also come. The course was a great success and I was invited to come again the following year and do a tour of the whole of the state of Victoria, giving workshops as we went. So, in spite of the *débâcle* stemming from my outburst at Adelaide airport, the trip did have a positive side to it! Familywise, it was an all-round catastrophe, except for Corin, who proceeded to her next partner.

There was a lot of sibling rivalry raising its ugly head during our stay in Adelaide. Bruce fell in love with a young teacher and enjoyed his first real love affair, but then was miserable to have to leave to go back to Canada.

We finally said our goodbyes and the four of us, Tessa, Sorrel, Bruce and myself, flew to New Caledonia, where I had arranged to do a week's work. New Caledonia was then (and possibly still is) a French colony, administered as though it were a part of France, much like Tahiti. So, from an English-language milieu, we were then immediately transferred to a French-language one.

We were put in the best hotel on the island and looked after in quite lordly fashion, while I worked with the children and the teachers. The powers that be were very pleased with how the work had gone and on our last day, which was a Sunday, said they would provide their best pilot to take us around the islands, wherever we wanted to go. So we were taken to the airstrip all keyed up, looking forward to a beautiful tropical day! The car had not yet stopped on the edge of the strip when we saw a plane landing, fire belching forth from one of its engines. The pilot managed to get out and the fire was soon put out, but the scene terrified Tessa and she said that on no account would she fly on these tiny planes, whether it was the world's best pilot or not!

So she got a lift back to the hotel. Sorrel, Bruce and I thought we would not be put off by the fiery landing and were introduced to our pilot, who asked us to get in the plane and we soon took off. We asked the pilot to decide where to go and he agreed. After about half an hour of flying over the incredibly beautiful island atolls, he landed the plane on an extremely narrow strip, trees on either side, with just enough room for the wings. We were all glad that Tessa was not with us because she probably would have fainted during the landing operations!

The beach was a few minutes' walk away. We donned our swimming gear and got ready to have a really relaxing time in these beautiful tropical surroundings. There was a little sandy bay surrounded by coconut trees, just as you might see in movies about romantic tropical islands! The water was crystal-clear, with many different kinds of brightly coloured fish swimming about in the shallow waters, with no fear of intruding persons! There was live coral on the sea floor, waving about with the slight motion of the water occasioned by the soft breeze that was blowing. I started swimming in the shallow water with my head submerged but with my eyes open, looking down towards the sea bottom. It was like swimming in an aquarium! We swam about in the bay, exploring various lagoons, but careful not to tread on dead coral. When coral is petrified, the edges become extremely sharp and one step can make your feet bleed. I had already learned my lesson by bitter experience in New Guinea, so I was able to warn Bruce and Sorrel.

We spent the whole day sunning ourselves, going in and out of the water admiring the aquatic fauna and drinking in the sheer beauty of the surroundings. When the pilot came to fetch us, we sadly said goodbye to our tropical paradise and flew back to the main island, pleased that, even if for such a short while, we had been able to commune with nature in such an intimate way in such wonderfully scenic surroundings.

The next day it was departure day. I was well paid for my labours in French francs and we took leave of our hosts in a whirr of propellers. The plane took us to Fiji, from where we could travel by normal jet and duly arrive back at Sherbrooke and take up our Canadian life again.

A year or two later, from Winnipeg, we made another trip to Australia, to honour my promise to do the tour of Victoria with my mathematical games. I had arranged to stop in Tahiti on the way, to give a ten day workshop, for which I had been invited by the local branch of the teachers' union.

We were lodged in one of the good hotels on the seashore, where the bedrooms were all little thatched shacks, but very comfortable. Tessa was not that keen on pretty little lizards running up and down the walls, but I enjoyed their antics! There was a pier near the dining area, from which I used to dive into the sea every morning to have my swim before breakfast. The teachers with whom I was to work knew that I had to have some children to show them how to handle the mathematics in practice and a bunch of beautiful-looking Tahitian children were provided for me to play with each morning. Each morning the children eagerly awaited the opening of the doors and I noticed that each time there were more children! It must have got around on their bush telegraph that there were fun things to do in school, in spite of the fact that it was their vacation time.

Even the local TV station became interested in the proceedings and persuaded me to work with some very young children in the studio. The work itself was a piece of cake, as we used logic blocks and they are always good for showing some unusual, spontaneous ideas occurring to children! But I had not reckoned with the heat of the lights in the studio. The children certainly enjoyed themselves and there was a notice about it in the local press, but at the end the poor things were all dripping with perspiration!

After a couple of days' of work, I happened to mention that it might be nice to have a look at the island. The next morning the manager of the hotel informed me that 'my car' had arrived. Apparently, our hosts had rented a car so we could drive round the island. On some days, 'after work', Tessa and I would get in the car and drive. There was only one road, the one that went round the island on which Papeete stood, so we soon explored that! We also found a museum in which Gauguin's paintings were housed, those he painted while he lived in Tahiti. Tessa was rather against even looking at them, as Gauguin had been so selfish in his private life and had neglected his family and come to live in his 'tropical paradise'! But the paintings were still very worth seeing. It seems as if it often happens that the expression of a creative urge in the form of artistic works impedes a similarly creative expression in dealings with one's fellow men. Why this should be so, I suppose we shall never know. We can just say with Hamlet, 'There are more things in heaven and earth, Horatio, than are dreamt of in our philosophy!'

All the children could speak French fluently, although their mother tongue was Tahitian. Several of them asked me if I could speak Tahitian and I had to admit, ashamedly, that I could not. One of them brought me a Tahitian reader one morning, plus a grammar and dictionary, and asked whether I would like to learn it. I was extremely touched and promised that if ever I came to Tahiti again I would be speaking to them in Tahitian!

I still have these Tahitian books. I sometimes look at them, but at this point in my life it is now very unlikely that I will visit Tahiti again; the little girl who gave me the books is now grown-up and probably has several children of her own.

When it was time to go, we were given a great send-off at the airport. Many of the teachers and children were there to say goodbye and we were given *leis* made of shells, since they said that flowers would fade. Among these Tahitian parting gifts was a crown, also made of shells, which was placed on Tessa's head as we were leaving to go through to the departure lounge. So we said goodbye sadly to Tahiti, possibly never to return and boarded the Air New Zealand jet bound for Auckland, where we would change planes for our journey to Australia.

The mathematical tour of Victoria was great fun. I spent about two days or sometimes three at each place and worked with both children and teachers, as was my established custom by now. We did some work on logic; we handled some geometrical ideas, using squares, triangles, tetrahedra and cubes, as well as employing many stories and dances that embodied mathematical notions. I left behind a set of notes on the work, which was distributed afterwards to all teachers who had participated in the workshops.

We also renewed our contact with Corin, Nigel and Jancis, so we were able to keep abreast of what was happening to their growing families: our grandchildren.

Chapter Twenty-One
About our Personal Life during the Sherbrooke Years

We tried to reach out both to the French-speaking and the English-speaking people in Sherbrooke with whom we came into contact. I believe our house was one of the very few houses in Sherbrooke, possibly the only one, to which people from both language groups would be invited at one and the same time.

Backing on to our Rue Adam house was the rear of the house of a francophone family. The man who lived there was on the university faculty and one of the children went to the École Eymard so we had two reasons to come into contact with one another. We became friends and the friendship has continued right up to the present, although as we live a long way apart now, it is not easy to keep it up.

One of the circuit judges had married a Hungarian lady, who very much wanted to be friendly with us. She threw some wonderful parties and we were always invited. The judge was keen on hockey, so if ever it was hockey night in Canada, you could not tear him away from the TV set! But his wife largely ignored the hockey and we always had a good time. One of her friends was the widow of another judge, so we became friends with her as well.

We became very friendly with Mr Wilson, the owner of the furniture store from which we had bought nearly all our furniture. His wife had recently died and he was left to bring up four young children alone, so very soon he succumbed to the charms of a lady, who then became the children's surrogate mother; she really made a very good job of it, much better than my own stepmother made of her job with me when I came to England!

In the winter we used to go out cross-country skiing together, often in the Orford Provincial Park, where there were some really scenic cross-country ski trails. On one of them was a log hut, in which a fire was kept going nearly all the time, so we often took off our skis there and munched the sandwiches which we had in our backpacks. Bruce used to come with us, but he went on the downhill slopes and was soon very good on the expert slope, which took him from the summit of Mount Orford to its base!

We once invited Corin and the twins over from Australia, when she was between husbands. The twins were about ten years old and they had never seen snow before. When they arrived, they went quite mad in the snow. One of the twins tried to lick the drainpipe and of course could not take his tongue away, for it was too cold and froze it to the pipe! He never did it again! Fortunately, there was not much damage done. We took them cross-country skiing with the Wilsons, which they thoroughly enjoyed, they could not wait to tell the kids at home about it. Since all the children we knew spoke only French, we could not find any playmates for them, but, being twins, they had each other so we had a great family time together.

Corin's current husband was a common-law husband, the only one in her entire string of husbands whom she did not officially marry. He was a chiropractor and taught Corin some of the tricks of the trade, enough so that she could set up as 'zone therapist'. They were very much in debt. I am not sure whether it was merely on account of unwise living, or from the purchase of equipment for the chiropractic work, but, be that as it may, Corin said to us that if they could only pay off the debts their relationship would build up and would that not be wonderful? We agreed to pay off their debt, to the tune of several thousand dollars, but very soon after this Canadian trip, after Corin and the twins went back to Australia, both the relationship and the practice collapsed and another relationship appeared on the horizon, in the form of the next husband! Whether we were right or not in trying to pull our daughter out of her difficulties, we shall never know. All we know is that we were several thousand dollars poorer!

We also made friends with our doctor, who used to invite us to a gorgeous Christmas Eve party in his old Victorian mansion, where we met all the anglophones of Sherbrooke of any note! Nigel came over while he was training to be a doctor, because he wanted to do his internship in Sherbrooke. He did this at Sherbrooke Hospital, the only English-speaking hospital in Sherbrooke and worked very closely with our doctor friend, who had a very beautiful young daughter whom he would dearly have liked Nigel to marry. He even offered to pass on his practice, as he was getting near retiring age, but it was not to be. Nigel eventually married Paula, an Australian girl, with whom he had two children, a boy and a girl and they lived in Montreal while he did his specialising in psychiatry. So we developed a closer tie with Nigel and his little family than we had been able to do previously, having lived so far from each other.

Just about every year we invited Rosalind, Tessa's mother and her second husband Jock, to come and stay with us in Canada. We even took them with us to Victoria once, on Vancouver Island, when I was invited there for an educational meeting. Jock was the kind of person who enjoyed everything: it was a pleasure to do things with him. Rosalind was a little more particular, but I think we managed to satisfy her needs also, which extended to a heavy consumption of alcoholic beverages, in particular beer! Since we were not bothered with financial problems, we could make her happy with a generous supply of whatever she wanted. Jock had worked in France and his first wife had been French, so he could get on well with the local francophones, who liked him right away as he was not as stuck up as some of the Sherbrookois thought most anglos were! He liked any kind of work, even though he was not a young man any more. He painted the rails of our Romeo and Juliet balcony, which possibly we did not appreciate enough, as he went to a great deal of trouble to do it perfectly.

We rented an apartment in an apartment block in Montreal, to have a place in town as well as in Sherbrooke. This was partly so that we could go to the Quaker Meeting, the only one in the Province of Quebec being in Montreal. Our first apartment was right next to the meeting house, so that whenever we spent the weekend in Montreal we could pop over next door for the meeting. Later we changed apartments and rented one halfway up Peel Street, so we were nearer the shops as well as a stone's throw from the mountain. Having a town place also allowed us to have more cultural life: we could go to the theatre and to concerts. I remember taking Tessa to a comedy called *Encore cinq minutes* and at another time taking her to the Place des Arts to see *Hamlet* in French! I believe 'To be or not to be' was rendered as '*Existence ou non-existence, c'est là le problème!*' or something of the kind. Since *Hamlet* was the compulsory Shakespeare play I had had to do for my school leaving examination, I remembered the plot and most of the soliloquies and it was really amusing to hear it all in French!

Each of our apartment blocks had a swimming pool in the basement, as well as a sauna, so every time we took any of our guests to Montreal they could benefit from these added conveniences, which were not available in Sherbrooke. Our visitors from England liked to dine in Vieux Montréal and soak in the very French atmosphere, invariably ordering snails and frogs' legs for their meals and pretending to enjoy them! Of course the 1967 Exposition was an attraction: it was great to tour the world yet remain on a small island in the middle of the Saint Lawrence River! Some children from the Toronto French school who had been not only learning everything in French but also learning to speak Russian were once taken to the Exposition. They went to the Russian Pavilion and they all started to speak Russian with the organisers of the pavilion. The Russians were amazed. How could such young children learn to speak Russian without ever having been to Russia? It seems to say a lot about the competence of the Toronto French School in the teaching of languages! I was in touch with the Toronto French School because I had been asked to give a workshop there at one point about how to handle mathematics at elementary level, at the same time Papy was there to tell them about his 'arrow method' for the upper school.

So we led a kind of multiple life, some of it in Sherbrooke, some of it in Montreal and some of it around the world doing various jobs I was asked to do. We had some pretty severe winters during our stay in Sherbrooke, the thermometer dipping below minus thirty at times. One winter we had a total of four metres of snow and the path to the front door was like a canyon, the snow reaching higher than our heads. It tended to be Bruce's job to do the snow clearing, until he went on strike to get a snow-blower, which we eventually provided for him. He also used to build his own igloo just outside the back door, take a candle and do his homework in it, sitting quietly inside his construction! It was amazing that the candle provided enough heat so he could stay in the igloo for quite long periods without getting cold.

Towards spring we would start going to the *partis de sucre* or 'sugaring off' which were parties held in celebration of the beginning of the sap running up the maple trees, which were tapped and the juice boiled and made into syrup. We would pick up handfuls of snow, pour on the boiling syrup and then eat the resulting substance, which was something like toffee. These parties were held in the *cabanes à sucre* in the depths of the woods and were always accompanied by singing, telling jokes and sometimes dancing. Tessa felt a little out of such activities as she could not easily communicate in French and she was not particularly thrilled with snowscapes.

I remember once coming back from Europe in April, having sat out in terrace cafes in Italy and France, we were greeted by a snowstorm on the way to Sherbrooke from Montreal when Tessa burst into bitter tears, asking why in the world I had brought her to such a godforsaken place! It seemed that Tessa was much more vulnerable to culture-shock than I was and I had to try and be understanding and realise that what, for me, might have been an amusing detail (such as a snowstorm), for Tessa would loom large and would approach the dimensions of a catastrophe! I am not sure how good I have been in exercising such empathy with her during all the many years we have been married, but I have certainly tried!

When Nigel came to Montreal to study psychiatry at McGill University, he was already married to Paula and had a little boy, so they bought a very old house in Westmount, which looked very romantic but it lacked a lot of the comforts one would expect in a house these days. We often used to visit him there when we came to Montreal, taking the little boy skiing in the nearby park and generally trying to be friendly. Paula was a bit of a 'good-time girl' and wanted to have fun, so we paid the fees for them to belong to a club where they could meet people and do sporty things.

One of their problems was that Paula's mother kept coming over from Australia and working on Nigel emotionally, for she seemed to want power over her. Paula appeared to succumb to these machinations and Paula's growing mother complex really started to come between them. At one point they decided to have a holiday in Australia; at least it was meant to be a holiday, but Paula had no intention of coming back. When Nigel discovered this, in Australia, he was very hurt and came back alone to Montreal. He was very much in love with Paula though and, some time later, decided to go to Australia to try and solve the problems they were having. He suggested to Paula that they should go on a kind of second honeymoon and get together again in more loving ways. Paula then declared that she would go but that her mother would have to come as well! A honeymoon with your mother-in-law! This was more than Nigel could take and he told Paula to go with her mother but that he was not coming on a three-person honeymoon! Strangely enough, Paula accepted this and she went on holiday with her mother.

This is where a young girl of the name Karin came in. She had been wanting to seduce Nigel for some time and now was clearly her opportunity. The state of mind in which Nigel must have been would have made it quite easy for a sympathetic-sounding young girl to break down any resistance, if in fact there had been any. So Karin and Nigel got together and they flew back to Montreal together, while divorce proceedings were started.

Nigel went back to his work. The two took an apartment on the corner of Peel Street and Sherbrooke Street and settled down to a common-law marriage. They discovered that Karin could not stay indefinitely in Canada unless she were legally married to Nigel and, since the divorce from Paula did not seem to take very long to come through, Nigel decided that he should marry Karin, particularly as she was already pregnant with their first child. They invited us to the simple wedding ceremony, which was to take place in a part of Vieux Montréal.

We were ushered into a smallish room and the judge came in who was to perform the marriage rites. She took all the particulars from Nigel and Karin and started to go through the ceremony, but in French.

I had to interrupt and I said to the judge, *'S'il vous plaît, madame, Karin ne comprend pas le français!'*

She was not unduly disturbed and restarted reading, but in English.

The vows were taken, the ring was placed on Karin's finger and then the judge said, 'You 'ave to poot maiden name *pour* Karin, see, nowadays with much marriage, divorce, when you catch sixty, you don't know 'oo you are!'

So Karin never became Karin Dienes but remained Karin Demasius, as she might not have known who she was after all the marriages and divorces the judge had foreseen were to take place in her lifetime!

Our personal life was continually influenced by people coming to see the centre from different parts of the world, as far from each other as Afghanistan and Japan. We nearly always entertained them in our house, often putting them up, so we got to know them personally as well. Tessa was very embarrassed when she burned the rice when we were having a Japanese professor to lunch. But when he said, 'Oh, this is just how I like it! My mother always used to serve the rice like this!' Tessa was relieved, although I doubt if she realised that this attitude on the part of our guest was merely a demonstration of traditional Japanese courtesy.

We had one visitor who had to have extremely 'natural' things. When he was offered wine, he asked for water. When we offered him a rather sophisticated salad, he insisted that he ate only grass. So Tessa went outside to collect some grass, washed it and put it before our guest. I think he ate a few blades, but, not having four stomachs, as cows have, I suppose he could not digest more than a few blades at a time! All in a day's work, eh?

I had established a close relationship between the centre and a Hungarian cultural institute, whose president came to Sherbrooke one day to sign a treaty of collaboration with us. We invited him to lunch to celebrate the occasion, but there was not much in the cupboard at that particular time. We were on good terms with the King George Hotel, where we often sent our visitors to stay or at least to eat, so Tessa called them on the phone and asked if they could send up a good meal, with bottles of good vintage wines, so we could have a festive lunch. The wine had to be 'lent' on account of some law and we would repay the debt by buying wine on another day and returning it to them. But the food they sent was out of this world! The cook had obviously outdone himself, not only in the cooking but also in the serving of the meal, which was delivered in a van just before our illustrious guest arrived. He complimented Tessa profusely on the lunch, admiring her excellent cooking ability! Needless to say, we never let on that the work of art that had been provided for his benefit had been put together by a professional cook!

One of our 'official' visitors was my brother Gedeon. He was well qualified in linguistics and I had a certain Père Richer on my staff who was a linguist and who had learned Hungarian. I wanted Gedeon and Père Richer to work together on analysing our mother's notes on development of language, when Gedeon and I were very young. Apparently, my mother had taken great pains to compile the whole sequence of development of language from the first utterances in babyhood till language appeared useful in communicating. Père Richer had gone through the material, but we wanted his work checked by another linguist who had command of the Hungarian language and who could be better than my own brother? He was travelling in the United States anyhow, and I only had to pay his fare from Chicago to Sherbrooke and then back to New York, plus a small fee for his work. Of course Gedeon did not realise, coming from Central Europe, that what I wanted him to do was some real work. He thought it was just an official excuse for getting him there. He has been cross with me ever since for not inviting his wife Maya as well. I was not made aware of his resentment until years later.

We had a party in his honour one evening and he hardly spoke to anyone. Gedeon and I never were on very good terms during our adult life – he had not taken to Tessa, whose ways he could not understand or appreciate much – and this invitation did not improve things. These days we are at least on speaking terms, but it does not seem that we can ever be very close, which I regret very much.

This resentment problem reminds me of a funny little scene I once witnessed at Vienna airport while I was checking in for a flight to London.

The person in front of me was overweight and the person at the check-in said that there was a certain amount to pay. The man in front of me was furious and said, 'Look! I have travelled round the world with this luggage and have never paid any excess. This is absurd!'

'I am sorry, sir,' said the check-in man as politely as he knew how, 'but, according to our regulations, there is excess to pay on your baggage.'

'But I don't even have the money!' said the angry man.

'It is not my fault if you travel with insufficient funds!' replied the check-in man.

The angry man finally did find some money and angrily paid it. When I had checked in, I saw him sitting in the departure lounge, looking very glum and cross.

I moved over and said to him, 'I couldn't help witnessing your problem with the luggage. But look at it this way: the money you paid over, you will probably make again in your business in a matter of minutes. But the minutes you are wasting being unhappy about this will *never* come back! Why not just enjoy life?'

He seemed to brighten up and he said to me very cheerfully, 'By Jove, you are right! Have a drink on me!'

At this, he pulled me over to the bar and ordered a couple of whiskeys, which we both drank quickly since our flight had already begun to board.

I often think of this incident when I find myself in danger of wasting precious minutes of life in being cross or disgruntled. I often wish that my brother Gedeon could do likewise!

During the Sherbrooke years Bruce and Sorrel grew out of school and began to think about organising their lives. The Lennoxville CEGEP and Bishop University, also at Lennoxville, provided an obvious avenue for their further education, as they both wanted to continue on the anglophone path. Gradually, they flew the coop and started having rooms and apartments of their own. They still had their own rooms in our house, so the umbilical cord was not broken too suddenly.

Bruce thought he would do mathematics and eventually did a year of mathematics at McGill in Montreal. At the end of his first year, a conversation between Bruce and myself ran something like this, 'Dad, I don't want to go on doing mathematics!' said Bruce.

'Why not?' said Dad, trying to put on a poker-face.

'Well,' Bruce replied, 'it is too trivial!'

'How come?' replied Dad.

'Well, you see, Dad, mathematics is just the collection of all the obvious things. If they were not obvious, they would not be true,' was the surprising reply.

'Are you sure you don't want to go on with it because it is too hard?' I suggested.

'No,' replied Bruce. 'It is because it is too easy!'

He was not sure what he wanted to do instead, so he decided to work for a while and see. His work consisted of filling up boxcars at railroad depots with all sorts of heavy cargo. To this day he has back problems as a result of this noble work. During the boxcar period he got interested in photography and out of his possibly meagre earnings bought himself a camera and other photographic equipment. So for a while he was a filler-upper of boxcars as well as an amateur photographer. Then he got so interested in photography that he decided to apply to take a degree course at Ryerson College in Toronto. The exam was a 'take-home' one and none of the questions had obvious answers. This appealed to Bruce, as he foresaw a lot of non-obvious possibilities on the horizon in the field of photography, as opposed to his Russellian conception of mathematics based on tautologies. He was offered a place, rather surprised at the fact, since only a tiny proportion of applicants were accepted.

We decided that we would support this project and said we would pay his fees. So this was the start of Bruce's first learning period. This learning was to last a very long time and eventually finish with an undergraduate and then a graduate study of psychology.

Bruce is the only one of our children who has remained a Quaker. In fact he has been extremely active all his adult life in the Quaker Youth Movement, attending many yearly meetings, both in the USA and in Canada and drawing spiritual strength from this three hundred year old movement.

Sorrel's academic career was not successful. She managed to get a few credits, some in psychology, some in art, but to this day she has not managed to put together enough credits to obtain a degree. Tessa puts this down to our peripatetic life, going from one place to another, resulting in Sorrel's having often to change schools, change friends and even change countries and cultures. The Quaker influence did not seem to be enough to give her a foundation of moral strength to grapple with life's problems. Some years ago she did become a Christian and was baptised, and from this source she has been able to draw more strength. Whether she will ever achieve any academic distinctions is now, for her, a less relevant question than before, as she thinks it is more important to concentrate on things eternal than on the ephemeral things of this world.

Chapter Twenty-Two
The Squeeze

When I accepted the job at Sherbrooke, I was given to understand by the then dean of science that the appointment was permanent, namely with tenure. Apparently, in the offer of appointment it said that my appointment was in line with the tenure arrangements in force at Sherbrooke University. At the time of my appointment nobody had tenure: in other words, the tenure arrangements were simply that there were no arrangements, *ergo*, logically, I had no tenure but did not know it. Had I known it, I would have done things quite differently and some of the things I am about to describe now would, no doubt, not have happened.

I was also not aware of the politics involved in the North American grading system. Apparently, grades are not given for merit, or otherwise adjudicated on how well or how badly a candidate has performed, but in a 'politically correct' way, which means that all graduate candidates get As, otherwise the result counts against the instructor involved. To have B grades anywhere in your records is simply unacceptable to candidates. And, since candidates are the clients and since clients are always right, a B grade is out of the question. When we were introducing masters and doctors degrees in psychomathematics, naturally nobody told me about these things, for it was tacitly assumed that everybody knew. In fact, most likely it would have been politically incorrect even to talk about it!

I had also been promised two assistants to work alongside me in the practical and theoretical research for which the centre was created. I was also unaware that such promises did not have any real basis, either logically or morally. It seemed that in many cases promises were made not in order to be kept but in order to obtain the services of a certain individual.

One of our doctoral candidates was a person whom I had brought over from Germany. This person had made arrangements with another colleague at the German university where they both worked that they would take it in turn to come to Sherbrooke. I was not aware of this, and, as my invitee asked each year if he could stay on, I continued to reappoint him each year. I only learned much later from his German colleague that this was breaking their arrangement. I apologised to him, pleading ignorance of the arrangement. Then, when he produced a rather third-rate essay as one of the requirements for our doctorate, I thought I was being quite generous in giving him a B grade. He was furious and resigned from his doctoral programme and, unbeknown to me, registered at Laval University in Quebec City for a doctoral programme. The other candidate also resigned and then the dean told me that we might as well shut down the doctoral programme and continue only with the masters programme.

We were eventually able to grant two masters degrees in psychomathematics, one to an Argentinian lady and one to a Quebecois nun, who both wrote very interesting theses, probably suitable even for a doctorate, but, since the doctorate had been abolished, all they could get was a masters.

I was getting a fair amount of federal money for running the centre, some from the National Research Council, some from the Canada Council and even some from the Quebec Provincial Government. I was asked at one point if I could use some of these funds for other purposes which they had in mind, letting me know that such action would probably be the best thing I could ever do for the centre. I was not well-versed in politicking and I had not understood the veiled threat implied in the suggestion and I politely declined. I was even asked if I could use some of the ISGML funds for their purposes. I informed the university that the funds for the ISGML all came from private subscriptions and voluntary contributions of members from royalties earned through publications of work done through the ISGML, and I thought it would not be fair to use these funds in any other way than in the way the subscribers believed they would be spent. Such a British attitude seemed quite alien in the prevailing Quebec culture, but again, simple-minded as I was, I had not caught on to the checks and balances applied to ethics in this part of the world!

I had one serious talk with the rector of the university, in which I expressed my opinion about some of the above, in particular on the issue of the two assistants. I was never one to mince words, not being schooled in diplomacy and the rector was quite horrified that anybody below him in the hierarchy could question anything whatsoever that had been said or done by anybody who was above. Obviously, such an interview and plain speaking were a diplomatic mistake and no doubt that I would have been much more careful if I had realised that I did not have tenure!

I sent off a telegram to one of my colleagues in Germany, asking if there were a possibility of my coming to work in Germany, if the centre was going to have difficulties. I got a positive reply almost immediately, so we began to think in terms of winding up the centre and moving back to Europe. My application took over a year to process and it had to go through three ministries: the Ministry of Culture, the Ministry of Education and the Ministry of Finance. The first two ministries approved my application, but the Finance Ministry turned it down, no doubt on account of the generous German arrangements for retiring, which they would have to apply in my case, for not very many years of service before I had to retire according to law. So the German plan was aborted.

The German assistant whom I had invited over from Germany went to the Mathematics Department, where he assumed the direction of the educational aspects of mathematics. This was obviously a hint that I should think of packing my bags.

During this latter period, the dean took over my best secretary, no doubt by offering her a higher salary and intimating that the centre's days were numbered.

The relations between the university authorities and myself gradually deteriorated, so it was really time to look for something else. Tessa and I decided to sell our dream house and move into an apartment. A new apartment block was being built near the municipal ski slopes and we persuaded the developers to turn three apartments into one, which we would rent. So we sold the house and moved into our apartment, which was on the top floor, with two terraces and a very nice view over the town. There was a swimming pool and a sauna, as in our Montreal apartments, so, even though we were sorry to lose our beautiful house, the move was not altogether a comedown. We thought it would be easier to give notice with a rental than to sell a house, if it appeared to be suddenly necessary to move.

The university actually called in the police to investigate me! One day I had a call from the police asking me who the lady was who had been in my car when I left the house and yet who was no any longer in it when I arrived at the university? I wonder if they thought I was engaged in some sort of subversive activity! I have no idea what people from the university might have told them. The mystery person was Sorrel and it happened that I gave her a lift to the swimming pool!

They also went through my university accounts with a fine-toothcomb, but they could not find a cent that was not properly accounted for. They clearly wanted to have some plausible excuse for getting rid of me.

The work in the schools was going very well, in fact too well. It seems that it is enough merely to do a few things really well for enemies to arise who try to put you down. The children at the École Eymard were doing so well that it got members of the Commission Scolaire quite upset. It became a saying that *On ne parle pas de l'École Eymard!* because it was too embarrassing, as, by comparison, it showed how badly the other schools were working. There was even talk of closing it down. So things were really getting bad for me all round, except that I was still getting funds from the federal government, enabling us to go on with research until the very last days of the existence of the centre!

Having failed to find any real excuse for getting rid of me, the dean wrote me a registered letter in which he informed me that my contract was not to be renewed. I was not at home to receive the letter and, by the time it arrived and I signed for it, it was a day too late for giving me notice for that year, so, legally, they had to keep me for that year as well as the next one. I went to see the dean and we had it out about the 'tenure' position. I told him that such misleading information in a letter of appointment was really not following normal professional ethics, but the dean was adamant that legally they could simply give me a year's notice, emphasising also the university's need for economy, the budget having been overspent by large amounts so cuts were having to be made all round to make ends meet.

I had to think over whether it was worth going to law about it and Tessa and I came to the conclusion that, in the separatist atmosphere in the province of Quebec, I would not have much chance of winning; in any case, working under such conditions would be most unpleasant. The centre was to close, my ex-assistant had been appointed head of the educational section of the mathematics department and if I stayed I would have to work under the person whom I had helped train and who had become in a sense an enemy. So there was nothing for it but to leave; but where?

I had been preparing things in Bologna, in collaboration with Pescarini, for setting up a mathematics education centre that was to be funded by the Emilia-Romagna regional government, in which Pescarini was responsible for education. I had spent quite a lot of time preparing things, such as establishing the kind of personnel we would have and had even chosen an old Italian stately home which we had started to convert into the place from which the new centre would operate.

At the same time I was offered a five year contract by Brandon University, in Manitoba, where they wanted me to work on the mathematics education of native teachers and children. The proposed remuneration was very good in both cases and Tessa and I spent a lot of time discussing the pros and cons of the two possibilities. But then something happened which made the Italian choice less desirable. In less than two years there were to be elections in Emilia-Romagna and, although I was verbally assured that, if I came, my position would be permanent, legally, they could not offer me a contract for more than a year. Since I had just been deceived by such promises of permanence, it was obviously not a good idea, or so we both thought, to be let in for another similar experience. So I accepted the Brandon job, we gave up the apartment and decided to move to Winnipeg.

By this time Bruce was in Toronto, taking his degree in the photographic arts, Sorrel was vacillating between taking courses and odd jobs and became involved with a young up-and-coming musician, so only Tessa and I moved to Winnipeg. We kept on the Montreal apartment for Sorrel's sake and, in fact, she did live there for a short time, her musician friend using it as his office.

Being squeezed out of Sherbrooke was the first time in my professional career when things had started to go down rather than up. When such things happen in life, one has to go through a so-called agonising reappraisal, which is a kind of post-mortem. A reappraisal is, of course, concerned with one's own role in events and, if anything were to be learned from the unfortunate turn of events, any blaming of outside factors or other actors in the events would be out of the question.

I recall a conversation with John Williams during the 'squeeze period', during which I said to him, 'What can the dean possibly do in such a situation?'

'He will do,' replied John Williams, 'exactly what suits him best.'

'Do you mean that ethics simply does not come into it?' I enquired, somewhat surprised at what I had just heard.

'No,' he said simply, 'only self interest.'

John Williams was obviously a better predictor of events using the self-interest model, than I was, using the ethical model.

So this was obviously one of the lessons I had to learn, albeit somewhat late in life, but then perhaps better late than never. Everyone looks after number one. Self-interest is the key motivator of action for most people.

Then I would ask myself the question, 'What advantage does the existence of the centre confer on those directing the university?'

The centre was a research centre. We were not there to attract a body of students. But the bread and butter of the university is the number of students, since the grant received is directly proportional to the number of students who register there. So the centre was not attracting any money to the university. It was attracting federal grants, but the university had no power to use these funds; I had that power and I would not relinquish it. So, money-wise, for the university the centre was a non-starter.

Since I was employed by the university, basically the university would expect some advantage to itself as a result of hiring me. What were these advantages? Many people from different parts of the world came to visit the centre or to work in it for shorter or longer periods. In Europe, when Sherbrooke was mentioned, people immediately thought of it as the place where the Psychomathematics Centre was operating. In the beginning this might have been seen as an advantage, but it soon became a disadvantage, because it stole the thunder of whatever else other members of the university might have been doing. Besides, there were too many foreigners. I was even asked once why I was always asking Hungarian, French and German workers to come and work at the centre and that I should advertise in the province of Quebec for assistants. When I did so and showed the résumés of the Quebecois candidates to the dean, asking him to compare them with what I could get from Europe, he reluctantly agreed that I was getting better value for money by bringing in my staff from Europe.

We did run a large number of *cours de recyclage* (retraining courses) for elementary school teachers, for which University credits were awarded. In the beginning I had to give most of these, since there were no others who were competent to run such courses. Gradually, my assistants and others acquired enough mathematical, psychological and educational skills to run the courses themselves, so the advantage for the university of having me there gradually decreased. I was working myself out of a job.

I tried to weigh up all these things in my mind. Should I have appointed an inferior Quebecois to appease the dean? Should I have realised that promises made by a dean were not really binding on the next dean? In general, should I not have been more 'service-oriented', in the narrow sense of 'service to Sherbrooke University' instead of in the idealistic sense of 'service to the world'? One can be idealistic up to a point, but, by not recognising the way the world really works and how to play the 'power game', one loses the chance of doing the idealistic thing for the world. Life is full of compromises; the sixty-four thousand dollar question is where does one draw the line?

So what could I learn from all this?

Certainly, in the future, I should be more 'service- oriented', being directed to the situation there and then, not so much to the world. I should constantly ask myself the question: what does my employer get out of me?

Another lesson could be that, in spite of my many achievements in my worldwide efforts to improve the learning of mathematics, I should never regard myself as in any sense better than my fellow creatures. Thinking you know best could be a policy leading to disaster. A healthy respect for views that I did not share might be another lesson I should have learned from the Sherbrooke episode.

So Tessa and I took the train from Montreal to Winnipeg, taking our cars with us. By this time we had disposed of our Holden and Tessa was driving a Mercedes and I a Chrysler, which we took with us on the train. The ticket for the cars included two free tickets for passengers, so we availed ourselves of this. It was a long but pleasant trip. We had a room to ourselves on the train and there were occasional stops where we could get off the train and look around. There was plenty of time to ponder what had happened and to think of ways of making things better in the future.

On a previous visit to Winnipeg we had looked around and bought ourselves a nice-looking house in one of the suburbs. We had arranged for our furniture to be moved so we could move in and start our new life in 'the West'!

Chapter Twenty-Three
Winnipeg

Part of the Faculty of Education of Brandon University was organised for training native teachers, who could then teach native children. Since this part of the faculty was doing a service to the native Americans, there were plenty of funds available from Ottawa and that was partly the reason they could offer me financial conditions which represented an improvement over those I had enjoyed at Sherbrooke. The centre that was working on these problems was located in Winnipeg, as from there manpower could radiate more easily into the various reserves in the north, where most of the work was taking place. At the time of my appointment to this post, which was a full professorship, about ten per cent of Manitoba's population was native. Winnipeg also had a fair-sized native population, so the work would have to be divided between service to urban and rural natives. In my letter of appointment there were no particular tasks mentioned; it said only that it would be the responsibility of the dean of education to determine what I should do. I had had some very cordial conversations with this dean before accepting the job and we seemed to see eye-to-eye in matters educational, so I did not think there should be too many difficulties ahead.

I soon got in touch with some of the Winnipeg elementary schools and had discussions with the respective heads about the kinds of things I could offer. One school seemed an obvious one, in which there was a large percentage of native children; another was a private school known as the Sacré Coeur, a French-language school, to which I added another French-language school in one of the francophone communities outside Winnipeg. It did not take too long to get things going, the positive feedback from the children was the factor which eventually persuaded most teachers that I must be doing something right.

The centre was run by a bunch of very left-wing operators, who appeared to have taken their views straight out of a Marxist textbook. The rooms used for offices and/or meetings at the centre seemed to be a shambles, whether by accident or on purpose, I shall never know, although I am inclined to believe the latter. I remember that when Tessa and I first came in she was certainly horrified. It was a cold day, so she had a fur coat on, which immediately did not fit with the 'left-wing' atmosphere. When we asked where to put our coats, as there seemed to be no allotted place, we were shown a pile of coats strewn around on the floor. So, like 'true socialists', we threw our coats into the pile!

Another capitalistic crime we were committing was Tessa's driving around in a white Mercedes. This simply did not tally with 'going native'. I overheard a little later, while I was coming in to the centre, something to the effect of, 'She thinks she is great, driving around in her white Mercedes!'

We invited members of the centre to our house to parties and they all came, with their wives and enjoyed themselves and we tried to make them feel at home, but we did feel as if we were in very strange company. One lady, when she appeared at the door, exclaimed, upon seeing the inside of the house, 'Oh, how elegant!'

True, it was a nice house, furnished with taste, but we did not realise that this was also a socialist crime! I wonder how these fake-working-class people would have managed in our cottage in Moreton or in Pitt, drawing water from a 150 foot deep well and cooking on a range. There was an NDP Provincial Government in power at the time in Manitoba (which is a rough equivalent of the British Labour Party) and I suppose, in order to make headway, one had to swim with the times.

In spite of all the above, we did manage to make friends with a number of people working on the native education project. Some of them would even come and visit the classes in the schools in which I had become involved and they liked what they saw. Eventually, it was decided that I should do some teaching myself. There were three plans: one was to take the course at Portage le Prairie for Metis teachers; another was to work on the Peguis Reserve for natives on that reserve; we also thought it would be good to get something started at the Island Lake Reserve.

Portage did not present any logistical problems as it was halfway between Winnipeg and Brandon and I could often combine it with going to Brandon, where I would have discussions with the dean and stay the night in rather a nice motel with an indoor swimming pool. I realised that the language the Metis spoke at home was French, so at the start of the course I would ask, *'Voulez-vous que le cours soit fait en français ou en anglais?'* ('Would you like the course to be given in French or in English?'). They had never had such an offer made to them before, so they excitedly agreed that we should have the course in French. And so we did, except in certain cases when I had to use some technical mathematical terms, which they knew only in English!

When it came near exam time, one of the students came to me, looking very sheepish and said, *'Monsieur! Est-ce qu'on peut avoir l'examen en anglais?'* ('Sir, could we have the exam in English?').

I discovered that, although they spoke French as their first language, their ability to read it or to write it was next to zero! So of course we had the exam in English.

I had learned my lesson about the grading system and all the students passed with flying colours!

The Peguis Reserve was a different kettle of fish. It was, so to speak, at the end of nowhere and the only way to hold a course there was to stay there. The only way to get there was by car, which in the dead of winter, was sometimes a hazardous undertaking.

We arranged with a member of staff who had a house in Peguis that we could stay in his house and he would come to Winnipeg to do his work while I went to Peguis to do mine. Tessa took a poor view of being left alone in Winnipeg, so she always came with me when I had work to do in Peguis. The road to Peguis was not very much used, and, in the case of a snowstorm, it was one of the last to be cleared by snowploughs. I recall one trip up to Peguis, when there was a severe snowstorm and the road was covered so thickly with snow that it was impossible to see where the road ended and the nearby fields began. Unfortunately, between the road and the field there was a ditch. The only way to avoid the ditch was to drive straight on and keep the telegraph poles at a constant distance from the car! Strangely enough, we managed to avoid the ditch until we entered the actual reserve and I suppose then that I must have let my attention slip and we found ourselves at an angle, the right-hand side of the car much lower than the left-hand side! That was the only way we could tell that the right-hand side must have been in the ditch! Some friendly natives immediately appeared, as though from nowhere and, putting their Tarzanesque muscles to work, within seconds they pulled us out of the ditch! This must have been quite a feat, since I was driving the Chrysler, which was not exactly a light car!

I had brought some multi-base blocks with me and the native teacher trainees really had a ball playing with the material. Since, theoretically this is what they should have done to start with (i.e. according to my own theories), I let them do it. It paid off, because soon they were quite willing to face the challenge of solving problems using the material.

We were told that it was really safer for Tessa to stay in the house, as some of the natives might not be reliable and she might be molested. So she spent the time preparing the meals we were to eat and reading all sorts of books she found on the shelf about native life in the old days, their legends and their culture.

The Island Lake project never really got off the ground. When the dean realised that I could speak seven languages (my original five, plus Spanish and Portuguese), he asked me if I would like to learn Cree, in particular Island Lake Cree. I instantly agreed, as I thought it would be fun to learn an Indian language and to communicate with the natives in their own tongue. I asked if I could have some tapes and an accompanying booklet and then I would do my best. I was flown up to Island Lake once, just before the freeze-up. The airstrip is on an island and it is necessary to use boats to get to the reserve. During the freeze-up or during the break-up, this is not possible, so at those times Island Lake is isolated from the rest of the world. In the summer they use boats, in the winter skiddoos, or they simply walk across the ice to and from the airstrip. During the winter there is also a 'winter road', made each winter for trucks, which trundle across the frozen ice on the lakes and cross various islands on roughly made roads. It is cheaper to make a winter road each winter and supply goods by truck than to use air transport!

When I was up at Island Lake, I visited the school, where there was already a native teacher teaching the children. Some results of the native education project were already visible in Island Lake. I asked the teacher if she would like me to do something with the children, to which she rather shyly agreed. I then played some multi-base games with the children, which they greatly enjoyed but which somewhat nonplussed the teacher, as she had been trained very traditionally and possibly rather quickly. This was probably a mistake on my part. I do not know whether it had anything to do with the project never getting underway, but it is possible that the teacher in question might have said something negative about my intervention. It seems that one has to tread very carefully where native sensitivities are concerned.

I paid a visit to the local chief and we had a good talk about many things. He told me that Island Lake was 'dry', but asked if I would still like something good to drink. I declined, but I asked him about a model canoe with two natives in it, which he had carved, placed on one of the shelves. He said that he would sell it to me. I believe we agreed on a price of around seventy dollars but I am not quite sure now. He packed it up and I took it home with me. In our home in Devon now, we have a large bookcase, with curved tops, which make the tops of the shelves look like waves! The native canoe, with its 'native' operators, still adorns the top of this large, tall bookshelf, sailing on the wooden waves, next to a New Guinea canoe, which has a sail, in the middle of which there has to be a hole, to let the evil spirits fly through.

I waited and waited for my Island Lake tapes, but there always seemed to be some difficulty in getting them made. I even had a talk with a Berlitz school in Winnipeg and offered to help them make a Cree Berlitz, but, in the absence of the tapes, I could not do anything. The dean had to get very angry before the tapes were actually produced and I remember trying to learn Cree on one of my trips over to Italy, but, by that time, the Island Lake trail was cold and nothing ever came of the project.

It is a pity it went that way, as I had even got some natives interested in a project for making parabolic mirrors out of aluminium foil, which would rotate with the movement of the sun; water would pass through the focus, which would thus be heated. I thought that if I could get the natives interested in such things, apart from getting hot water cheaply, it might get them interested in scientific ideas and possibly even in mathematics, so that what I was about to teach their children would make some sense to them.

Sorrel came to see us once in Winnipeg. She had been to stay with my cousin Aga, having made friends with one of Aga's children in Montreal and she had been trying out the working girl's lot by working for a while in a hotel in Banff. She stopped off with us for a short time, but she had no intention of settling down with her parents again and she made her way back to Montreal, where she had many friends.

Nigel also appeared at one point during our Winnipeg stay. He was going through a difficult time and seemed very unhappy. He liked our Chrysler and enjoyed driving it around and, when he left, he wondered whether he could drive it back to Montreal. I said that he could not only drive it back but that he could have the car; after all, we had another one. Tessa was not so sure about this idea of giving the car away and to this day she thinks that she eventually stopped driving because at that particular point we were left with only one car and she could use the Mercedes only when I was not using it; although I offered to buy another car, she declined, thinking it was a waste.

Bruce also appeared, together with a young girlfriend. They were hitch-hiking, heading west. They had a notice they held up when they wanted to get a ride, which merely said, '*elsewhere*'. It seemed the notice did attract quite a lot of cars and they were never very long waiting at the roadside. They stayed with us for a few days and then, on one of my trips to Brandon, we took them to Brandon, filled their bags with goodies and dropped them on the TransCanadian Highway and watched them walking away towards the west, their heavy packs weighing them down, until we saw a car stop to pick them up.

When the children have flown the nest, parents are quite lucky to see their children cross their paths, even if they do so only occasionally. With a shrinking world, both parents and children become more mobile, which means that they are likely to be far away from each other, but the shrinking world also makes it possible for them to meet, which in the olden days would not have been possible, not when the world had not yet shrunk to its present space-time dimensions.

After about a year in our 'elegant' house, we began to feel that we ought to move, partly because it was difficult to leave a house for my trips to Europe and partly because our basement had flooded once and Tessa thought it was not safe to stay. Anyhow, we started looking round, both in Winnipeg and in Brandon. The estate agent we contacted in Brandon became a friend, for we seemed to have a lot of things in common. She showed us no end of houses, but, compared with what we had in Winnipeg, the Brandon houses seemed very poky.

I said to Tessa once, after the estate agent had shown us a particularly poky house, 'This is at least nine on a ten-point pokiness scale!'

Our estate agent friend was tickled pink with the idea of having a scale for pokiness and she took it up with some of her other clients, who were also very amused. I am not sure if the ploy ever gave her a sale she otherwise would not have made, but it was fun.

We made friends with some of the members of the mathematics department in Brandon and I used to go there to give seminars about how to make up mathematical games. These meetings were always friendly and jovial and we had lots of laughs about the games, although I do not know whether any of them made use of them in their teaching. But, there again, it was a fun thing to do and we all enjoyed it.

There was a Quaker meeting in Winnipeg; not a very large one, but a fairly lively one, whose members also became our friends. It was good to be able to commune with people who were more likely to be sympathetic towards the same sorts of things and situations as we. After all, one's spiritual values permeate a great deal of one's everyday activities and one can feel more peace if one's friends share these values.

Our house-hunt eventually ended with our selling the house and moving into a penthouse in a very desirable block overlooking the Assiniboine River. Tessa was not at all sure whether she wanted to sell it, as it was a very nice house, with a huge 'rumpus room' and a very large study for me, also used as a second lounge for entertaining 'official' visitors. There was a comfortable L-shaped bathroom, in which we had placed an easy chair so that one of us could have a bath while the other sat there being the entertainer! One evening we invited the estate agent to dinner and we had a good dinner, with perhaps more wine than was good for us and we talked about the pros and cons of selling the house.

At one point, rather foolishly, Tessa said, 'Let's toss a coin to see whether or not to sell!'

We tossed a coin and it came out to 'sell'. The agent had a signed offer in her bag, which she pulled out for us to sign. We both signed, being co-owners of the house.

After the estate agent left, Tessa changed her mind. She called the estate agent to say that she did not want to sell the house. But the agent was smart, even though possibly somewhat unethical, because she had already been to the people who had made an offer, who very much wanted to buy the house, and would not let us off. So the toss of the coin had decided the fate of our house!

In the apartment block into which we moved, there was an underground garage for the car, and, on arrival at the building, all we had to do was give the key to an attendant, who would drive it down. If we wanted to have the car out, we called down and asked for it, and, by the time we had taken the elevator to the ground floor, the car was waiting with the engine running. There was also an emergency bell, which we could ring if anything untoward happened, and a security guard would appear in less than a minute. But, above all, we could always leave and go anywhere by turning the key in the front door and we knew that everything would be looked after.

When we had moved in, Tessa said to me, 'I wonder who will be the first person to call us in this apartment.' Our telephone number was the same as the one we had in the house.

No sooner had she said that than the telephone rang. I picked it up. It was a professor from Bogotá who was inviting me to come there, because they were naming a college after me and they wanted me to be there for the opening ceremony. We had already booked our trip to Tahiti and Australia and we were to leave within three days. It would not have been impossible to call at Bogota on the way to Papeete and still keep my engagement in Tahiti, but then we would have had to leave all our stuff unpacked and we would come back from the trip to a grand clear-up. So I politely declined the invitation. I do not know to this day what happened about the college; for all I know there is a college there named after me, unless it has been turned into an anti-terrorist centre for drug-busting or some such thing!

Jancis came to see us in Winnipeg with her children Raj and Rilka. She also brought her man. I am not sure now whether this man was a husband or a common-law husband. My recollections of him are by no means pleasant. We rented them an apartment across the passage from our own, so that they could keep themselves to themselves if they wanted to and socialise with us if they felt so inclined.

The 'man' invariably wanted something to eat when we had all finished a meal. He was very demanding about what he would eat and he always lounged in the best chair, never offering to lift a finger to help. It really does try one's patience to have someone like that around, but, since it was Jancis's 'man', we thought we would put up with him.

We were members of a club called the Winter Club, where you could swim, skate, have saunas and do gymnastics as well as sit and have drinks or even meals. We arranged for Jancis and the children to join as temporary members, so they could enjoy the privileges. It was hard to persuade them to go in the beginning, but, once they got the hang of skating, they really enjoyed it. I also took Raj and Rilka out cross-country skiing. One easy ski run was to ski down the slope into the Assiniboine River, which was of course well and truly frozen over, so the kinetic energy acquired on the downhill run took you right across to the other bank of the river. There was also a forest a few miles out of Winnipeg, where there were some trails made especially for cross-country skiing; this took you round from five to ten kilometres of quite scenic skiing. Another good area where I took them skiing was the local zoo. This was only a little way from where we lived and it was possible to ski there, ski around the zoo and ski back if one had the time.

While I was in Winnipeg, I was also responsible for some courses at the Université de Montréal, so at one point we were commuting weekly between Winnipeg and Montreal by plane and therefore got to know all the airline hostesses and crew personally, so they gave us special attention, which was very nice for us as the commuting became really quite tiring in the end. Our membership of the Winter Club allowed us to use the facilities of other clubs, in particular those of a very good club in Montreal, where we used to go after Friends' Meetings for brunch. Apart from this commuting, I arranged with Brandon to have three months off each year (with pay) for doing work in the projects I was running in Italy, Hungary and Brazil, so our world travelling continued unabated during the Winnipeg years.

One year we spent quite a long time at the Marchesa Incisa's San Guido property, where we invited Nigel, Paula and their first child Jamie. It was good to see them and we would take them down to the beach, only about a kilometre away, where Jamie could safely paddle in the sea while we exchanged stories of recent happenings. The experimenting I was able to do at the school was useful in trying to devise a new curriculum, which I was later to try out in my experimental schools in Winnipeg.

For the Brazilian trip, which turned out to be the last one, Tessa flew to Montreal first, to be with Nigel for a few days and I was to meet her in the transit lounge at New York's Kennedy Airport before boarding our Varig flight to Rio. It was wintertime and the roads in Montreal were very icy. Tessa and Nigel and a friend of his were walking up Peel Street, rather faster than Tessa could manage, when she slipped and, as it happened, broke her ankle. But she managed to crawl after the two young men without too much fuss.

'Nigel! I think I have broken my ankle!' said Tessa, expecting them to slow down.

'Impossible!' said Nigel. 'With a broken ankle you could not be walking!'

This 'expert' medical opinion turned out to be wrong, as, when we arrived in Brazil and had Tessa x-rayed, there had indeed been a fracture!

Somehow Tessa managed to get on the plane for New York. I was frantically trying to find her, looking everywhere in the Varig departure lounge, when I saw a Varig employee pushing someone in a wheelchair. This was Tessa, her legs up in horizontal position, all bandaged up! Of course I had no idea of what had happened, but Tessa soon enlightened me, expressing a poor view of these young people who did not treat their elders as they should!

For this trip I had insisted that I could not give four consecutive weeks of workshops and had insisted on a week's holiday between the first and the second two week periods. This was good for Tessa, for our hosts took us to a lovely place way up in the tropics, with a gorgeous garden and swimming pool. I enjoyed the swimming and walking around in the tropical surroundings, while Tessa was convalescing, trying not to move too much.

At the end of this period of workshops we were due to go to Italy, to discuss possible future work with the Giunti-Marzocco publishing house, which had taken over the rights from OS. We arrived in London and both Tessa and I felt ghastly. Somebody told us that we might have 'the amoeba', which is what you get in the tropics from unwashed salads and fruit.

We went to have check-ups by a Harley Street specialist in tropical medicine, who tested us both. He said that the tests came out negative, but that this did not necessarily mean 'no amoeba'; it just meant that the test did not detect any! As a scientific researcher, I knew the difference very well! The specialist forgot to tell us not to drink any alcohol with the medication he had given us, which we had to take just in case we had 'the amoeba'. The next day we took our plane to Pisa and, since we always travelled first-class in those days, we naturally had a lot of drinks on the way. We were met at Pisa by a representative of Giunti-Marzocco, who drove us to Florence and dropped us at our hotel.

Both Tessa and I felt absolutely ghastly by then. We did not have any dinner: we just went to our room and went to bed. We felt so ill that we each thought this must be what death was like and did not want to disturb the other! If one is going to die, one might as well do so quietly, without disturbing anyone else.

But we did not die. I was to meet Giunti at nine o'clock in the morning. I still felt terrible then, but no longer dying. I dismissed the idea of a taxi. A walk would do me a great deal of good, I thought, and I really must be alert and on the mark, as we were going to discuss contracts! I knew Florence very well, so I walked over to Via Gioberti and fortunately acquitted myself quite well during our first meeting. Arrangements were made for future co-operation and Tessa and I could relax and have a little holiday before returning to England and eventually to Winnipeg.

After I had spent three years at Brandon University, there was a change of deans. The one with whom I was friendly left and an American was appointed, who was determined to make a business out of the university. Efficiency became the key word and he called together people working on the native projects and told them that they would have to become more 'efficient'. Apparently, among other things, this meant that every native reserve would get somebody flying in for not more than six weeks to teach the would-be teachers all the mathematics and accompanying pedagogy they would have to know for teaching children from grades one to eight. When I heard this I was horrified. I thought one term was short enough to do the job, but just a few weeks was absurd.

I had a meeting with the new dean and said to him that, in my experience of the past three years, the natives who came to be trained as teachers knew practically no mathematics so how did he imagine that they could be taught everything in such a short time?

'We have to be seen by the public to be producing the teachers,' said he. 'Otherwise the money will dry up and I won't be able to pay your salaries!'

'But particularly the natives we are trying to educate,' I would reply, 'need even longer to learn these rather abstract ideas that mathematics mainly consists of!'

'If you were to say that in public you would be branded a racist!' replied the dean.

'I don't mind what they call me,' I retorted. 'Surely we are here to do a decent job, not to pretend to be doing one!'

'Well, well, well!' the dean would reply very smoothly. 'I can see you are not a politician! The job you have to do is what is dictated by the circumstances: you must take it or leave it!'

'In that case,' I replied, 'I have no alternative but to leave it.'

Thus ended my short career at Brandon. I just could not bring myself to do a 'cheat' job and pretend to be training teachers when in fact I was not doing anything of the kind.

I got on to the telephone to Germany, England, Italy and Montreal, trying to see where another 'job' could be found. I was almost immediately offered one at Trois Rivières and in a few days' time I was told that the contract would be in the mail that day. I said to the person at the other end of the phone not to trouble, as I was coming over to Montreal and we could then discuss matters in person. Tessa and I flew to Montreal and stayed with Nigel. In the morning I phoned Trois Rivières and they said they would phone me back but they did not. So we went to Trois Rivières and, when I wanted to see the professor with whom I had been dealing, I was told that he had gone to Quebec. Clearly, somebody had thrown a spanner into the works. I was eventually told that it was all a mistake and there was no job for me at Trois Rivières.

Fortunately, by this time my European enquiries had borne fruit. I was offered a temporary job at the Goethe University in Frankfurt, I was also offered some months of workshops in Haringey, North London, a contract for three years with Giunti's publishing house in Florence, as he had taken over from Dr Abbele, who had unfortunately died of a heart condition and a three year contract by a South Devon Trust (The Howe Green Trust), to do more or less what I liked which would be good for South Devon education.

It seemed obvious that we would have to go back to Europe. It was not obvious whether it should be Devon, London or Florence, but that could sort itself out in good time. So we went back to Winnipeg, gave notice for our penthouse apartment and arranged to sell some of our furniture and goods and to ship the rest to Europe, together with our white Mercedes, which we would ship by train to Montreal. We would take the Russian ship, *Pushkin*, to England. We got very good prices for our furniture and other goods. This was probably due to our location. It was also a good thing that we had the emergency button, as, when one is selling things, one does not know who might be coming into one's home. Tessa did have to push the button once and the security guard got rid of our awkward customer within a minute of the button's being pushed.

We also advertised our Mercedes, just in case somebody turned up who would give the right price. Perhaps unfortunately, somebody did turn up. They wondered what was wrong with the car since we were selling it so cheaply! We were selling it for what we gave for it when we bought it. In any case, this person bought the car and we cancelled the freight booking on the train and on the ship. The next day we found ourselves on the train for Montreal.

Nigel and Paula came to see us off at the quayside in Montreal. Nigel had tears in his eyes: he really was unhappy that we were going.

'Why do you have to go, Dad?' he would ask.

'I have not been able to find any more work here,' I replied. 'You have to go where you can work!'

'Why don't you set up a business?' he asked.

'I would probably go bankrupt,' I replied. 'I am not a good businessman. It is a great pity, but I do have to go!'

So Tessa and I boarded the *Pushkin* and we soon cast off and sailed down the Saint Lawrence River, leaving our very sad son weeping on the wharf.

Chapter Twenty-Four
To Europe Again

So we were on our way across the ocean in a Russian ship, the last time we were able to use a ship for crossing any ocean. The voyage was calm and the food and wines excellent. My only problem was that I developed a very bad backache! I went to the ship's doctor, who could not speak English or French, or any of the languages I knew and, stupidly, I told him in Russian that I could not speak Russian very well, after which he went into a long tirade about my back, not a word of which could I make out. However, he gave me some pills, which I assumed to be painkillers, which I took and, indeed, the pains did subside a bit for a while after each pill.

We arrived in England and went straight to Rosalind's house in Putney, as I had accepted a contract to work for a few months for the Haringey school system and we had nowhere else to go. Rosalind was quite taken aback that we thought we could just come and stay with her and could not cope with the idea, so we got in touch with the North London Quakers, who directed us to a certain Mrs Strange, who had some rooms to let in Muswell Hill. This was not too far from where I was to work and the 134 bus also came near, which would take Tessa directly to Putney in case she wanted to spend the day with her mother while I was working. We made ourselves at home in Mrs Strange's house and I commuted to Haringey to work with the local teachers and children.

I was working quite a lot with six year old children, who were definitely pre-operational from the Piagetian point of view in practically anything which could be called mathematical. I played the following game with them: I got them to place a large number of paper plates in a row, it did not matter how many. Then I asked a child to put one bean in each plate. Then I asked another child to put one bean in each plate. Then I asked a third child to do the same and we went on putting beans in all the plates, until the children lost count of how many beans there were in each plate, but they knew that every plate had the same number of beans in it.

Then I asked a child to cover up the first plate with another plate and to put a bean in every plate that was uncovered. Then I asked another child to cover up the second plate and to place a bean in each of the plates that were still uncovered. We went on like this until all the plates were covered and not a bean could be seen.

'Where are there more beans, on this side or on the other side?' I asked the children, pointing successively to the beginning and to the end of the series of plates.

'Over there, at the other end!' cried all the children together.

Then I pointed at two plates next to each other and asked them, 'Which of these two has more beans?'

'That one!' said several children, pointing to the one which had been covered up later.

'How many more does this have than the other one?' I asked.

The children started thinking, then one little boy said, 'It must have one more, because, before we covered them up, they had the same number, then we covered this one, then put another bean in that one and then we covered that one up as well!'

Some of the children followed this reasoning but others did not. The next day I repeated the experiment but only with those children who were quite sure why two plates next to each other were always such that one had one more bean in it than the other one. Then I picked up two plates, with one plate in between the two.

I asked, 'How many more does this one have than that one?'

Nobody could give an answer to this difficult problem.

One boy suggested, 'Let's empty the plates, then put the beans from one plate next to the beans in the other plate, then we shall see how many more this plate has in it than that one!'

We did this and the children noted that the difference was two beans. We tried again with two more plates with one left in between them. We went through the same procedure and checked that the difference was still two beans. Then I picked two plates, again with one untouched between the two, from near the 'end' of the series. I asked, 'What about these two?'

'There must be at least three more beans there, or even four!' suggested one child.

'Why?' I asked.

'Well, you see, there are so many beans up there it must be more!' answered the same child.

We went through the same procedure and the children were truly amazed that the difference was still only two beans. But there was no child who was able to say that the next plate had one more, the next one after that had one more still, so that must be two more altogether. Their problem clearly had nothing to do with knowing that one plus one makes two, as they all knew that. But they did *not* know that one more and again one more was two more!

Then I asked them this, 'Supposing you made a series of plates, but right across the school yard, ending up at that door over there, would the difference still be two beans if you picked up two plates with just one plate between them?'

The children said that they were not sure, and the only way to find out was to do it!

Next week I came back to the same class and several of the children came running up to me, saying, 'Dr Dienes! Come and look at our series of plates! We have made one that goes right up to that door, like you said! And the difference is *still* just two beans!'

I had had no idea that the children would go to that much trouble, but it was certainly enough proof for me that they were in the pre-operational stage.

'Supposing the series went right up to the church on the top of that hill, would you still get a difference of two beans?' I asked.

'That would take much too long to find out!' replied the children.

It takes this kind of detailed work with very young children to show teachers what children know and what they do not know. Knowledge should not be thought of as the ability to give the right answers to questions invented by educators, but as a potential to tackle problems that arise in one's environment. Such potentials grow naturally as children grow, as well as arise out of actions being internalised, so that appropriate actions can be taken in given circumstances.

The population of Haringey is racially very mixed. I have already alluded to the fact that children of African origin often seem to learn more effectively through embodiments which include the use of rhythm. Rhythm can be introduced through bodily movement, song or language, each of these areas having its own rhythmical aspects.

While working with children in Haringey, I often resorted to the use of rhythmic exercises, since, in nearly all the classes with which I worked, the majority of children was black. Starting with a very definite pattern such as:

Short short long long/short short long long

the children would start to do:

step step step hop step step step hop

Then they would try to make words that fitted the pattern, such as:

Do you like this/way of playing

and they finally realised that you could sing:

Oh my darling/Oh my darling/Oh my darling/Clementine

to the same rhythm and step and hop at the same time.

The road is then open to the children examining the songs they know, getting the rhythmic patterns out of them, so that they can sing them and dance to them with even more gusto.

Some seven year olds were surprised to discover that the nursery rhyme *Pat-a-cake* consisted of exactly forty-eight elements and could be sung by touching all forty-eight blocks in a set of logic blocks one after the other while singing the rhyme.

I have given the above examples to give a rough idea of the kind of way the work progressed and how it was that I was always able to find ways to motivate the children and, through the children, to motivate the teachers.

Unfortunately, my back problem had not been cured by the Russian doctor on our ship over, so I had to pay frequent visits to a chiropractor, to continue doing the work. On many days I would have to remain standing all day, as sitting down and standing up or getting on and off the floor entailed too much pain, so I could not concentrate on what I was doing! But the treatment did work after a time and my back problems largely disappeared in time.

We were expecting our furniture to arrive towards Christmas, so we thought we would have time to think about where to put it. But, unfortunately, we had notice that our things were at the docks and the removal company wanted instructions about where to send it! I made some enquiries about the possibility of continuing the work in Haringey, but the response was non-committal. There was the work planned in Italy and in Germany, but Tessa did not feel that she could make the adjustment to live permanently in either of those countries. This left Devon by a process of elimination.

So one Thursday we took the train to Totnes and presented ourselves at an estate agent.

'We want to buy a house and we want to buy it now!' we said to the agent. 'So don't show us any houses with anyone living in them.'

He took us round and we looked at a house on the edge of Dartmoor, but when Tessa saw a mousetrap on a windowsill that was the end of that house, as far as she was concerned.

'I am not going to a house with rats!' she announced firmly.

We saw another house in the main street of Ashburton, then we were taken to a very romantic-looking little thatched cottage near Newton Abbott. Tessa had hardly gone inside when she exclaimed loudly, 'I must get out! I must get out! This is an evil house – we cannot possibly come here!'

So that was the end of that one. Much later we went back to visit that house and we discovered that a murder had been committed there.

Then we saw a very small house in Totnes. I could not see how we could swing the proverbial kitten in it, but Tessa thought it would be all right. In this case I definitely vetoed that one, on account of its size.

We were just about to give up when the agent said, 'Just a minute! I think I have the house that you will like! Come with me!'

We were taken up a steep drive, at the top of which stood a very large, Victorian-looking mansion and beyond it, standing on its own, was a little house, more like a bungalow, but it was really a split-level house. We had a look. It had a gorgeous view at the back, overlooking the Dart Valley and Dartmoor, with Haytor beckoning on the far horizon. It was full of windows; on the inside, the walls were wood or brick. It looked very nice inside, although somewhat insignificant outside. We were in a hurry to put the furniture somewhere so we said we would have it.

We went to see a lawyer straightaway and said that we wanted to buy the house right away, as we wanted to move in immediately. The removal firm wanted to charge fifty pounds a day for storage, so every day we waited increased the price of the house! The lawyer said she would try her best and, indeed, she was able to do all the searches before the weekend and we paid for the house, writing a cheque out on one of our Swiss accounts.

The next weekend we moved in, with all the furniture that had arrived from Canada. The 'mansion' next door had been, in fact, subdivided into three parts. The part nearest to our house was occupied by a divorced lady with her five daughters and this lady, whose name is Chris, was extremely friendly, helping us with everything; even the girls came in and helped arrange the books on the shelves. We soon became fast friends!

It happened that an old friend from the Dartington days, Pauline, was living in the vicinity, a little way out of Ashburton. I had asked her by letter from Canada if she could enquire whether there were any schools in which I could start work and she came up with about half a dozen schools, all situated somewhere between Plymouth and Torquay. So I was able to make some arrangements with the respective principals to start working on various projects after Christmas. When we finished with Haringey, Tessa and I flew out to Florence and established links with some of the Giunti staff, as well as with a number of key persons, such as an inspector of schools and some teachers, who would form the nucleus of a team writing up a new mathematics series for Italian elementary schools. I also contacted the Goethe University in Frankfurt, where we went for a day or so for some talks. I gave one or two seminars to the mathematics and the education faculties and, when they realised that I could comfortably work in German, they invited me to come for a semester.

So after Christmas, the situation looked like this: firstly, we had bought a house in Devon and were under a three year contract with the Howe Green Trust to do three months' work a year for three years, for which I was paid a reasonable honorarium. Secondly, we were given a furnished house to live in for the time I would be working with the Giunti-Marzocco publishing house. There was also a three year contract for this work. This was to be renewed for an eventual total of eight years. Lastly, we had rented an apartment in Frankfurt and I had a contract for just one semester to work with the students in the Mathematics Education Department, directed by Engel.

This involved a certain amount of commuting. I had to put in time for the Florence work while Goethe University was not open, but I had to do the Devon work and the Frankfurt work simultaneously. I arranged my lectures for Tuesdays and Wednesdays at Frankfurt. After the afternoon class on Wednesday, we would drive to the airport and get a flight to Gatwick, then a train to London. This would be well in time to catch the night train to Plymouth. The train was bound for Penzance, but the Plymouth carriage was shunted to a siding at Plymouth and the night passengers were woken up with a cup of tea at about seven o'clock in the morning. At about seven thirty, one of the teachers would be waiting for me at the station, to take me to the first school. Thursday and Friday would be spent in various schools. Saturday and Sunday, we would go to our new house in Totnes and recover from the toil! On Monday morning there would still be some work to do and one of the teachers would take us to the station at midday to catch the London train. We would get off at Reading, get the bus to Heathrow and the plane back to Frankfurt, so that on Monday night we would already be in our apartment in Frankfurt. This went on for two terms.

The work in Frankfurt was bread and butter work; in other words it consisted mainly of giving courses in mathematics education to future teachers who had come to Goethe University to train. I got on well with the students, as I knew how to make mathematics and its teaching interesting and we had a lot of good laughs together.

When we first arrived, Tessa and I had come up by car from Florence and had trouble finding the person who was supposed to meet us. Eventually, we found each other and we were shown our apartment, which was not up to the usual German standard of cleanliness, so, before we could settle down, we had to spring-clean the place! Sorrel was with us at this time, as she had come over from Montreal, where she was living at the time, for a break in Europe. So the three of us together soon made the place habitable and we settled in for our German experience.

In order to keep myself from having more back problems, I would go to the nearby swimming pool and have a swim on most days when there was time to do so. The shortest way to the pool was through a park; once Tessa came with me to the pool and she was horrified at what she saw.

She said to me, 'Don't you realise that this is the place where all the drug addicts meet? Look at them all, looking dazed as though they were in some other world! And look at all these syringes, just strewn about on the flower beds!'

Indeed, I must have been very unobservant, for I had not noticed these rather obvious aspects of the park across which I often walked to the pool. So I promised Tessa that I would take another route to the pool in order to avoid any danger from crazed drug addicts!

We did find a Quaker meeting in Frankfurt, but the average age of Frankfurt Friends seemed to hover around eighty or so. Many Germans had joined the Quakers after the war, as they realised that they were there to help, following Jesus' suggestion that anything you did for the poor, the hungry and the imprisoned you did for Jesus. But they did not seem to have evolved a way of attracting younger people to the meeting. I would guess that by now the Frankfurt meeting probably no longer meets for lack of members. We did, however, find a very good Episcopalian church, part of the 'European diocese', there being another part of the same diocese in Florence. It appeared to us that the minister there, as well as the people who attended, really had their heart in what they were doing for God, so we joined them!

Before we left, we decided to buy a German car, so what could be better than a Mercedes? If you took it out of the country in less than a month, you got the tax back at the border. And if you ran it on the continent for three hundred and sixty-five days, then you could import it into England without paying the car tax. The problem was that a car with oval plates, denoting foreign ownership, could only be insured for a maximum of three hundred and sixty-five days. So when we left Frankfurt, having got our tax money back at the German border, we drove the car to Florence, where we used it for a year. On one of our journeys from Florence to England, we arranged to get on the night ferry at Roscoff, bound for Plymouth exactly on the three hundred and sixty-fifth day of the life of the car. When we arrived in Plymouth, the car was three hundred and sixty-six days old, so exempt from tax. We provisionally taxed and insured it. So we had a Mercedes in England, with English registration plates, on which we had paid no German tax, no English tax and no customs duty. So we sold our other car and continued to use the Mercedes for many years to come, even though it was a little awkward driving it in Britain, the driving wheel being on the left-hand side.

During the English Easter break, I was invited to take part in a conference held in the old part of Dartington Hall, where different people from various parts of the world came together to discuss creativity. It was a very interesting set of meetings, to which I contributed my bit about how I thought, through the interdisciplinary approach, you could make mathematics learning into a creative activity. I gave an address as well as some workshops, during which the participants could get the feel of creativity themselves by creating situations of a mathematical kind, using some the Sherbrooke materials, which I had brought along.

We stayed at the hall during the week of this series of meetings, although sometimes we would go back to our house and to Chris, who very kindly looked after Tessa when she caught a touch of flu, possibly on account of all the commuting she had been subjected to!

This is when computers were beginning to come in, so I said to myself that, if I were to continue to be of any use in bringing mathematics education forward, I would have to learn to use a computer, in particular learn to program in BBC Basic, which was a modified version of the 'conventional' basic, with bits of Pascal incorporated into it, which was beginning to be used in the schools in Britain. So I bought myself a BBC computer and taught myself to program. This went really against the grain as far as my own personality was concerned! Computer programming is extremely finicky. You put a colon instead of a semi-colon and the message 'syntax error' comes up immediately. Another frequent message is 'no such variable'. Sometimes the computer is kind and says things like 'comma missing in line 230', in which case you go to line 230, insert the comma and hope that the program will run! I now have a small laptop, what the computer freaks call a 'notebook' and it is IBM-compatible, so I can write programs in IBM-adapted BBC Basic or in Pascal, the latter being even more finicky than any Basic. You have to 'declare' all variables and state what 'type' each one is. Woe betide you if, while using one of your variables, you use rules that do not comply with the type you have called it. It then 'will not compile'.

I was soon able to put together some fun programs, trying to transfer a lot of my games on to the computer screen. I also tried to write programs which had to be used alongside concrete materials, the user having to solve problems with the materials and then using the program to check his work.

I finally settled down to visiting three or four South Devon schools, which were brought together as the South Devon Interdisciplinary Working Group. I would meet the teachers in these schools regularly on my Devon visits from Frankfurt and later from Florence. I developed the 'birthday story' for introducing the idea of measurement in a playful way and both the teachers and the children hugely enjoyed the fun. The story consisted of a child walking to his friend's birthday party, with all sorts of things happening to him on the way, as a result he would be carrying different sets of things on him in different parts of the road. By knowing precisely what he was carrying, it was possible to deduce how far he had got on the way to his friend's house. With the younger children, we role-played the whole thing, involving many children. Some had to be fish, others had to be goats that gobbled everything up and there were parts for birthday children and grandmothers who prepared birthday cakes. We then prepared 'measuring rods' on which we drew the salient events that took place on the trip to the birthday party and used such rods to introduce 'real' measurement.

With the older children, this led to the introduction of infinite but recurring sequences. We were led to problems such as the following: three children want to share out a litre of milk, but they only have containers which hold half a litre; one-quarter of a litre; one-eighth of a litre and so on. How do they do it?

Clearly, the half-litre bottles are no good, as one child would not get any milk. So they each get a quarter of a litre. But there is a quarter left over. The eighth of a litre bottle is again no good, as one child would go without, so each one gets another sixteenth of a litre. And so it goes on, each one getting:

$$1/4 + 1/16 + 1/64 + 1/256 + 1/1024 + \ldots\ldots\ldots$$

In other words, every other container is not used! So the sequence becomes something like:

no yes no yes no yes no yes no yes

Where will it end? Of course, in practice it will end when there is so little milk to share out that nobody will think it is worth worrying about. In theory, of course, the sequence can go on for ever. This is an amusing introduction to the intractable idea of infinity in mathematics!

I also tried to work out an amusing method for breaking numbers down into their prime factors. We used only the numbers 2, 3, 5 and 7 and their products, so we obtained all the numbers that occurred in the tables and of course a whole lot more. In order to get children to visualise what was happening, I got them to make 'roads' with cubes that were either 2 or 3 or 5 or 7 cubes long. Then such roads were put together into fields, but only by putting together either 2 or 3 or 5 or 7 roads, all equally long. Then equal fields were placed on top of one another, as long as there were either 2 or 3 or 5 or 7 such equal fields. These constructions were our 'boxes' or 'houses'. Then we made roads of houses, put close to each other but not touching, as long as you used 2 or 3 or 5 or 7 equal houses for making a road. Finally, we made 'fields of houses' as well as 'houses of houses', the first such esoteric construction being used for visualising the number 64! Then by taking these constructions to bits, children could see what kinds of factors the corresponding numbers had.

At one lesson one little boy said to me, 'Sir! I can make a house so that all its faces are squares and I can make houses so that only two of its faces are squares and I can make them with no square faces at all. But I have not been able to make a house with just four faces that are squares!'

I then explained this child's problem to the rest of the class and suggested that, for homework, they should try to make a house with exactly four square faces, this being an impossible task.

When I saw them the next day and asked how they had got on with their homework, there was deathly silence. Nobody had succeeded in making a house with just four square faces.

'Is there a trick in it?' asked one of the children.

'Sure, there is a trick!' I replied.

'I think I know!' said a little girl. 'There is no such house!'

'How come?' I asked.

'Well, it's like this,' replied the little girl. 'Two opposite faces must have the same shape. So you can have a house with two square faces opposite each other. Once you have a third square face, this must lie between our first two square faces and so must be the same size, because for a square all the sides must be the same. So rolling the house, holding it by the first two square faces, we shall always see square faces in between! So if there are more than two, there must be six!'

No doubt this little girl had constructed her first proof! Not many of the children were able to follow her argument, but what actually convinced them, apart from the fact that I told them, was that nobody in the class had succeeded in making a house with just four square faces!

So we had lots of fun enquiring deeper and deeper into matters mathematical and I was able to teach a small nucleus of teachers enough so they could hold workshops themselves and pass on their knowledge and understanding to other teachers.

There was another section, run independently, of Dartington Primary, where so-called retarded children were sent. Some of these unfortunate children were in very sorry states from the point of view of cognitive ability. Mathematics for them was not a practical option. But some of the children with measured IQs between sixty and seventy-five, were able to learn some of the games with the logic blocks, particularly those of the type: this one is to that one as this other one is to which one?

Most of them were able to put pairs of blocks together which differed from each other in exactly the same way as other blocks, already in pairs, differed from each other. This helped them to realise what such similarities really were, by having concrete representations of them and some of them were able to go on to solving similar problems with words, thereby flexing their mental muscles and starting to develop their intelligence, which they were not able to get a start on before.

I also offered my services to the Education Faculty at Exeter University and they used to invite me to talk to their teacher trainees, as well as hold seminars for members of the faculty. In fact, they made me an honorary research fellow of the university, which distinction I still enjoy; every time I go to England I still give them seminars and keep in touch with their teacher trainees. Of course, Devon is the place where I went to school when I came to England and I retain a certain sentimental attachment to the place and its people. I took great pleasure in retracing my steps of so many years before across the moors, along the romantic Dart Valley and along the beautiful South Devon coast. Tessa by this time suffered from rheumatism and her idea of fun was certainly not tramping over the moors, but I was able to go sometimes with members of the Totnes Quaker Meeting and sometimes with Pauline, our old friend, who lived on the very edge of Dartmoor.

At the time I write these lines, I am not really able to walk more than two miles or so and even that has to be a fairly level walk. For my last walk on the moor I took my three year old great-grandchild (accompanied by his father, one of Corin's twin boys and his wife!) on a walk, when we actually reached a tor, opposite Sharp Tor, whose pointed peak we could gaze at on the opposite side of the deep valley of the Dart, directly below us. I am not sure how many more walks on the moor I shall be able to do. Fortunately, I have a great-grandchild who is not one year old yet, so if the three year old gets too fast for me there will be another one I can go with!

Now let me return to the situation in Florence. I tried to persuade Giunti to use my computer programs, but by the time he got round to appreciating that the future of books is probably very uncertain, another publisher had acquired the rights for the BBC computer so this connection was never made. I did introduce computer programming to our Ravenna Project, where an Italian 'born teacher' was working with me to establish interdisciplinary situations in the schools. Most of the children, by the time they reached the age of ten, could write simple programs, as they had had some foundations in logic through the innumerable logical games they had played.

An ISGML meeting was held in Rome, Siena and Ravenna, the Ravenna section being largely devoted to the problems of how to take advantage of computers in schools. My one-time Stanford colleague and collaborator Pat Suppes was there, as well as Bob Davies, not to mention other old ISGML members such as Ricardo Pons, Angelo Pescarini, Ermanno Pasini, Brian Greer and Tamas Varga. This was followed two years later by the Siena meeting of the ISGML, which coincided with Siena University conferring on me an honorary doctorate. As far as I can see, that meeting will be the swansong of the ISGML, as no other meetings are planned and other international organisations are now responsible for diffusing information about the 'latest' in mathematics education.

The Italian work lasted without any interruptions from 1978 until 1986, the whole period being covered by contracts between the publishing house Giunti-Marzocco and myself. From the publisher's point of view, the aim of the contract was the writing and publication of a five year series of textbooks for the Italian five year course in elementary school. The books would encompass every subject that was to be taught, according to the Italian Ministry of Public Instruction. So I got together a team, consisting of an inspector and a number of teachers, with whom I would try to construct this mammoth series. There was a fine line between what was psychologically correct and what was commercially viable, so I had to learn to walk that tightrope. There was also a time limit, as books had to be ready by the spring so teachers could have a look at them and, if possible, adopt them for their classes. We also had to tread warily with regard to ministerial guidelines about the various curricula so our attempts at making the series truly interdisciplinary were only partially successful.

Since the series we were producing was breaking new ground, never before attempted commercially, Giunti thought it desirable that I should give a number of talks, workshops and/or seminars up and down the country, to whip up enthusiasm about the new publications. This meant travelling up and down the country in trains and although we always travelled first-class and in suitable sleeping accommodation on night trains, it was very tiring nevertheless. As often as not, trains were late or we had to wait for trains on cold platforms where there were no porters. I remember once waiting at Livorno for a Milan train which was so late that, by the time we got to Milan, the person who was waiting for us to arrive was just going to give up. And then he had to drive us to Lugano, in Italian-speaking Switzerland! Naturally, very few people were still left waiting for us, as they did not know why we were not coming. Such things prompted me to say to Giunti that the train rides were over and that we would have to go by car, our own car and that they would have to pay me kilometrage. We agreed on the same kilometrage that would be charged by a car rental company. After this, the journeys became less nerve-racking.

The books were continually revised, sometimes against my better judgement, so I had to keep my study group together fairly permanently and keep tabs on how the next edition would be altered. This also involved classroom trials, which took place on the whole at the Scuola Bechi in Florence, where I got to know the teachers and most of the children by name. They loved having fun with mathematics whenever I turned up. There was another school where I was experimenting with the music-mathematics connection and we all thought the work we had done in a grade one class was so interesting that we wrote it up in a paper, which was published. Later, it was also published in the *Journal of Structural Learning* in English translation.

It was my first attempt at putting music, mathematics, language and bodily movement together in one single learning sequence, without, however, subordinating any of the four disciplines to the others.

This is how it was done. I started off by playing a Scottish tune to the class on my recorder and just asked them to listen. Then I asked them to close their eyes and listen and relax and notice what they could 'see'. This I did several times until the children started telling me something about the imagery evoked by the tune.

I drew pictures on the board and wrote short descriptions of what the children were bringing me. I then asked the children if they could make a story out of all the 'pictures'. When a short text was agreed, I asked them if we could sing our text to the tune. Clearly, this was not possible, as there was no temporal-rhythmical match between the music and the words.

Now came a study of the words, breaking them up into syllables, distinguishing between long and short vowels, single and double consonants and trying to match these properties of the words with corresponding properties of the music. This resulted in a couple of verses about a bucolic scene in the country, involving a shepherd and his flock, flowers, birds, a stream through the woods and such. We then learned to sing our song, giving a performance to one or two of the other classes. Later, half the class would sing and the other half would act or dance to the tune being sung. Some of the interpretations were very much liked by the children and they wanted to find a way of remembering them. So we discussed the possibility of turning the whole thing into a collective dance. There happened to be just sixteen children in the class and this allowed for four moving squares of four children each to be the basis of the collective dance. There was much discussion about how the squares should move: whether the children should face inwards or outwards and whether they could change between the two ways while still holding hands. The end result was a carefully choreographed dance.

The above sequence had the following important elements: Firstly, carefully listening to a simple piece of music and observing one's own visual and emotional reactions to it. It was also an exercise in creative writing, constructing the story arising out of the listening experience. Thirdly, it was a lesson in the internal construction of words out of syllables, vowels and consonants; becoming aware of the rhythm of language. One also became aware of the construction of the musical piece in question, musical phrases, bars and notes, long and short. Then there was the problem of matching the rhythm of language to the rhythm of music. Then there was learning to sing in tune, following the notes played on the recorder and singing the appropriate notes. Then the children had to learn to adjust their bodily rhythm to musical rhythm to carry out the dance. This involves realising that there is always a phase difference between the two (the foot coming down has to coincide with the musical stress, but the foot must come up before it can go down!). Then they learn a great deal about the geometry of the square by moving the child elements (vertices) of each square as well as by moving the whole square around on the floor, not to mention the square of the four squares.

It is obvious that there is much mathematical content, starting with the subdivision of the tune into phrases and then into bars and ending with the study of the properties of the square. There is musical value in the exercise; first in learning to listen, then in composing words to a tune and then learning to sing it. There is also much linguistic value in the sequence. The creation of a story from given elements is interesting, not to mention the part in which the words themselves have to be analysed and dovetailed with the music. The movement aspect is also very important, providing an opportunity for co-ordinating mouth and movement, learning to move while you sing, not to mention the learning of the disciplined movements necessary to carry out the agreed dance. It remains for others to expand and to study the possibilities opened up by the kind of experiences that synthesise the above four disciplines into one activity.

So our return to Europe started off by my doing four jobs: one in North London, one in Devon, one in Germany and one in Italy. The North London job ended within a month or two, the German job ended within a year. So my European activities were to continue in Devon and in Italy, with a sprinkling of work from time to time in Hungary.

During one of our sessions on the Marchesa Incisa's estate, we heard that my mother was not well. We telephoned Budapest to find out whether the trouble was serious enough to merit our coming over, but my brother was away in Russia and his daughter-in-law, who answered the phone, thought there was no immediate danger.

When we arrived at Heathrow, the captain passed us a telegram, which informed us that my mother had passed away. He also said that we could immediately fly over to Hungary; we did not even have to enter the terminal buildings. Tessa and I discussed this possibility briefly, but decided against it. After all, if my mother were already dead, there was not much point in going.

In a day or two we were down in Devon. It was a beautiful June morning and I went for a walk by myself in North Wood, which I knew so well from the days of my boyhood as well as from the time I had been teaching my six to seven year olds about nature by walking with them in these woods. I paced along the narrow path, following the River Dart, the path I had taken many decades before with David or Brenda and then turned into the northern end of North Wood and stopped to contemplate.

'If you can hear me, *anyuka*, I am with you now and I say goodbye to you from this beautiful part of the world. You know I have always loved nature, so let us part now. Let this be the end of our *'petite promenade'*, which we started together in Paris, remembering this beautiful Temple of Nature as the place of our last conversation! I thank you for everything you ever did for me and please forgive me for any wrongs I have done to you. Goodbye and may God be with you!'

I walked out of the woods, having performed the last rites for my mother as best I could.

Chapter Twenty-Five

The Italy–Devon Period

During the period from 1980 to 1986, we used our Devon house as a home base while the major part of my professional work was done in Italy. The three year contract with the Howe Green Trust in Devon naturally expired at the end of three years and for various reasons the Dartington Hall trustees were reluctant to renew the contract or offer another one. This might have been due to the following factors.

Dartington Hall School was, in a sense, winding down. The school had been 'inwards-looking' for some years, not realising that what it stood for in the 1930s had, for the most part, now been achieved by the state school system. The main points that distinguished Dartington Hall School from other schools in the beginning were the fact that it was co-educational and there were no rules relating to being a boy or a girl; that relationships between staff and children were based on friendly interaction rather than imposed authority; that there was freedom of speech and discussion about any subject (in other words there were no taboos); and that, within reason, children could choose the subjects they wanted to study. By the late 1970s and early 1980s all the above had been achieved by the majority of state schools, in particular by the village school in Dartington village. Parents began to realise that they could get what Dartington Hall had to offer for a high fee, for no fee at all in the state system. There had also been some scandals, which for obvious reasons I cannot mention and many parents started to take their children away from the school until it finally had to close. I did visit some of the senior classes before it closed and tried to interest the pupils in various amusing approaches to the study of three-dimensional geometry, but the general Dartingtonian attitude of 'Well, we know it all!' was not helpful in getting anywhere with them.

After the closure of the school which had been operating at Foxhole in the 1932 building, a new type of school was started in what had used to be the boarding houses for the juniors. This was run by one of the teachers who had taught at Dartington Hall and he clearly was prey to the 'know it all' syndrome. I had a number of talks with him at his home on the estate and we were on the point of organising something together when I lent him videotapes of the mathematics learning programmes I had made years earlier in Adelaide. Having looked at one of these, he declared that my approach was 'much too directive' and cancelled the arrangements. It was no use to point out to him that, in a thirty minute programme, if you wanted to pass some information to the public, you could not leave the children to 'discover' everything – you had to guide them to some extent. Whether my failure to co-operate with the 'new' Dartington had anything to do with the Howe Green Trust not renewing my three year contract, I shall never know. I will let the reader have his or her guess!

Notwithstanding the above-mentioned failure, my work in the few selected South Devon schools went on as scheduled – well beyond the three year period of the contract, in fact – until there were enough local teachers who could themselves hold workshops and I was getting less and less necessary on the scene! I also gave some seminars at Exmouth College to teacher trainees, as well as keeping up my contacts with the School of Education at Exeter University, where I continued to give seminars to the faculty and lectures and/or demonstrations to the students.

So as the years rolled by, the main thrust of my professional work became more and more oriented towards Italy. Tessa and I used the car for commuting between Totnes and Florence, usually taking the ferry from Plymouth to Roscoff and then driving through France, usually taking the Mont Blanc tunnel, making the journey each way into little holidays. In fact, we probably tried all the possible ferries and many routes across France, also across Germany and Switzerland, whenever we took a ferry from the south-east of England. On some of these journeys I would sometimes stop and do some work, by giving some seminars and lectures in Heidelberg, Scwäbisch Gmünd and other places.

The lectures and seminars I was asked to do by Giunti took me to practically all towns of any size in Italy, including the islands of Sicily, Sardinia and Elba. Some of these were courses lasting a whole week, others just one-shot affairs, the purpose being to let the local teachers know that there was something interesting brewing in the textbook area. During these activities, Tessa and I became acquainted with much of Italy and its people, most of our experiences proving very positive, although there were times when things went wrong.

We were once driving down towards Cagliari, the capital of Sardinia, from the north of the island. We thought that we had better confirm our hotel reservation, so we stopped at a public phone and did just that, saying that we were definitely coming and the hotel manager assured us that a very good room was reserved for us.

When we arrived, the manager coolly informed us, 'We are fully booked.'

I reminded him of the booking we had made and the telephone call that was supposed to confirm it, but the manager would not move. Then I had an idea.

I said to him, 'I don't believe you have paid your camorra (protection money for the 'mob')!'

'*O Signore*, I am so sorry! We have the best room in the hotel for you and we will pay up tomorrow!'

'Too late!' I said in a nonchalant tone. 'We must go. You will be hearing about this!'

'But, *Signore*, it was all a misunderstanding! Please, do come back!'

But we walked out haughtily, leaving a completely terrified hotel manager gesticulating at the front door while we calmly drove away. We had no problem in getting a very good room at another hotel. Possibly the news of our arrival had already spread through the area!

The above was merely a theoretical brush with Mafia affairs, but on another occasion we came much closer to the reality of the mob. We were supposed to drive from Cosenza, in Calabria, to a small town on the Aeolian coast. Some people we had made friends with in Cosenza told us that we should not drive there alone. We could drive in our car and they would follow, just to make sure we were all right. They took a young student with them, we supposed for helping us lift the luggage. When we arrived at the hotel, which was in a somewhat isolated position, our friends came with us into the building, while the student stayed in their car. There was again some problem about the rooms. At this point our friend gave a significant look in the direction of their car. The student was sitting there and we could all see the glint of a rifle he was holding at the ready, just in case! When the manager saw this demonstration of force, he changed his tune and offered us a room, but our friends were of the opinion that it was better to go somewhere else. So, while the student kept us covered, we walked back to the car, got in smartly and drove away.

'When we heard the name of the hotel you were going to,' said our friend, 'we knew we must come with you. That is where the Mafia meet and any suggestion of there being no room invariably means that there is a meeting that night. So we must go to another hotel!'

We thanked our friends for having thus protected us and, indeed, we were glad to have such friends who knew their way around these things! Of course, for all we knew, our friends might have been involved in a rival mob, but, there again, we shall never know! On another occasion, I was going to give a talk in a small town in Liguria; we had been told to leave the autostrada at a certain point and a police car with *carabinieri* would be at the exit to meet us and take us to the hotel. We thought this was a little strange, but then, when in Rome you do as the Romans do! It only occurred to me at dinner that evening, which was a kind of gala dinner with all the local dignitaries present, that there was something strange going on, because the same two policemen were there, mingling with the guests, but in civilian clothes. When I asked the person who had invited me to give the lectures about all that police protection, he just replied vaguely that one never knew what might happen and it was better to be on the safe side!

When I was working in Bari, we were told on no account to leave the hotel, except in a taxi or accompanied by the local authorities. Of course, Bari is a port and ports are notoriously gangster-ridden places. However, the work in Bari was very interesting, being intended to show teachers of retarded children how one could awaken what intelligence there appeared to be in such children and how one could begin to develop it. So my games on logic were very popular with the teachers, as, when they tried them with the children, they immediately realised that something in them responded to the challenge and the grey matter began to tick over faster!

But our worst experience of this kind happened in Sicily. I had given a short course to teachers in Catania and in the morning it was time to go. For some reason, on account of some strike by the petrol garages, we had to take a certain route so we could fill up and this took us right through Catania. We had stopped in a traffic jam in the middle of a square when two people came up to us, waving some cards. Tessa thought they were police and opened the door. I knew at once what they wanted – they were reaching for Tessa's bag, which was under her legs.

I shouted to Tessa, 'Look out! He is trying to take your bag!'

While one of them got hold of the bag, the other one restrained Tessa. During this 'restraining' some of her hair came out, but it all happened in seconds and, before you could say Jack Robinson, they were away on their motorbike. A few things dropped out of her bag as they ran, but the bag and its contents went with them. The bystanders told us it was a good job we had not resisted or sounded the horn, as the thieves would surely have stabbed or shot us.

I had just been paid several million lire for the work I had done in Catania, which we had split between us, for just such an eventuality. Many people came to commiserate but not one had lifted a finger to try and stop the bandits. We were too upset to go to the police station to do what they call *la denuncia*, but we did this when we eventually got back to Florence. Several days later a parcel arrived for us in Florence, with Tessa's documents, which I supposed the bandits thought were too hot to keep and make use of. They must have stuffed all that in a mail box, as the parcel came to us from the Sicilian police. The parcel contained the credit cards and other documents of another couple, to whom no doubt the same thing had happened. Since they were US documents, we took them to the US Consulate in Florence, who were at first reluctant to accept them, but Tessa just put them on the counter and left. Several months later we had a very touching letter from the owners of the documents, one of whom was a Hungarian and was going to give a lecture in Palermo!

We found out later that at about that time there was an article in *The Times* of London, describing precisely what had happened to us! Apparently, certain areas are reserved for such activity by the Mafia and woe betide anyone who interferes! So we understood why nobody had moved against the robbers.

The connection I had already established with Serena Veggetti was further strengthened through the two of us beginning to collaborate on a longitudinal study lasting five years, this being the length of time that an Italian child remained in elementary school, comparing the children in the Scuola Bechi in Florence with a control school in Rome. We examined aspects of learning such as anxiety and creativity as well as acquisition of mathematical skills. We have only recently worked out all the statistics and written a paper, which will appear in one of the European journals.

Serena was working part-time at Siena University and part-time in Rome at Sapienza University. For three years I was myself attached to the Psychology Department of Siena University, giving the course on the psychology of mathematics learning to students who were training to be teachers. As a result of this work, I wrote up a number of articles for the journal known as *Riforma della Scuola*, some on the abstraction process and some on creativity, which caused quite a little discussion amongst those engaged in tackling allied problems.

Quite an important day for me was the day I was awarded my honorary doctorate by Siena University. Ushers dressed in medieval costume announced the beginning of the ceremony by playing ancient bugles. Then I was asked to make my presentation in front of the rector and the deans of the university, which was about the conditions that needed to be fulfilled in order to generate intrinsic motivation for the learning of mathematics. After this, I was robed and I was pronounced *dottore ad honorem in filosofia*. Then there was a festive reception, at which we discussed some of the points I had raised in my address. It is true that I had done much for mathematics education in Italy but I was, nevertheless, truly grateful that this was recognised in such a splendid way by one of the oldest universities in the world.

Siena is about an hour's drive from Florence and it can be a very pleasant drive. There is a 'superstrada' which takes you there in record time, but there is an older road which is much more scenic. It takes you through the Chianti classico country, where around every corner you are driving by yet another winery. There was one which became our favourite during our years in Florence and that was the Castello Verrazzano, the owners being descendants of the Verrazzano who had first explored the Hudson River in the New World! We made friends with the manager of the winery, whose name was, appropriately, Napoleone. He always opened a bottle of wine for us and garnished it with all sorts of home-made goodies every time we came by. True, we always purchased a crate or two of this very excellent wine, but the welcome we always received was truly appreciated. The castello itself is built on the top of a high hill, where all the cellars are located and where all the ageing of the wines takes place. The road up to the castello takes you through the most romantic densely wooded country and the vines only become visible when you reach the summit, when you can feast your eyes on the rolling Tuscan landscape stretching as far as the eye can see in all directions.

There was another interesting stopping place on the Siena road: it was a tiny church! There was literally no room for more than about ten worshippers, but the priest kept the church really beautiful. As you entered, you felt immediately the loving care that had been bestowed on the place. As you stood in front of the simple altar with the cross of Jesus over it, you were enveloped in a mysterious twilight, broken only by the few candles which had been lit previously by worshippers or by the priest himself. The priest was a great lover of animals and just by the church there was an area where there were a number of animals, such as goats and peacocks, ambling about quite freely. The priest always had time to talk to us. He would tell us about his flock and how he enjoyed looking after them and showing them God's ways in his little church. Once, when we were passing, we admired his peacocks, so he went straight into his small cottage, brought out a whole bunch of peacock feathers and told us that we could have them! These same feathers still adorn a corner of our lounge in our Devon house!

While we lived in Florence, we often used to go up to Fiesole, especially when we had visitors. In Fiesole there are a number of interesting attractions. One of them is the Roman amphitheatre, which we could not use anymore for playing lions and Christians, as all our children had grown up and had flown the nest! But our adult visitors enjoyed the amphitheatre as well as the adjoining museum, in which all sorts of ancient Etruscan artefacts were on display. On top of the hill behind the amphitheatre is a very old monastery, where often we would go and contemplate, particularly in one of the small cloisters, which was beautifully tended, with flowers growing in it almost at all times of the year. We could think of and feel all the people hundreds of years ago who must have prayed, meditated and contemplated there, searching for the true meaning of life in some mystical unity with the divine. We could wonder whether we had come any further in that search than they had been able to achieve!

Then we would go a little beyond Fiesole and turn off the road on to a forest track, rutted with potholes and strewn with boulders, but taking us through the most amazing vistas of rolling, forested hills. All of a sudden we would come across a very modern-looking villa, but, even though modern, it somehow fitted in with the rolling landscape. We got to know the person who lived there. He was an architect and sculptor of some renown. In fact, he had sculpted one of the presidents of the United States of America, I now have forgotten which one, in a size several times lifesize. The statue was taken, with great care, to the United States and now it stands in the middle of a square of some Midwestern township, again I am not sure exactly where. Of course it does not matter who the president was, nor exactly where his statue now stands proudly overlooking a square. What is important is that some people felt impelled to have this president commemorated and by a sculptor who, no doubt, had known him and would give the statue true authenticity.

There was a terrace, overlooking the forest, from which there was a very big drop to some rocky ramparts below. The sculptor told us sadly that his wife had committed suicide by leaping from the terrace to the rocks below. As we looked at the rocks so far, so deep, below the terrace surface, we tried to imagine what this person must have felt, living in almost total isolation at the very end of a rutted track, even though bathed in natural beauty, perhaps deprived of the love and attention she craved from her fellow human beings or perhaps just from her husband.

On the occasion of one of our visits we found the house empty. The statue had gone, the unfortunate person who had taken her own life had gone and even the sculptor had gone. The garden was beginning to look neglected, wild flowers taking over where a formal garden had once greeted the curious visitor. We sat silently, contemplating the scene. There was nothing to say but much to feel. I wondered about the things we hold dear. I wondered about the worthwhileness of ideals when the end is just empty desolation. Is artistic creation worth the suffering it causes? Was Gauguin right to leave his family and go and paint in Tahiti? Had I been right in going to Australia against my family's wishes, in order to do something for the world? Was 'the world' not the person right next to me – in fact, for me, was it not my childhood friend, my wife Tessa?

The part of the Florence work I found most pleasant was the work with the children at the Scuola Bechi. I worked with two classes, which two teachers taught together as a team, but, since Italian classes are not very big, the two classes together numbered just over thirty children. In Italy, whenever possible, the same teacher takes the children through from grade one to grade five, after which they leave to go to the scuola media. So the teachers and I got to know all the children very well. We knew about their family situations (some of which were not too happy!) and the children considered me a friend. Whenever I arrived, they all rushed to embrace me and every single one had to be kissed separately, both at the beginning and at the end of each lesson! A scene totally unthinkable in either Canada or in Britain!

I used music a lot in the lessons, using special arm positions for the different notes. Very quickly, I was able to move my arms and the children would respond with the tune I was giving them through my arm movements. We used a lot of Italian folk songs, but also Hungarian, Irish and Scottish tunes, to which they would make up texts in ways which I have already described. Once we had an Australian visitor, who sang *Waltzing Matilda* to them and it was not very long before we had an Italian version of *Waltzing Matilda*, naturally nothing to do with the Australian text, because the words were invented by the children to express what the tune was telling them in the form of imagery.

The children were quite used to recognising imagery in their own minds generated by music to which they were listening. This was because often I would play some short pieces of music on a tape recorder and ask them to respond to it by movement and mime. I had learned this trick from Dr Klára Kokás, who had been to the Sherbrooke Centre several times and showed us in practice how these things were best done. I recall one time when I played a somewhat sad-sounding piece to the Bechi children: after a while only two children remained active, as all the others had started to watch what the two were doing. It was all impromptu and very moving. The boy was lying on the floor, 'dying' and the girl, whom we took to be acting the mother, was showing all the anxiety and grief that such a situation demanded. When the music stopped, the two children stopped moving, the girl bending over the boy in utter desolation! There was deathly silence and nobody moved or said a word for several minutes. Gradually, we all came to life and were able to talk about what had happened. The expression of emotion is all part and parcel of Italian life so the kind of experience I have just described fitted well into the Italian culture. It is much more difficult to elicit emotional responses in Anglo-Saxon countries!

The 'expression sessions' were often followed by a discussion of what had happened and how it could be described, so that the same or similar scenes could be enacted another time. This brought in the need for some elementary choreography, which was a good cue for some geometrical explorations. Apart from the conventional parts of the syllabus, I was able to amuse the children with all sorts of games involving mathematical groups, rings and fields, finite geometries and matrix transformations, which they all lapped up like kittens lapping up their milk.

When they passed on to the scuola media, they still kept wandering back to see us, for they found the higher grades very formal and very boring.

One of them came back one day and said, 'You know, our mathematics teacher doesn't really know about positive and negative numbers.'

'Why do you say that?' we asked.

The child then explained how the teacher had given quite the wrong 'explanation' about some properties of integers.

'And what did you say to the teacher?' I asked.

'Oh, nothing,' replied the child. 'He can't help it if he doesn't understand!'

This remark not only shows that the child in question had a full grasp of the concepts in question, but seems to indicate a certain moral maturity.

Once the children were discussing me with one of the teachers while I was not there, as this teacher later told me.

One of them said, 'It's much more fun when *il professore* teaches us!'

'Why do you say that?' asked the teacher without getting cross.

'Well, you see,' answered the child, 'with *il professore* you never know how things are going to turn out!'

'But,' chimed in another one, 'if we are in any difficulty, he seems to appear, have a look, make one remark and walk away and we know directly how to go on!'

The first scene shows the absolute thoroughness of understanding the children had achieved, the second illustrates what I might call the light touch in teaching. By this I mean: 'give away as little as possible.' It also shows that uncertainty of outcome is an intrinsic motivator! Not knowing how things are going to end is the basis for motivating readers to buy thrillers, so why not take a leaf out of their book and use it to motivate children to explore mathematics?

One year I spent very little time in Florence and the team then responsible for the revision of the textbooks completed their task without letting me see the result. Some of the changes I would not have agreed with and I asked Giunti about that.

His reply was this, 'Why don't you write a series in which you put everything just as you think it should be done? We could call it *A scuola con Dienes* and publish it as a kind of extra.'

I agreed to do this, and, with the aid of the teachers at the Scuola Bechi and the school in Ravenna, where I was working with my other 'born teacher', I put together a five year course in mathematics for Italian elementary school. We immediately started implementing it at my experimental schools, where the teachers could appreciate what it was all about. But it took several years of prodding to get any response out of Giunti, who eventually declared that, according to several teachers to whom he had submitted the manuscript, it was much too difficult and certainly would not sell. It could possibly be transformed into a series intended for teachers but it was not suitable for children. This manuscript is still in the hands of Giunti, and, according to our contract, he has every right to do or not to do anything with it, since anything I produced while working in Florence became Giunti property. I suppose the moral obligation to keep to the promise was of no great importance.

I also produced a book of mathematical games. This I had written originally in English, but it was translated into Italian and carefully revised by myself to eliminate any mathematical or psychological mistranslations. It was soon published in Hungarian and became a favourite book in Hungary for making mathematics teaching more lively. In fact, the Petö Institute for children with cerebral palsy is using these games to this day, as a result of some visits by myself to the institute.

I recall playing a game with some eight year old children at the institute in which, by looking at one side of a pack of cards, you could 'guess' the number on the other side; the 'guessing' involved making little piles of matches and putting them in a box, depending on the number of red and blue flowers on the card. When the actual number always came out correctly as a result of the 'match trick', one child said, 'How is it that it always comes out right?'

'It's obvious,' said another child. 'It's magic!'

The next day, as I was coming to see them again, I heard one of the children say, 'Look! The magician is coming!'

The magic of mathematics can so easily be introduced to very young children so they can enjoy the power of what it can do! I still feel that, in spite of a lifetime of efforts, I have only made a small dent in the problem!

The 'magic trick' worked like this: You pick a card and put it on the table with the flower-side facing up. Then you put five matches on every red flower and put all these matches into a box. Then you put two matches on every flower, whether blue or red. After this you remove any sets of five matches from the card. Then you put the matches that remain on the card into the box. Then you count the total number of matches in the box. This will be the number on the back of the card. You turn the card over and check.

If r denotes the number of red flowers on the card and b the number of blue flowers on the card, then the number on the back of the card is given by the mathematical formula:

$$\text{Number} = 5r + (2r + 2b) \text{ MOD } 5$$

where MOD 5 is short for: 'Divide by 5 and keep the remainder'.

Naturally, the formula would strike terror into most people not familiar with the language of mathematics, but the magic trick with the matches was easily understood and carried out by the eight year old, cerebral palsy sufferers in Budapest!

To come back to the Italian version of the games book, this book, after many years, is still sitting on somebody's desk, unless of course it has already found its way to the wastepaper basket. I have several times written to Giunti and the reply has always been that it was being looked into. I have a separate contract with Giunti for publishing the games book, but it seems that this promise is not being thought of as of much importance either.

The textbooks now being brought out by Giunti are produced by another publisher allied to Giunti and they sell much better than the Dienes series did. The new books have the outer appearance of being modern, but, in reality, they are pandering to the conservative teachers, who of course form the largest part of the market. I know that some teachers who have preserved the Dienes books prefer to spend their own money duplicating them and use them in their classroom, rather than the new ones, which they consider to be a pale shadow of what was available before. I suppose the almighty lire is the deciding factor and a poor, retired professor cannot do much to promote something that is good and exciting for children, if there are other ways in which more lire can be accumulated!

For Christmas 1985 we invited Bruce to come over to Devon, because he was in a very bad way. He had been working for Katimavik, a programme started by the government for helping young unemployed people to survive;

as soon as the Tories won the election, this scheme was cancelled and, from one day to the next, Bruce was out of a job with no notice and a big VISA bill owing! The house he was living in was to be sold, possibly on account of his having kept it so well that it had become more saleable. To top these misfortunes, his long-time girlfriend had deserted him so he was ready to do desperate things, such as going to Africa to bury his sorrows!

So Bruce came to Devon and we had Christmas together, after which he came with us to Germany and later to Florence. Coming through the San Gottardo tunnel, we found a great deal of snow lying at the southern exit. Bruce helped put the chains on the wheels, otherwise we would have been stuck in next to no time. We tried to cheer him up by taking him everywhere with us. We went on one of our routine trips to Ravenna, when, instead of going via Bologna on the autostrada, we took the road across the Appenines. As we got to the pass, again we found deep snow and the chains had to be put on again. It was snowing heavily and Bruce took quite a long time doing the job and got soaking wet. By the time we arrived at our hotel, he was feverish so we put him to bed. We stayed a few days, while I did the work for which I had come, but Bruce never ate a thing, only drinking large quantities of fruit juice. He was barely able to get out of bed when it was time to drive back to Florence. Fortunately, by then the snow had melted and we had a much easier return journey.

We talked about the possibility of buying a house in Wolfville, which we could jointly own, but, naturally, we would pay the whole amount of the price; he would then have somewhere to be and we could come over at times and share some of his life. Soon after these discussions, Bruce announced that he had better return to Canada. Neither Tessa nor I saw the reason for the hurry, but I suppose the point was that in Italy he was in a kind of limbo and it would be difficult to sort things out in such a situation. So he left.

A few weeks later we got a telephone call from him. Bruce said there were three houses on the market that would be suitable, within the price range we agreed on, but he thought there was one that was more suitable than the others. Should he go ahead? Whether it was right or wrong, we shall never know, but we told him to go ahead. We had several more calls with him, during which we discussed price; finally, we agreed on the price of one hundred thousand dollars and sent the money over. This was another *alea iacta est* (the die is cast)! We had bought a house in Nova Scotia, Canada, but we did not want to sell our Devon house, so we embarked on another commuting period: instead of commuting between Devon and Italy, we would be commuting between Devon and Nova Scotia!

At the end of the third contract with Giunti, he informed us that we would have to give up the little house that had been placed at our disposal for all these years. So we had to pack up and we knew that the Florence experience was over.

We packed up the car with what we could take with us and arranged for another van to bring more things to Totnes at a later date. We had news that Rosalind, Tessa's mother, was dangerously ill so we had to hurry back. We had booked the car on the train from Nice to Calais, so as to be quick, but when we arrived at the station we were told that we had to take everything off the roof rack and put it inside! But the inside was already full to capacity so we had to cancel the train trip and get on the road! We made it to Aix en Provence that night, where we had a brief stay and early the next morning took the autoroute north. By a miracle, we arrived in Calais that evening, managed to get on the ferry and then drove to Sussex, where our daughter Corin was then living. The next morning we made a beeline for Suffolk, where Rosalind was hospitalised in Colchester. We went to Diana's house and together we drove to Colchester.

We were in time, but Rosalind barely recognised us. We stayed with her most of the day, trying to talk to her and comfort her. She kept clutching at Tessa's hand, possibly wanting to tell her something, but she never succeeded in doing so.

In the evening, we decided to leave, as the doctors told us there was no immediate danger. Tessa was unhappy about leaving, especially since it was partly on account of an arrangement I had made to work the next day with the teachers at the school where Russell, Corin's youngest son, was a pupil. I was doing this work regularly, in return for the school having Russell there for no fees. While we were travelling on the bypass, I asked Tessa whether we should turn round and go back. She thought not.

When we got back, we were told that the work at the school had been postponed. And the next morning, we had a call from the hospital, informing us that Rosalind had died. Of course we were able to go back for the funeral, at which they played the twenty-third psalm and Tessa and Diana sadly said goodbye to Rosalind for the last time. So this was the end of another chapter in our lives.

In spite of childhood experiences, Tessa remained very attached to Rosalind all her life and her death was a great blow. The ending of the Florence connection was another blow, for we had been looked after really well there. The little house we were given was furnished with everything we ever asked for, only the best hotels were used on our trips and my pay was always religiously deposited in one of our Swiss bank accounts. As for work in England, there was the work in Russell's school, to keep him there for no fees. There was still my link with Exeter University, where the School of Education kept asking me to give seminars and lectures, but there was no financial remuneration for these. The work in the schools in South Devon had pretty well come to an end, as those who wanted to know how to handle the new maths thought they knew enough, so there was not much for me to do in this line of work. On the other hand, Wolfville was the home of Acadia University, where there might be some openings for me and the local school systems were virgin territory! So why not start from square one again?

Chapter Twenty-Six

The Devon–Wolfville Period

We bought our house in Wolfville in a street called Grandview Drive, which lies in the southern section of the town, at an elevation where it is possible to see the grand view over the Minas Basin, which is the eastern end of Fundy Bay. We have one northern window in the main bedroom and, if you strain your neck a little, you can see a part of the Minas Basin, with the cliffs of Blomidon rising up out of it in the far distance. It is not as romantic a situation as we have in Devon; it is, in fact, a rather typical suburban house of a Cape Cod type, surrounded by other houses and their gardens, all of them of fairly decent construction and style. There is a sun deck at the back on the western side and a covered veranda at the front, which, logically speaking, is on the eastern side. Many of the houses around are occupied by people who work at the university, which is only a quarter of an hour's walk away. The centre of town, where all the shops are, is about a mile away and the gymnasium, which has an Olympic-sized swimming pool, is just at the bottom of the hill. The house itself is reasonably laboursaving, all heating is electric, there is a built-in dishwasher and we soon acquired a washer and drier. Rosalind had left us some furniture (as well as a little money) in her Will, some of which we had sent over here, together with some of the antique furniture from the Devon house, so, when we eventually arrived in June, Bruce was there to welcome us, already installed in the house, with most of it already furnished.

During the first few years of this period we tried to stay in Canada for less than one-half of each calendar year, as in this way we were not liable to pay Canadian taxes. After a while, when we had begun to feel that we really lived here and went to 'visit' England, rather than the other way round, we stopped that procedure and became Canadian residents again, thus paying full taxes. It seemed unethical to get out of paying one's share by using a technicality! I contacted the Psychology Department at Acadia University. Some of its members were indeed interested in the cognitive problems thrown up by our attempts to learn such abstract subjects as mathematics. So I was elected an honorary research fellow in the Psychology Department, which distinction I retain to this day, although most of the work I have done has been either in continuing education or in the School of Education of the university, giving courses to teachers in training as well as in service courses. My attachment to the university has the advantage of letting me use the Internet, which I can contact through my own modem from our own house. This keeps me in touch with Serena Veggetti in Rome, as well as with a number of other collaborators.

I contacted the Kings County School Board and suggested that I could do work in some of their schools, on a voluntary basis. It was suggested that Coldbrooke School would be a place where go-ahead ideas on teaching mathematics might go down better than elsewhere. I made contact with the principal, who introduced me to some of the teachers. This school had classes from kindergarten through to grade eight, so there was a great deal of scope for working with a number of different age groups. I started visiting the classes and working with the teachers about two or three times a week.

So here I was, starting everything afresh, back at square one!

During this time I organised a number of courses, from continuing education at Acadia University as well as from the teachers' centre in New Minas, a nearby village. In one of the continuing education courses, I offered to visit the classes of any participant who wished me to do so at least once during the course. Most of the participating teachers wanted me to come, so I travelled extensively over Kings County, giving demonstration classes.

One course was so oversubscribed that I had to do it in two sections. One section was given on the campus here in Wolfville and on another day of the week the other section was given at the Windsor Elementary School. In this way I got to know a large number of teachers, most of whom also asked me to come to their schools, so I got to see a fair number of schools and learned something about how elementary schooling was organised in Nova Scotia. There was one particular teacher at Coldbrook School by the name Doug Holland, who got really hooked.

He really enjoyed the way I was dishing up what he had thought was 'difficult mathematics' and, after some thought, he said to me, 'Is the mathematics you are showing us what you would really like to do with us?'

I thought this was very observant of him, so I replied, 'I would like to do a lot of fun things, but I have agreed with your principal that I would keep to the curriculum!'

'Never mind the curriculum and never mind the principal!' he retorted. 'Why don't you just show us the things you would like to show us?'

'All right,' I agreed and I was as good as my word.

We played all sorts of fun games, incorporating many esoteric mathematical notions, but never actually using the correct terms. I always used playful terms, arising out of the activities in which we were engaged. The correct terms could come later! Why spoil a perfectly good game with a lot of technical terms? One of the fun games was the Sixteen Game. This was played with sixteen cards, a four by four board and a small object that was moved round the board according to the card played, each card representing a move for the object. For the benefit of any reader who also wants to play the sixteen game, here are the moves:

HOR: move two spaces either to the left or to the right.
VERT: move two spaces either up or down.
DIAG: move two spaces diagonally.
KNIGHT 1: do a knight's move but stay within the top half or within the bottom half of the board.
KNIGHT 2: do a knight's move but stay within the left half or within the right half of the board.
POINT: travel in a straight line to the centre, then go on along the same straight line and stop when you are as far from the centre as you were at the start (point symmetry about centre of board).
POINT H: point symmetry about centre of top half or bottom half, depending on which half you are in.
POINT V: point symmetry about centre of left half or right half, depending on which half you are in.
POINT Q: point symmetry about centre of top left, top right, bottom left, or bottom right quarter, depending on which quarter you are in.
POINT 1: point symmetry about centre of left three-quarters or of right three-quarters of the board, taking a path parallel to a side of the board being forbidden.
POINT 2: point symmetry about centre of top three-quarters or of bottom three-quarters of the board, taking a path parallel to a side of the board being forbidden.
REFHORPT: reflection between the first and second rows, or between the third and the fourth rows, depending on

whether you are in the top half or the bottom half of the board.
REFVERTPT: reflection between the first and second columns, or between the third and the fourth columns, depending on whether you are in the left half or in the right half of the board.
REFHOR: reflection about the line separating the top half of the board from the bottom half.
REFVERT: reflection about the line separating the left half of the board from the right half.
ZERO: stay where you are.

Now pick any of the above sixteen moves. Make your object travel across the board from any chosen spot, called the starting spot, to where it must go according to the rule of that move. Your object will thus reach the end spot. Find three cards, so that by carrying out the three moves one after the other, you can get your object from your chosen starting spot to the end spot. Do this in five different ways, thus using up all the fifteen cards.

At one point we pushed sixteen tables together to make a board and let the children walk on the tables to carry out the moves. Several children could do the walks simultaneously and after each move, assuming no errors are made, any two children would remain separated by the very same move, meaning, for example, that if Jack and Jill could change places by performing a certain move, they would still be able to change places after any number of collective moves, using the same move as before.

Another interesting arrangement poses an even more challenging problem. This is what you do: Place one card on the table, then place the remaining fifteen round it in a circle, but in such a way that, in any five successive cards, the effect of carrying out the moves of the first one, the

KNIGHT 2	REFHOR	POINT
VERT	POINT H	POINT Q
POINT 2	REFVERT	HOR
ZERO	REFHORPT	POINT 1
KNIGHT 1	REFVERTPT	POINT V

DIAG

second one and the fifth one should have the same effect as carrying out the move indicated by the central card.

For any doubting Thomases, here is one solution:

Central card = VERT, around which you place the others thus:
ZERO ➡ DIAG ➡ REFHORPT ➡ KNIGHT 2 ➡ HOR ➡ KNIGHT 1 ➡ POINT Q ➡ REFVERT ➡ REFHOR ➡ POINT 2 ➡ POINT 1 ➡ POINT H ➡ POINT ➡ POINT V ➡ REFVERTPT and back to ZERO.

The geometrical aspects of the game should be clear even to a non-mathematical reader and the fun aspect will be appreciated by actually playing the game! Cards could be made which are more descriptive than mnemonic, so it would be quite obvious what kind of move each card told you to do.

Here are some suggestions:

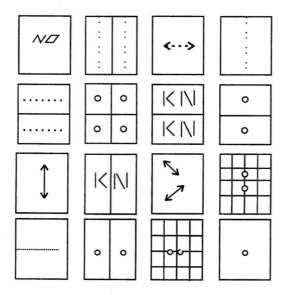

the dotted lines indicate 'mirrors' in which the movable object has to be 'reflected' and the Os indicate the points about which the 'point symmetries' must be carried out. When the card is halved or quartered by means of solid lines, it means that the movable object must not cross the solid lines. This of course does not apply to the two cards with the Os in the centre of each three-quarters of the card! I have given this much detail of this game to enable the reader to construct the game and play it and thus enter into some of the situations I have tried to create for children in the various parts of the world described in these pages!

Trips to Europe continued unabated, with a regular rhythm of two trips a year, one in late spring or early summer and the other during November or December. The first one of these trips took us to Spain. We went to England first, where we stayed in our Devon house and socialised with Chris, our neighbour, who had become a really great friend. We had similar tastes in music, art and literature and we had both raised or were raising five children, which also contributed to the bond established between us. Chris very kindly kept an eye on the house for us while we were away, forwarded our correspondence and made sure the gardener kept the garden in shape. And whenever we arrived, if it was winter she had the heating on in our house, had aired the beds and got rid of the woodlice, which kept creeping in through all sorts of impossible entrances known only to small insects!

I also worked with the teachers and children at Greenfields School in Sussex, to 'pay' Russell's fees there After visiting our daughter Corin, who had changed her name to Jasmine on becoming a Scientologist and renewing our relationship with her twin sons and with Russell, we were ready to fly to Spain. The first port of call was Barcelona, where we were greeted by our old friend the ex-priest, Ricardo Pons, and I was introduced to the teachers with whom I would have to work for a week. It was a long time since I had worked in Spanish, so I was a little apprehensive about communicating with the teachers.

'Look here,' one of the teachers said to me. 'Why don't you speak to us in Italian, but slowly? Every Italian word is either a Spanish word as well as a Catalan word and, since we are all bilingual, we should have no difficulty in communicating. You obviously understand us when we speak in Spanish so you speak in Italian and we shall speak in Spanish!'

As the course progressed the Italian was used less and less and the Spanish more and more, until, by the end of the week, I was able to give press interviews in Spanish; obviously, I was not going to have any more language problems for the rest of the trip. This was just as well, as the next port of call was a small country place near the Portuguese border called Caceres, where nobody would have understood any Italian! The hotel in which we were placed was very pleasant. There was a very scenic courtyard bespattered with palm trees, where we had our meals to the sound of soft, Spanish music. We found it a little hard to get used to the arranged schedule. I had talks to give in the mornings and then at nine o'clock in the evening. After the evening sessions, it was not until around midnight that we sat down to our evening meal in our romantic courtyard. A film crew was using Caceres as their location for shooting a film and the producer, the director and a number of the actors were staying at the same hotel, which was very interesting but it meant that we were never in bed before about two or three in the morning. By half past nine the next day, I would have to be fresh enough to interest teachers in some abstruse aspects of the problems of mathematics teaching.

Another knotty problem was to do with transport. There was no airport at or even near Caceres, so we had to travel by train from and to Madrid, which meant a trip of several hours in each direction. Needless to say, there were no refreshment cars on such local trains. So, in order to get a drink, I had to get off the train, buy a drink at the bar and rush back to the train as it was already moving off with Tessa in it! Another hardship was that, at least on the way back to Madrid, we had to change trains. Getting off a Spanish train with a certain amount of luggage is only surpassed in difficulty by getting on to the next one. One had to negotiate steep steps to reach the ground while holding on to a rail with one hand and a heavy case with the other. Imagine doing the reverse, when, instead of being aided by gravity, one is definitely impeded by it!

We spent only two days in Madrid. By this time I was quite fluent in Spanish, remembering to lisp my Cs and Zs, unlike South American Spanish! I had made some transparencies for my talks, which went off without a hitch. Apparently, my spelling and grammar on the transparencies were also correct, as many people complimented me on how well I could speak and write Spanish and asked how long it took me to study it. I had not the heart to tell them that I had never studied it, but had managed to pick it up along the way!

The day before we were to take the plane back to Canada, we just missed an ETA bombing at one of the intersections in the city, which our car had to pass, as it was between our hotel and where I was giving the lectures. Tessa and I decided that it was indeed time for us to leave! Our flight was very comfortable, as we were travelling business class (one notch down from our previous first-class trips on our many world trips!) and the hostesses were very kind and helpful. We broke our journey in Montreal and went to visit our son Nigel, who had moved out to the depths of the country near Lake Memphromagog and had a thriving psychiatric practice there. He and his wife already had three children, two boys and a girl, who seemed to like the wilds of Quebec. Tom, the oldest, was very keen on nature, as his grandfather had been. He would take us into the nearby forest, part of which was their land and showed us many rare plants and flowers. After a few days with Nigel's family, Nigel took us to the airport and we flew to Halifax where Bruce met us at the airport and drove us back to our little Cape Cod residence in Wolfville.

Bruce, being a very active Quaker, got together a fair number of people from around the 'Valley' who were to form the nucleus of a new monthly meeting. This happened quite soon and we transferred our membership to it. The meeting in Wolfville is a small one but lively. The hour of worship is nearly always silent, unless Tessa or myself become minister! The meeting takes part in interchurch activities and engages in various projects to help those less fortunate than ourselves.

During our first year in Wolfville we were in the house with Bruce, sharing it with a student, so as to have some company for Bruce of someone nearer his own age! He was still smarting from being abandoned by his girlfriend, who in the meantime had married someone else. It was a Quaker wedding and we all went to it. It was held outside in the open. Bruce was glad that we all went since he was keen not to have any hard feelings. He even asked the new husband to do the work of finishing our basement, where he had already built a darkroom; in this way we had a proper storeroom, as well as an extra room for our use.

Bruce suffered another tragic event: a girl he was contemplating marrying committed suicide. She was apparently schizophrenic, but, with medication, was able to live a normal life. At one point, however, she decided that she was not going to be dependent on medication, that she could handle life on her own. It turned out that she could not and this was a great blow to Bruce. Eventually, he began to be friendly in a romantic way with another friend he had known for quite a while, whose name is Gwen. One Christmas we invited them both to spend some time with us in Devon.

One problem for Bruce was that Gwen wanted to spend time in India before deciding whether she wanted to throw in her lot with Bruce. This seemed hurtful to Bruce and very strange to us. But then, modern ways are different from what our ways used to be when we were young and we thought we should be tolerant about these things.

As it turned out, when Gwen returned from India, she decided that she was ready to marry Bruce. Gwen is from British Columbia, where her family still live. So they decided to have two weddings: one in British Columbia and one in Wolfville. The Wolfville one was a Quaker wedding and all Bruce's and Gwen's friends were present – it was a great crowd! By this time they had decided not to live in our house, as they wanted their own pad and they rented a large farmhouse not very far from Wolfville, next to a small lake. It was a beautiful modern house, with a view over Wolfville and what they call the South Mountain, the lake in the foreground. We often went to see them there; in summertime we swam in their lake and in wintertime we used skis for the last part of the trip to the house.

Bruce decided to do psychology, for he could not see himself as a photographer for the rest of his life. So he took courses at Acadia for two years, until he got enough credits to have the equivalent of a degree, after which he was accepted as a graduate student for a masters cum doctorate programme at the University of Illinois at Urbana, where he is at the time of writing, having already obtained his masters and now he is about a year away from completing his doctorate. So Bruce and Gwen now live in Illinois; they come back to Wolfville in the summer and sometimes for Christmas. This year we are expecting him for the long weekend of American Thanksgiving and hopefully also for Christmas. So our move to Wolfville as a 'rescue Bruce operation' has, in fact, resulted in his rescue in the shape of a marriage in which they are both happy and in the shape of a career in psychology to which he is looking forward.

The fact that we are left high and dry without him in Wolfville is beside the point! There are a number of quite beautiful areas very near Wolfville, which I was determined to explore. One of these is Lumsden Dam, which is an artificial lake, but one can hardly tell that it is one, except when going very near the actual dam. In the summertime the water is very warm and many people from Wolfville and other places nearby come to sun themselves and swim. I used to make a point of swimming right across the lake and back, which meant a total of over eight hundred metres, but, when I later developed angina, I was told quite severely by the doctor to swim in the pool only when there was a lifeguard in attendance! Another beautiful walk is the one up to Cape Split which is a tongue of land sticking into Fundy Bay, surrounded by precipices (assuming you are on top!) except at one point, where you can access it. It is about a two hour walk from the nearest road, through a forest track and, needless to say, it is uphill all the way! The good part is that it is downhill all the way on the way home, when you are likely to be more tired. Bruce and I and some friends made the trip several times and I really enjoyed both the forest walk and the glorious view from the top, with water nearly all round, the northern shore of Fundy Bay looking at you mysteriously from the dim distance.

Another fun excursion from Wolfville is what they call the 'three pools'. These 'pools' are surrounded by rock faces connected to each other by waterfalls. To reach the upper pool a fair amount of scrambling over rocks is necessary, some of it over somewhat hair-raising slopes, where one slip would mean a fall of more than one hundred metres. But the upper pool is worth all the effort.

Bruce and I once went up there by ourselves and, finally having reached our destination, I said to him, 'Let's jump in the pool and swim over to that waterfall and have a shower!'

'Yes, let's do that,' replied Bruce enthusiastically.

There was nobody else around so we both stripped naked and jumped in. The water was quite warm and we did manage to swim across to the waterfall and even climb up into it.

'Why not take a picture?' I asked him.

Bruce swam back to where our clothes were, took the camera and swam with it, holding it up high out of the water and, while treading water, took a snapshot of me standing in the rushing waterfall in my birthday suit! We have this picture in one of our albums, but we often turn two pages over 'by mistake' if we are showing the pictures to a more conservative friend!

For several summers running, Nigel and his family would come over to Wolfville for a holiday. They would drive through the Gaspey and Northern New Brunswick, making a holiday of the journey as well. They would rent a little, family cottage at a nearby motel, where there was a pool. The place was also quite near Evangeline Beach, which they liked to explore. We used to go together to the South Shore and show the children the lovely sandy beaches there, as well as some of the picturesque fishing villages. In the evening, the whole family would crowd into our little house and Tessa would lay out a spread on a large table on the sun deck, where they did not have to be so careful. By the time these trips took place, there were four children! As I am writing these lines, their respective ages are twelve, eleven, ten and eight, the ten year old being a girl, the others boys.

Travelling about on holiday with such a brood must have taken some doing! When we did our mad holidays with the children, it was easier for us, as our children came in two 'batches', the first three came within five years and then there was a six year gap before the second batch of two came along.

We also used to go and visit Nigel and his family in Quebec. The journey took about two days, as the journey each way is 1200 kilometres. Sometimes we would take the route via Fredericton and Rivière du Loup and other times we would drive through Maine, which was much more scenic, as it took you through the Sugarloaf country. If it was summer, we would walk in the woods; if it was winter we would put on our cross-country skis.

One day we got the very disturbing news that Nigel had cancer of the oesophagus and had to be operated on almost immediately, otherwise he would live only for a month or two. With the operation, the surgeon gave him a year. This meant more frequent visits to Nigel's family. They had moved into Granby because of the children's schooling, where they had bought an old but rather picturesque house, with quite a lot of land, but of course nothing like the country place they had. They kept the country place on for weekends and holidays and Nigel had part of his practice there and another part in Granby. While he was in hospital for his operation, Karin was with him just about twenty-four hours a day and of course we had to cope with the children. We shopped and cooked and we saw to it that the children went to school and generally looked after them, to free Karin completely.

When Nigel came out of hospital, after some convalescing, he seemed much better. He was very keen on the spiritual force behind prayer and was convinced, even though medically he realised he was a terminal case, that he could be healed by seeking help directly through prayer. Every evening Nigel, Karin and all the children had a serious prayer session, during which they prayed for deliverance and healing. Nigel was able to continue with his practice, albeit on a reduced scale. Once he was called back to Sherbrooke to the hospital for a few days, yet he managed one day to get a day pass, go to his country house, treat a psychiatric patient and then drive himself back to the hospital! But he clearly became less and less able to do his work and we began to have to help him and his family not only by coming to help with the children but by putting considerable sums of money into his bank account. We all tried to be positive with him: he never wanted to hear of any possibility of not being healed!

During these visits we also took the opportunity to see the friends we had made in Montreal and Sherbrooke. We never missed seeing the Wilsons, whose family had of course grown up and they were having grandchildren to visit! We talked about the old times when we went out cross-country skiing. They remembered the twins from when they were only about ten years old and could not believe that they were both grown up, had jobs, were married and were raising families! We could already brag about our great-grandchildren, whereas all they had to show us were mere grandchildren!

I arranged eventually with the School of Education of Acadia University that I should be regularly responsible for the mathematics education course for one class, the professor taking the other class having attended a number of my continuing education courses, so he was in sympathy with my ideas. We decided to 'team-teach'; in other words, we put both classes together and worked with the whole lot as one group. We even hired some mathematics students who would do tutorials with them in the afternoons and evenings, so they could have enough hands-on experience with materials which we provided. We also made some videotapes for the students to watch during these tutorials; my colleague, Norm Watts, and myself making the videos during the weekends, with the children of our various friends and acquaintances.

I also made a contract with the media services of the provincial government and with their collaboration made five quite exciting sections of videotapes, using the concept of ratio as a recurring theme. I managed to include some of my fun games in these videos, such as the one about the roads, fields and houses as well as the one about measurement based on a child's walk to his friend's birthday party. So I felt that at least some of these fun activities would be left behind after I was no longer on the planet. I tried to schedule all these activities to leave time for any work in Europe, which, as it turned out, was to be had mostly in Italy. The mathematics education course would finish some time during April and we could take a plane to England and then use our car to travel to Italy.

The work in Italy was divided between work in Emilia-Romagna and work in the south. There were two books available for such courses, arising from some previous work I had done for the region, published by the Bologna publishers Cappelli. Then there was also the *Piacere della matematica*, also brought out by Cappelli, which gave an account of many amusing ways of dealing with the cube, as well as work on intrinsic motivation, using the game so much enjoyed by the Follonica children some years previously.

In addition to these, the AIAS of Cosenza (Associazione Italiana per l'Assistenza agli Spastici) having translated the notes I was using for my Acadia students, published them in a restricted private edition. So there was enough fairly recent literature available, which I supplemented by sending other materials from England, Canada and Hungary, so there was no shortage of hands-on activity available to the participants of the courses. At first the work was concentrated in Puglia, in particular Bari and Bisceglie, where there was much interest in how to improve the intelligent functioning of children and adults who showed signs of retardation. A little later the Cosenza AIAS became interested in having me over in person, which they did several years running. This included my giving lectures to diploma students who were studying at the AIAS Centre in Cosenza, as well as courses in neighbouring areas.

The first year Tessa and I were put in a hotel for the duration of the work, but later we were offered an apartment in the building where the courses were to be held, together with an offer of taking us shopping to buy anything we liked in the way of food and wine. So we opted for the apartment, as a steady Italian diet was not likely to be as good for us as what we wanted for ourselves. Tessa was taken shopping once or twice a week; she chose what she wanted and AIAS paid the bills. All I had to do was to take the elevator and go down to the basement where my students were waiting for me and, while I was working, Tessa would prepare the midday or evening meals. If she did not feel like cooking, there was a little restaurant up the road where we could eat very well and, again, AIAS picked up the tab. My honorarium was also very good and they paid the Italian taxes due on such honoraria. So it seemed a very good arrangement, which lasted for several years.

The only trouble was that the courses became too popular and the local operators, who were also supposed to give in service courses to teachers, did not like it. So the provveditori, who were the government's appointed watchdogs for the area, decided not to give permission for teachers to be absent from school to attend my courses. Some teachers countered this by ringing in sick and still came. But, since you can't beat City Hall, as they say, the number of actual customers who were allowed to come dried up and as of now, no further AIAS visits are envisaged.

During this work, we made friends with some of the participants, who would sometimes take us to their homes or out to meals in nice places. One such family was the family of a doctor, whose wife attended one of my courses. They had a little boy, with whom we also became very friendly. One day this family took us to a place quite a long way out in the country and we met some of the doctor's friends, who treated us to some very interesting things to eat. We really enjoyed the relaxed atmosphere which is so common in Italian households, especially after lunch, when a certain amount of good wine has been consumed!

The following night I began to feel really sick and noticed that my heart had gone into a serious arrhythmia. I could not even take my own pulse, it was so irregular. This was accompanied by strange feelings to which I was not accustomed. Was I having a heart attack? Tessa immediately phoned our doctor friend. It was not long before he arrived, together with a cardiologist, who examined me. By this time the arrhythmia had subsided and they both assured me that any danger was now past and that I could go to sleep. But they said that I should come to the hospital and have myself checked over. The problem was that the next day I was supposed to be going to give a week's course in another area of Calabria.

In the morning we packed up the AIAS car with mathematics materials and drove to the hospital. I was given every possible test and told that what I had felt had not been a heart attack or it would have shown up on the electrocardiogram, which it did not. Our doctor friend thought it would be wise to stay in hospital for a couple of days for observation, but I said that I was feeling fine, returned to the AIAS car and we drove off!

When we finally arrived, several hours late, the teachers were still wistfully waiting, sitting in a room, hoping that I would turn up! I thought it would be churlish not to do at least a little with them, so, notwithstanding great objections from Tessa, I went in and gave them an hour's pep talk on what they were going to learn during the coming week. The course was a great success and we made up the difference stemming from my late arrival by having an additional final session on Sunday morning. Every single teacher turned up for this last meeting and we were showered with gifts of all sorts to take away with us! So my 'fake' heart attack, happily, had not stopped me from giving this service to this small group of teachers, who were all teaching in little country schools and were so keen to do a good job for the children in their care!

Our doctor friend had invited us so many times to supper at his own place as well as to restaurants that we thought it was time we returned the invitation. So Tessa and I asked them over to our top-floor apartment in the AIAS building for a meal. The meal was prepared meticulously by Tessa and was enjoyed by all. Giovanni, the little boy, said that he would do the dishes, which he did and very nicely too, but neither he nor we realised the catastrophe that would result from his kindness!

Tessa had placed her dentures in a piece of tissue paper and, the morning after our guests had departed, she could not find this piece of tissue paper, so we thought maybe Giovanni had thrown it away. We went to look in all the rubbish, even that down below on the street, as our factotum had already removed what we had placed outside our front door. I found nothing. But of course I am probably the world's worst finder of things when it comes to looking for something. Tessa is sorry to this day that she did not check herself whether the tissue paper with her dentures in it was or was not in the stuff that had already been removed from outside our door!

We were directed to a dental technician, who said that he could make new dentures for Tessa but of course at some horrendous price. We had no option but to accept! There were, in fact, only a few days before we were to depart from Cosenza, so time was of the essence. We were taken to the technician by an AIAS car during a time when I was not working.

While we waited, we were asked what we would like to drink. I replied, '*Una birra per la signora e un Cinzano rosso per me*' ('A beer for the lady and a red Cinzano for me'). We were amazed to realise that the drinks were on the house!

The technician turned out to be a very amiable person, and, after several visits to him, we became very friendly and he invited us for a meal to his house. He had a beautiful house a little higher up in the hills from where the AIAS Centre was, standing on several acres of ground, over which he took us for a walk. A little way away was a very old-looking house, in which his mother still lived. He said to us, '*Sono nato in questa casa, e mia madre non vuol andare altrove. Vuol rimanere sempre nella vecchia casa!*'('I was born here and my mother doesn't want to go anywhere else. She just wants to stay in the old house!').

We had a chat with his mother, who was happy enough to live as she had always done, doing things for herself in the old ways, without any modern conveniences. How wise of her son to let her stay there and be contented.

The technician had a little boy of about eight, who seemed very grown-up. He showed us a picture of a pretty little girl, saying, '*Guardate, la mia fidanzata!*' ('Look, my fianceé!').

'*Una bella bambina!*' I ventured to reply. ('A lovely little girl').

'*Niente bambina!*' he said quite sternly. '*Una ragazza!*' ('She's no little girl! She is a girl').

Then he took us to the top of the hill and pointed to a house on the other side of the valley, saying, '*La sua casa sta lì, in quel bosco!*' ('Her house is there, in that wood').

He was very serious about his girlfriend and was quite certain that he was going to marry her! Of course we should not have been surprised, as it seems that a lot of Italian children, starting in grade one, each have a *fidanzata* and the teacher often has to wipe away bitter tears when such relationships break up, as they inevitably do! I had seen that happen a lot in the Scuola Bechi in Florence in my experimental classes!

Although we had to pay several million lire for the dentures, which were beautifully made, we had made some new friends, which was a good feeling.

One of our Totnes friends, Vlin, had an Italian friend, Nicola, who used to come to Totnes quite frequently, but lived in Matera, in the south of Italy, in the area made known by Carlo Levi's famous book *Christ Stopped at Eboli*. So sometimes, on our way down from Ravenna to Cosenza by car, we would stop and stay with Nicola in his flat in Matera, which overlooked the strange old buildings, constructed on hills, of this ancient town. During one of our sessions in Cosenza, he called us and invited us to the wedding of a friend of his. Weddings in Italy are enormously long and elaborate affairs, we had been warned. The wedding was to take place on a Sunday, so we could come from Cosenza by car on the Saturday to Matera, which was a three hour run, if you are lucky not to be stuck behind a slow truck on the part of the trip which is not an autostrada. We duly arrived on Saturday afternoon and, as usual, had to totally empty the car, otherwise its contents would no longer be in the car by morning!

Nicola then took us for a walk around what they called the *Sassi* or the Rocks, where there were caves in which people used to live not so long ago, together with their animals. There was even a cave church, with quite old frescos painted on its walls, still used for celebrating mass on Sundays. Having done all the above touristy things, we settled down to a meal. It consisted of lots of wine and a dish of wild asparagus, which Nicola's wife had been out to collect while we were seeing the sights of Matera!

The next day was the wedding. The wedding dinner was so abundant that I could not understand how any guest could possibly eat even a small fraction of it. A normal Italian meal consists of an *antipasto* (*hors d'oeuvre*), a *primo* (soup, pasta or seafood), the *secondo* (the main dish), the *dolce* (the dessert) and the *frutta fresca* (fresh fruit), all accompanied by different wines for the different courses, finishing with coffee and liqueurs. In this more festive repast, every course was triplicated, not in the way of 'either… or' but in the way of 'as well'! So by the time they started serving the main course, we had already eaten six courses! After picking at the first of the three *secondo* dishes, we really had to excuse ourselves, saying that we had a long way to drive and that I would have to be working the next day!

Through Nicola and his wife, the news got round that I was giving courses in the south so I had some enquiries whether I could do some in Matera. The enquirers were the organisers of a co-operative established by teachers themselves to improve their teaching, so they could not offer as good terms as AIAS but I thought it was a worthy cause, so I agreed to do a few days at the end of the Cosenza work, before driving back north. The course was to be held out in the country, in an old baronial hall which had been converted to educational uses. The road leading to it was dusty and rutted and I began to wonder if the car springs would stand the punishment when the huge building loomed up before us, with enormously high walls around it, supposedly to protect the place from bandits. We were shown to a kind of dormitory, where Tessa and I had to make our beds, youth hostel style and cook our food for lunch, the ingredients for which had been left in the kitchen by the 'co-operators'. In the end it was all worth it, as the participating teachers were extremely enthusiastic and enjoyed actually making the materials, of which they could instantly make use. The teachers taught during the day and would come in the late afternoon and evening to participate in the course, which sometimes did not end till nine o'clock in the evening.

Several of these trips to the south were accompanied by work in the Romagna area as well as in Rome. The Rome work was organised by Serena Veggetti, some of it at the university, some in a nearby school her daughter Livia attended. I learned afterwards that their favourite game was what they called the *gioco del pulman* or the bus game! It was played something like this: we arranged eight chairs in four rows of two, which must have been what gave the children the idea of a bus and sat a child on each chair. The 'bus' was divided into halves in three different ways, each half having a name. A diagram of the bus is shown here.

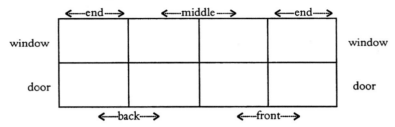

Each child had three 'names', depending on where he or she was sitting. For example the one on the top left corner of the diagram would be called 'back window end' and so on. At the start of a game all the children were either sitting or were standing, then there were three commands that a ninth child would give, to make some of the children stand up and/or sit down. These were:

'KEEP', 'ADD' and 'EXCHANGE',

to which you had to add just one of the six words

back; front; door; window; middle; end.

For example 'ADD front' meant that those 'front children' who were not already standing would have to stand up. Or 'KEEP door' would mean that out of the children now standing only the 'door children' should remain standing. 'EXCHANGE window' meant that any 'window children' now standing would sit down and any 'window children' now sitting down would stand up. Any 'door children' would stay as they were, whether standing or sitting.

A given group of children was selected, who must be left standing at the end of a series of commands, while all the others not so chosen must be sitting. The game consisted of getting just the chosen children to stand in the least number of moves. This was a logical game, as any mathematician will at once realise.

For those not versed in ideas mathematical, it is perhaps worth saying that the 'ADD' command introduced the logical function of 'either... or', since the result of the command was that any standing child was either already standing before the command or has just stood up as a result of the command. The command 'KEEP', on the other hand, introduced the logical 'and', because the result of a 'KEEP' command was that, if you were standing after the command, you must have been standing before the command and you must have remained standing as a result of the command. The 'EXCHANGE' command is a little harder to explain, as it represents the logical function 'If and only if not... then', which is rather a mouthful if you are not used to such logical contortions in your everyday life!

The work in Romagna was organised on two occasions by the Garzanti Foundation. This foundation was not bound by ministerial guidelines about honoraria, so was able to pay me as well as AIAS. The third time it was organised by the teachers themselves, so the hotel accommodation was less luxurious, but the teachers' interest in the work made up for that very well! On all three occasions there was work in the local schools, with the children and the teachers, thechildren's positive reactions to my suggestions always being the most powerful argument in any ensuing discussion with the teachers.

On each of these occasions Tessa and I visited our old friend Ermanno Pasini, enjoying his excellent wines and his wife's brilliant cooking. It was on one of these occasions, when we were on our way to Budapest from Cosenza, that we found a telephone message from Karin, Nigel's wife, waiting for us at Pasini's house: Nigel was very ill and we should come right away.

We cancelled the Hungary part of the trip, drove straight away to England and flew to Canada. We went to Wolfville first and called Nigel on the phone. Apparently, things were not as bad as they had thought: there would be time for us to drive over in our car. We stayed for two weeks and tried to look after the four children so Karin could concentrate on Nigel.

As it turned out, these were the last two weeks of his life, as the cancer from which he was suffering, having started in the oesophagus, had spread to the stomach and there was not much anyone could do, except pray for a miracle, which we all did. The miracle did not happen. One evening Nigel ate some fish, which he had been told was not suitable, while he was celebrating his fiftieth birthday. He had all his presents from the children, but soon after he had great pains and was taken to hospital. He died the next morning.

The custom in Quebec is to have a 'viewing period', when friends and relatives can pay their last respects. We came of course and so did the children, who were told that Daddy was now in Heaven, which they truly believed as they had been raised as believing Christians. One of the children, Dan, remarked, 'It's a pity Daddy had to die now, just as he was beginning to like me!'

Nigel had a Catholic burial, at which we all said goodbye to him.

Nearly all our later trips to Italy were also accompanied by a trip to Hungary. At first we used to drive through Yugoslavia, as both Pécs and Szeged were in the south of Hungary and it was shorter than driving through Austria. On one of these trips we managed to sell a number of garments we had picked up at a church sale in Florence, but of course we had to use the dinars for buying something else.

We stayed the night at Szabadka (Subotica) and asked at the desk, 'What time do the shops open?'

'At five in the morning!' was the surprising reply.

This was indeed a 'workers' republic' if work started that early! We rose very early, for later that day I was due to give some lectures at Szeged University and went to spend our dinars, saving enough to pay the hotel bill. We found some interesting-looking rugs, which we bought, thus having turned some rather indifferent clothes into hand-made rugs from Bosnia! We packed everything in the car, paid the bill and drove to Szeged.

Sandor Klein organised some of the work on these visits, which included working with teacher trainees at Szeged and at Pécs Universities, as well as work in some of the local school systems, not to mention work with my 'saint' in Budapest, who always welcomed me as the original teacher of all the good things that came out of the Trojan Horse of new mathematics!

It was on these occasions that I also worked at the Petö Institute for cerebral palsy children, now world-famous for achieving improvements with such children, which the medical profession regard as impossible. During one of these visits we were taken to see my mother's birthplace in Szekszárd. In that town, where my mother was born in 1879, a school had been named after her, and every year, on my mother's birthday, the 25th of May, the children do something to celebrate the occasion. I could never be there for those days, because it was only then that I could do my work in Cosenza, but in early June I could be there, a little late, mingle with the teaching staff and talk to the children.

The principal and the mayor of Szekszárd and I had a meeting at which we decided that it would be a good idea to acquire my mother's house of birth, as it was then for sale. The idea was to start a centre where all the Dienes books and materials would be collected together and eventually it could be used as a cultural centre for spreading abroad some of the contributions of the Dienes family to general culture. This would include my mother's work on dance, Gedeon's work on linguistics and my own work on the psychology of mathematics learning and mathematics education in general. The house has been acquired, but, as yet, not enough funds have been subscribed to make the centre operational. I have offered to put all my mathematical games at the disposal of the centre, which they could then market and sell and the profits could be used to help run the centre. At the time of writing it is not yet certain whether the above will become a reality, but we are all hoping that soon there will be such a centre operating from the house where my mother was born.

During our visits to Devon I was always keen to swim at least two or three times a week. The pool was less than a mile away, so it was not too hard to accomplish. During one of my swims, after the seventeenth length, I began to feel bad, but, rightly or wrongly, I finished twenty lengths, which was my regulation swim. I could hardly get out of the pool. I dressed with difficulty, drove back with even more difficulty and collapsed on the divan. Tessa called the doctor, who diagnosed another arrhythmia, possibly angina. He put a nitro patch on me and told me to keep quiet and rest.

The angina diagnosis was eventually confirmed, but in a couple of weeks' time I was deemed fit to travel back to Canada, which we did. I saw a cardiologist in Kentville, who kept me in hospital, just in case and I was treated very gingerly as a 'cardiac patient'. It took quite a long time for the doctors in Nova Scotia and the doctors in Totnes to work out a regime which would keep my blood pressure down as well as handle the angina. I was given an exercise programme, which I keep to to this day, involving three swims a week, which can be anything between one hundred metres and five hundred metres and a walk round the university nature trail on 'non-swim days'.

Soon after my first angina attack, Sara, who had been living in Montreal, decided to come and live in our house, so there would be somebody present in case of emergencies; also, she could be of help to us doing things that we were both beginning to find harder to do, such as lifting heavy objects. In the winter she has been coming with me on cross-country skiing trips, which in the winter replace the walks in my exercise programme. This was good, just in case my heart did not behave itself on one of the ski trips! We also made a ski trail at the back of the house, as well as round the house and obtained the neighbour's permission to run down at the back of their lot to the empty lot at the bottom of the hill. This gave us quite a nice hill and going up and down that hill about a dozen times was sometimes my ration of exercise!

A few months after Nigel died, Karin sold both their houses and moved over to Wolfville. She has a house about one kilometre away from ours, so we can visit mutually without much difficulty. In the winter some children have even skied over to our house! All the children go to a Christian school, about twenty kilometres away. Karin is very keen that the children should receive a thorough spiritual and moral foundation in their schooling, so she thinks it is worth all the trouble arising from the school being so far away. The children like the school, although they took a little while to learn to read and write in English, as they were attending a French-language school in Quebec.

This is as far as I can go now, since I cannot foretell the future. Tessa and I are still commuting between England and Nova Scotia and we have some hopes of more work in

Italy. Bruce and Gwen are in Illinois and we see them for brief periods. Sara is with us, helping us manage in this last lap of life's journey and we see Jasmine and her six descendants twice a year.

Postscript
November 1997

We have had some sad times recently. Russell, our daughter Jasmine's third child, was killed in a car accident. Diana, Tessa's sister, died of breast cancer. Life seems to be very easily extinguished and at my age now, of eighty-one, it makes me think that one has to begin to concentrate on 'things eternal', since, as it is very wisely said; 'you can't take it with you'! Each morning now when I wake up, I thank God that He has allowed me to wake up and to undertake the labours of another day and I ask Him to give me the strength to do His Will.

We finally sold our house in our romantic Devon and sent the contents to Canada. Tessa was very sad to do this, as she felt cut off from her homeland and could not feel Canadian enough to think of Nova Scotia as her real home! We were both sad to leave our neighbour and good friend Chris, but we promised that we would go on seeing each other in spite of the big pond between us! We did invite her to come over to stay with us, sending her a ticket as a big thank you for all the kind things she has done for us. We had a good time together, walking round the nature trail at times and generally chatting together, as friends would or should. She was even here for the 'last snow' and managed to snap a picture of my skiing down the hill at the back of our house!

Tessa and I have been finding it difficult to express our feelings of friendship towards Chris, since the 'stiff upper lip' idea is still very prevalent in England. We eventually wrote her what we thought was a good compromise between too much emotional content and the restrained British attitude; Tessa suggested that one way she could express herself was through dance! But how can you dance across the Atlantic? So Tessa thought of putting on a very dance-like dress and a long white chiffon veil and going out into our garden. She did just that and I followed her round with a camera as she performed her 'friendship dance'. Then we picked what we thought was the best of these and sent them to Chris!

Let me now just bring my story up to date by telling what has happened to us in the last few years. In 1995 we managed to do a European trip, which included some work at Russell's school in England and a visit to Chris and to our Devon friends, not to mention the momentous trip to Exeter University, during which an honorary doctorate was conferred on me by that university! Paul Ernest pronounced the eulogy, using some amusing facts taken out of these memoirs! It made me feel as if I were practically being canonised!

In 1996 Tessa and I went on quite an extensive European trip, which included Hungary, Italy and Greece. Together with my brother Gedeon, we made a TV programme of our life as children and of my later life, dealing mainly with my activities in trying to improve the mathematics education situation in different parts of the world. Ludovico Geymonat's son, Mario, invited me to give a talk at the University of Venice. I also renewed contact with Ermanno Pasini, who came to Venice to see me, as well as a teacher of the name of Sambim, with whom I used to work in one Venice elementary school. While Tessa and this teacher and I were walking through the narrow streets of

Venice, a young man rushed up to us and said to me, '*O Professore! Come mai La trovo qui a Venezia?*' ('Oh, Professor, how come I find you here in Venice?').

Apparently, he remembered me from when I was working with his class, when he was a small child!

I was very moved that I could walk through the narrow streets of Venice and be so warmly recognised!

From Venice we took the boat down to Greece. It was a beautiful trip, in much greater comfort than I remembered during my childhood crossings of the Adriatic Sea in fourth class! A Greek professor of the name of Vougiouklis was waiting for us at the port and took us up to the University of Thrace in Alexandroupolis, where we spent a few days. They got very excited about my proposals and wanted to invite me again for a longer period and do a whole course. We spent a day with Leonidas Kounguetsoff and talked about old times. He thought I should come and start a Psychomathematics Centre in his university! I think somehow that such dreams cannot now be realised, but I left enough material with the Primary Education Department and established an e-mail link, so something might be done even without my physical presence.

Another rather nice thing that has happened lately is this: when Paul Ernest got in touch with the publishing house QED of York about the possibility of publishing some of my recent writings, he got the response, 'Anything that the great man has written!'

So most of the new things I have thought up in the last few years, both practical and theoretical, have now been sent to York and I hope that in the not too far distant future some of these ideas will be made available to the interested public: teachers, parents and children.

So this is how we are in November of 1997, still living in our little house in Wolfville, Tessa, Sara and myself. We try to have as much contact with our grandchildren as we can, but the generations are very far apart and, sadly, it gets more difficult with every passing year. In the winter I still continue my cross-country ski runs, in the summer I take my walks along the nature trail through the forest and as often as possible I do my three to four hundred metres of swimming in the local pool! And we still go to a Quaker meeting, although sometimes we go to worship in the local Baptist Church, as we can do so with our daughter Sara.

I hope that, as I near the end of this earthly existence, my friends and colleagues with whom I have been in contact during my life can feel that I have given them service, trying to make life better for people, in particular for children, by liberating their thinking from the shackles of imposed ideas. May God forgive me for any wrongs I may have done them and I pray that He may bless all with whom I have been in contact during my life.